Climbing the Gamma Ladder
to Light-Speed and Beyond

Volume 3

If you have the desire to be set free from any worries that the speed of light may be a limit strictly imposed by Nature but are intrigued by the speed of light, then this book is a great match for you.

Essentially, I go to good lengths to explore *how* and *why* the speed of light, as an ultimate limit, enables far more exotic and wide-ranging travel than any otherwise-superluminal warp drives, wormhole shortcuts through space-times, and the like.

You, as a reader, will, hopefully, not be put off by my frequent discussion about GOD and divine providence in the context of exploring the perhaps

utterly infinite scale of the cosmic order and the bold, open-ended eternal opportunities for exploration. After all, if we assume that human civilization will in some way last forever, then such eternal duration may necessarily need to assume divine providence. Otherwise, we would almost certainly destroy our species with technology and unsustainable practices.

For conservative purists who are adverse to invoking the concept of a god in attempts at theoretical discourse, all I ask of you is to give this book a try. You will likely find the open-ended speculation a joy, in that you, hopefully, will come to see the glory in light-speed impulse travel in a brand-new light.

ESSAY 1) An Internal Structure of Planck Time Units?

Now, the reader of my previously published books may recall the concepts presented herein about how a spacecraft having run out of room to travel the accidental aspects of a cosmic order of travel may then begin to travel into the substantial principles of a cosmic order such as traveling into the prime matter. Accordingly, the spacecraft would have an infinite Lorentz factor and run out of future time to travel forward in a given universe, multiverse, forest, biosphere, and the like, and hyperspace of travel.

Now, a light-speed inertial spacecraft having nonzero invariant mass would travel an infinite number of light-years and an infinite number of years into the future in one Planck time unit ship-frame.

Now, each Planck time unit in the ship-frame is as if it's a discrete time tile, perhaps blurred by Heisenberg's uncertainty principle.

However, based on existential continuity, a Planck time unit may have a backward base, a medial bulk, and a forward roof.

So the location of the spacecraft at the base of a given Planck time unit in the spacecraft world line may be associated with an infinite spacecraft travel distance through space and an infinite temporal travel distance into the future.

Thus, for a suitably infinite Lorentz factor spacecraft, the spacecraft may travel infinite distances through space and infinite temporal distances into the future during the base of a Planck time unit ship-frame.

Another consequence of the above scenarios is that a spacecraft, accordingly, may travel an infinite distance through space and an infinite distance in forward time in a sub–time unit.

Accordingly, a light-speed finite invariant mass inertial spacecraft may travel infinite distances through space and forward in time in effectively super-instantaneous travel in the spacecraft reference frame.

The greater the infinite spacecraft Lorentz factor, the more super-instantaneous the ship-frame rate of travel is.

We may conjecture further that the base of a ship-frame Planck time unit may itself have a base that we refer to as the second lower base, while also having its own medial portion and roof, which we refer to as the second lower medial portion and second lower roof of the Planck time unit.

Likewise, we can conjecture a third lower; base, medial portion, and roof of the Planck time unit.

We can continue the conjecture to an arbitrary *kth* lower; base, medial portion, and roof of the Planck time unit, where *k* can be any positive integer.

Thus, there are potentially *k* distinct levels of super-instantaneous travel through space and forward in time in the spacecraft reference frame.

We may affix the following operator to formulas for gamma and related parameters to denote the above conjectured super-instantaneous ship-frame travel.

> [w(Arbitrary kth lower; base, medial portion, and roof of the Planck Time Unit where k = 1, 2, 3, … in k distinct levels of super-instantaneous travel ship-frame)]

ESSAY 2) Teleported Spacecraft Having Extensions into Mixing Parameters

Now, we consider the following electrodynamic field mixing parameters.

Electricweak, magnetoweak, electroweak, electric-strong, magneto-strong, electric-gravatic, magneto-gravatic, electrogravatic, electric-weak-strong, magneto-weak-strong, electroweak-strong, electric-weak-gravatic, magneto-weak-gravatic, electroweak-gravatic, electric-strong-gravatic, magneto-strong-gravatic, electro-strong-gravatic, electric-weak-strong-gravatic, magneto-weak-strong-gravatic, electroweak-strong-gravatic.

In all, we have twenty mixing parameters. We also have Higgs field augmentations, thus adding twenty additional parameters via appendage of the preface Higgs to the items in the above roster, and three additional parameters associated with electrodynamic-Higgs mechanisms. However, we have analogous supersymmetric mixing parameters thus contributing to other list items. It is prudent to note that we do not yet know what the most appropriate or most accurate supersymmetry theories are, so we simply double the number of Standard Model terms outright, which we placed at a total of 43, and add 3 more items for the additional electrodynamic mechanisms and then add the quantity, *k*, to the resulting value of 89, to yield a first-order consideration of *(89 + k)* mixing parameters.

Plausibly, each electromagnetic wave, a.k.a., photon, and each differential volume element of a magnetic field and an electric field may have the above-listed appendages.

So in the condition that a spacecraft would be entangled with a carrier photon and an originating particle, the spacecraft may, in a sense, have extensions into these mixing parameters and, perhaps, may be extended beyond the primary effective midpoint or averaged location of the carrier photon. The same may be considered with a classical feedback loop that would be used to complete and close the teleportation process. Such teleportive processes may extend a full radius of a cosmic light cone, at which the spacecraft may be re-teleported across another cosmic light-cone radius distance. The teleportation apparatus for this next leg of teleportation may be teleported along with the spacecraft.

The basic idea here is that the photonic mixing parameter appendages may extend beyond the pure electromagnetic aspects of a photon in ordinary 4-D space-time and/or extend slightly into other dimensions such as hyperspaces and the like. For hyperspatial extensions, the projection of the maximally length vectors extending the length of the mixing parameter distributions over the length of the ordinary photonic extensions may be greater than 1.

As another possibility, the purely non-electromagnetic components of photons may be tucked away in the appendages of the photon, but in ways not associated with other dimensions or hyperspatial extensions. Thus, a spacecraft might be entangled mostly or almost completely with any subset of non-electromagnetic photonic components and then transported in such manners that the spacecraft under teleportation hides within subsets of ordinary 4-D space-time yet removed from the ordinary aspects of said 4-D space-time.

We may include the above mixing parameter appendages of extension in the formulas for gamma.

ESSAY 3) Extra-Impulse Non-General-Relativistic Travel

Here, we consider a class of travel that is loosely related to impulse or inertial travel, as impulse or inertial travel is to general relativistic travel methods such as warp drives, wormhole travel, and the like.

Accordingly, the new class of travel is a kind of "maxed-out" translational travel through space and not necessarily of Keplerian velocity translational travel through space.

Actually, the maxed-out translational travel concept is derived from the possibilities of light-speed impulse travel at infinite Lorentz factors.

Accordingly, a spacecraft may attain so great of Lorentz factors such that the opportunities or states of ordinary impulse travel are no longer full of descriptions of the spacecraft travel but instead, a first level of maxed-out translational travel manifests.

The general idea here is that a spacecraft's invariant mass-specific kinetic energy will become so great, as will the spacecraft Lorentz factors, such that the spacecraft travel level overflows ordinary impulse or inertial travel energy states and topologies.

As the spacecraft Lorentz factor climbs to ever greater infinite values, ascending levels of overflows of ordinary impulse and inertial travel may manifest.

At first and via cursory study and observation, these levels of overflows may be undetectable. However, upon more rigorous study methods, the distinction among the ascending levels may become apparent.

The mechanisms of ascending levels of the travel mechanisms can be viewed as if a spacecraft Lorentz factor becomes so great that the impulse or inertial travel modes become full, resulting in an analogue of overflows into a higher level.

We may affix the following operator to formulas for gamma and related parameters to denote the relevant mechanisms of the ascending levels of impulse or inertial travel associated with progressively infinite Lorentz factors.

[u(Relevant mechanisms of the ascending levels of impulse or inertial travel associated with progressively infinite Lorentz factors)]

ESSAY 4) Levels of Infra-Now Travel at Light-Speed

Now, we can intuit that a light-speed spacecraft having attained the least infinite Lorentz factor can travel almost the least infinite number of light-years through space in one-year ship-time and an infinite number of years into the future in one-year ship-time.

However, the spacecraft, as such, already having the least possible infinite Lorentz factor, can likely still obtain greater infinite Lorentz factors to travel greater infinities of light-years through space and forward in time in a one-year ship-frame.

Eventually, the light-speed of the spacecraft will become so extremely impressed on the spacecraft such that time will cease on the spacecraft in the spacecraft reference frame. This is because nothing will be able to move in the spacecraft reference frame, thus resulting in time coming to a standstill.

However, there will still exist a now in the spacecraft reference frame for such maxed-out light-speed spacecraft.

Oddly enough, the entire universe, multiverse, forest, biosphere, and the like of spacecraft travel will be so relativistically aberrated to be located in front of the spacecraft that such realms of travel will pull on the spacecraft with infinite accelerative force via the force of gravity. Hopefully, the entire spacecraft is pulled on by the same accelerative force so that the spacecraft is not annihilated.

Once the spacecraft enters a permanent timeless now state, it may then progress to a timelike intranow.

Accordingly, the spacecraft would enter a state that, as we suggested, would be called an intranow, which is to now as now is to motion through time.

Perhaps in an intranow are some form of hard-to-comprehend dynamic processes.

Eventually, a spacecraft having maxed out a first level of an intranow will then move into a second level of intranow. We refer to these first levels of intranow as intranow-1 and intranow-2.

As the spacecraft continues to accelerate, it may max out an intranow-2 to move to a third level of intranow, or intranow-3.

We can surmise that a never-ending ascension of intranow levels becomes available, for which each level of intranow is associated with great and profound differences and experiences for the crew members relative to that of the preceding levels, each with their own fundamental dynamical analogues.

We may affix the following operator to formulas for gamma and related parameters to denote the intranow levels of acquisition by a light-speed spacecraft together with the option of taking arbitrary propulsive force, acceleration, and kinetic energy derivatives and integrals of first and higher orders of intranow acquisition rates.

> [w(Acquisition rate of kth to (k+n)th intranows and optionally the first and higher-order thrust force, acceleration, and kinetic energy derivatives and integrals thereof)]

ESSAY 5) Hyperinteger Lorentz Factors In Light-Speed Travel

Now, the speed of light is an all-important constant in the operation of our universe as we know it.

The speed of light shows up in special and general relativity, quantum mechanics, classical electrodynamics, and just about every other branch of modern physics and engineering.

In reality, everything accidental in our universe in its motion is relative to the speed of light. The speed of light is so important that we can enter into a great mystery through research and development and flying manned spacecraft able to travel at close to the speed of light.

A real godsend would be the development of spacecraft that can achieve the speed of light and associated infinite Lorentz factors. In doing so, we would be able to travel infinite distances in one Planck time unit ship-frame and infinite numbers of years into the future in one Planck time unit ship-frame.

Should we figure out how to travel at light-speed in translational impulse travel through space-time, we would, in a sense, become far more fundamental in terms of the accidental properties of our spacecraft and her crew members.

In theory, there should be no limit to how great of infinite Lorentz factors our light-speed impulse spacecraft could attain.

In theory, our spacecraft may attain hyperinteger Lorentz factors to travel hyperinteger light-years through space and hyperinteger years into the future in one Planck time unit ship-frame.

Hyperintegers are values beyond the possible infinite integers but on hyperextended number lines.

We will go still further to conjecture that light-speed spacecraft may attain Lorentz factors beyond the hyperintegers to levels most appropriately referred to as hyper-values.

Hyper-values would be so large or more appropriately, so far removed from even hyperintegers that hyper-values are not properly quantitative values, but instead, some things were far more abstract and located along a hyperextended-hyperinteger number line.

So any light-speed limit strictly enforced by Mother Nature is very good rather than something to fret about, and it gives us all the greater possibilities.

We may affix the following operator to formulas for gamma and related parameters to denote the wondrous ramifications and opportunities for light-speed travel of impulse forms in infinite real number Lorentz factors, then hyperreal number Lorentz factors, then hyper-value Lorentz factors, and perhaps so on.

> [u(Wondrous ramifications and opportunities for light-speed travel of impulse forms in infinite real number Lorentz factors, then hyperreal number Lorentz factors, then hyper-value Lorentz factors, and perhaps so on)]

We can, by corollary and symmetry, consider travel at light-anti-speed or light-anti-velocity, or perhaps anti-light-speed and anti-velocity of light.

As yet another set of levels, we can, by corollary and symmetry, consider travel at light-anti-anti-speed, light-anti-non-anti-speed, or light-anti-non-anti-velocity, light-anti-non-anti-velocity, or perhaps anti-anti-light-speed, anti-non-anti-light-speed, and anti-anti-velocity of light and anti-non-anti-velocity of light

As yet another set of levels, we can by corollary and symmetry consider travel at light-anti-anti-anti-speed, light-anti-anti-non-anti-speed, or light- anti-anti-non-anti-velocity, light- anti-anti-non-anti-velocity, or perhaps anti- anti-anti-light-speed, anti-anti-non-anti-light-speed and anti-anti-anti-velocity of light and anti-anti-non-anti-velocity of light, light-anti-non-anti-anti-speed, light-anti-non-anti-non-anti-speed, or light-anti-non-anti-non-anti-velocity, light-anti-non-anti-non-anti-velocity, or perhaps anti-non-anti-anti-light-speed, anti-non-anti-non-anti-light-speed and anti-non-anti-anti-velocity of light and anti-non-anti-non-anti-velocity of light.

As you can see, the number of terms increases by a multiple of 2 for every level relative to the previous level.

We might otherwise affix the following operator to formulas for gamma and related parameters to denote the above speed and velocity alternatives.

> [w(Of first class; travel at light-anti-speed or light-anti-velocity, or perhaps anti-light-speed and anti-velocity of light)]:[w(Of second class; travel at light-anti-anti-speed, light-anti-non-anti-speed, or light-anti-non-anti-velocity, light-anti-non-anti-velocity, or perhaps anti-anti-light-speed, anti-non-anti-light-speed and anti-anti-velocity of light and anti-non-anti-velocity of light)]:[w(Of third class; travel at light-anti-anti-anti-speed, light-anti-anti-non-anti-speed, or light- anti-anti-non-anti-velocity, light- anti-anti-non-anti-velocity, or perhaps anti- anti-anti-light-speed, anti-anti-non-anti-light-speed and anti-anti-anti-velocity of light and anti-anti-non-anti-velocity of light, light-anti-non-anti-anti-speed, light-anti-non-anti-non-anti-speed, or light-anti-non-anti-non-anti-velocity, light-anti-non-anti-non-anti-velocity, or perhaps anti- non-anti-anti-light-speed, anti-non-anti-non-anti-light-speed and anti-non-anti-anti-velocity of light and anti-non-anti-non-anti-velocity of light)]:[w(Classes 4 thru infinity)]

However, we likely instead should affix the following contracted operator for the sake of brevity:

> $\{[\Pi(k = 1; k = \infty\uparrow)]:[\Pi(h = 1; h = \infty\uparrow)]:\{[[(Anti),k]-[(Non),h]-light-speed]$ and/or $:[[(Anti),k]-[(Non),h]-velocity-of-light]\}\}$

As we learn more about the speed of light and why it is valid, we can learn more about these other conjectural parameters.

Some of these constructs may be operative in the mind-body link in Lorentz invariant and Lorentz varying manners, especially at the speed of light.

ESSAY 6) Light-Speed Travel but neither Spatially nor Temporally

Here we consider travel at the speed of light, but neither in spatial nor temporal direction or directions.

Accordingly, such a spacecraft would have a speed equal to that of light but without a velocity vector.

Such a construct is rather bizarre but fascinatingly exotic or far out.

Another construct includes the acceleration of a spacecraft without direction in space and/or time. Accordingly, such acceleration would have no vector.

As yet another construct, we consider a spacecraft with force propulsion, but where the force has no direction. Accordingly, such force would have no vector.

We may affix the following operator to formulas for gamma and related parameters to denote these vectorless otherwise-non-scalar quantities.

[w(Directionless light-speed, acceleration, and thrust force)]

ESSAY 7) Levels of Super-Classical Physics in Light-Speed Travel

For a spacecraft to obtain the velocity of light, and infinite Lorentz factors, the process of designing, assembling, crewing, and flying the spacecraft would need to have the utmost rational and rules-based procedures, and thus would be, in a sense, super-classical.

Just as quantum mechanics and the Heisenberg uncertainty principle looms strong in the quantum mechanical realm, a super-classical paradigm would lie on the other side of the classical realm.

Classical scenarios are somewhat predictable but not completely so, and in many cases, a butterfly effect can result in a huge change in the future direction of events from a small mesoscopic event such as perhaps the ways a butterfly flaps its wings over a period of a fraction of a second.

However, attaining light-speed and infinite spacecraft Lorentz factors would require superb predictability for how the R&D program and spacecraft missions would be carried out if for no other reason than that precision and utmost spacecraft predictability and rules-based operation would be required to enable the spacecraft to overcome natural obstacles and actually safely attain light-speed.

We may refer to the first level of super-classical events by the term super-1-classical.

In super-1-classical physics, there is a very large infinite number of ordinated steps for which the super-1-classicality becomes ever more extreme, such as, for example, increases in the infinity of the spacecraft Lorentz factor.

In a so-called super-2-classical realm, we consider scenarios such as the conjectural immortality of the human soul.

Accordingly, the existence of the human soul is not uncertain, just as, assuredly, each human person has his or her unique soul or existence.

Even under the circumstance where GOD would create two human souls with exactly the same traits and place these souls in identical environments, both souls would be unique. So if one soul were to be divinely annihilated, the other soul would be existentially unaffected.

We may consider a so-called super-3-classical realm and perhaps a super-4-classical realm and so on.

These realms would more properly define GOD or created realities GOD may create in the cosmically distant future. These created realities have ramifications for infinite Lorentz factor light-speed travel. This is so because the crew members may survive to experience the creation of said super-classical levels via infinite future time travel.

We may affix the following operator to formulas for gamma and related parameters to denote the above super-classical realities and ship-time, gamma, acceleration, and kinetic energy derivatives and integrals of the rate of change thereof.

> [w(Super-k-classical realms) and arbitrary orders of ship-time, gamma, acceleration, and kinetic energy derivatives and integrals of rate of change thereof)]

ESSAY 8) Super-Infinite Lorentz Factors

We can contemplate spacecraft Lorentz factors of arbitrary countable infinities.

We can also contemplate spacecraft Lorentz factors of arbitrary uncountable infinities.

We can also consider spacecraft Lorentz factors of arbitrary super-infinities.

However, perhaps very little, if never before considered, are spacecraft Lorentz factors that are so great that these Lorentz factors are qualitatively infinite and thus are larger than any quantity.

We may even go on to consider spacecraft Lorentz factors that are qualitatively super-infinite.

All these huge classes of Lorentz factors are lived out if possible in associated effectively superluminal velocities of spacecraft travel in space and forward in time.

We may affix the following operator to formulas for gamma and related parameters to denote the above Lorentz factor options for light-speed spacecraft.

> [u(Countably infinite, uncountably infinite, super-infinite, qualitatively infinite, and qualitatively super-infinite spacecraft Lorentz factors and effective velocities through space and forward in time)]

Now, we generally are familiar with the concept of infinity.

Note that Hyper4(a, n) is equal to a tetrated n, or a raised to the power of itself n-1 times. The latter value is symbolically written as n subscript a.

For example,

> 3 EXP 4 = 81, but 4 subscript 3 is approximately equal to 10 EXP (1,000,000,000,000).

Alternatively:

> 4 subscript 2 = 2 EXP 2 EXP 2 EXP 2 = 2 EXP [2 EXP [2 EXP 2]] = 2 EXP (2 EXP 4) = 2 EXP 16 = 65,536

For example, Hyper5(4, 4) is equal to 4 tetrated 4 tetrated 4 tetrated 4. This value is commonly referred to as 4 pentated 4.

Hyper 6, (4,4) is 4 pentated 4 pentated 4 pentated 4 and is also referred to as 4 hexataed 4.

Hyper 7, (4,4) is 4 hexated 4 hexated 4 hexated 4 and so on.

Aleph 0 is the infinite number of integers.

Aleph 1, according to the perhaps unprovable, and thus unfalsifiable, continuum hypothesis, is the number of real numbers that is greater than Aleph 0 by a multiplicative factor of infinity.

Aleph 2 is similarly greater than Aleph 1.

Aleph 3 is similarly greater than Aleph 2.

Aleph 4 is similarly greater than Aleph 3.

And so on.

In general:

Aleph n = 2 EXP [Aleph (n-1)]

The number Ω is commonly stated as the least infinite positive integer or ordinal.

Now here is a real zinger.

So we can produce this abstraction: [Hyper Aleph Ω (Aleph Ω, Aleph Ω)]

We can go to ever greater infinities.

So we can consider

Hyper [Hyper Aleph Ω (Aleph Ω, Aleph Ω)]([Hyper Aleph Ω (Aleph Ω, Aleph Ω)], [Hyper Aleph Ω (Aleph Ω, Aleph Ω)]) numbers

Dig a little deeper into infinities and you'll learn that there are countable and uncountable infinities.

Thus, we may consider this:

[Hyper-[Hyper Uncountable Infinities (Uncountable Infinities, Uncountable Infinities)] ([Hyper Uncountable Infinities (Uncountable Infinities, Uncountable Infinities)], [Hyper Uncountable Infinities (Uncountable Infinities, Uncountable Infinities)]

Dare we not say that there may still be greater categories of infinities?

How about infinities that are so extreme they are larger than themselves by extreme extents? These extremities of extents may be arbitrary. Needless to say, we can form the following operator function:

[Hyper [Hyper infinities that are so extreme they are larger than themselves by extreme extents (infinities that are so extreme they are larger than themselves by extreme extents, infinities that are so extreme they are larger than themselves by extreme extents)] ([Hyper infinities that are so extreme they are larger than themselves by extreme extents (infinities that are so extreme they are larger than themselves by extreme extents, infinities that are so extreme they are larger than themselves by extreme extents)], [Hyper infinities that are so extreme they are larger than themselves by extreme extents (infinities that are so extreme they are larger than themselves by extreme extents, infinities that are so extreme they are larger than themselves by extreme extents)]

Well, if you have ever fretted that light-speed may be an utterly inviolable limit imposed by Mother Nature, I offer a friendly gesture to say "Chill!"

At light-speed in the least infinite Lorentz factor, say the commonly stated value of Ω, which is the least infinite ordinal, you can travel almost Ω light-years through space and Ω years into the future in one-year ship-time.

At Aleph 1 Lorentz factor, you can travel about Aleph 1 through space in one-year ship-time and Aleph 1 years into the future in a one-year ship-frame. Aleph 1 is 2 EXP (Aleph 0).

At Aleph 2 Lorentz factor, you can travel about Aleph 2 through space in one-year ship-time and Aleph 2 years into the future in a one-year ship-frame. Aleph 2 is 2 EXP (Aleph 1).

At Aleph 3 Lorentz factor, you can travel about Aleph 3 through space in one-year ship-time and Aleph 3 years into the future in a one-year ship-frame. Aleph 3 is 2 EXP (Aleph 2).

And here we go again.

At Aleph Ω Lorentz factor, you can travel about Aleph Ω through space in one-year ship-time and Aleph Ω years into the future in one-year ship-frame.

At [Hyper Aleph Ω (Aleph Ω, Aleph Ω)] Lorentz factor, you can travel about [Hyper Aleph Ω (Aleph Ω, Aleph Ω)] through space in one-year ship-time and [Hyper Aleph Ω (Aleph Ω, Aleph Ω)] years into the future in one-year ship-frame.

At Hyper [Hyper Aleph Ω (Aleph Ω, Aleph Ω)]([Hyper Aleph Ω (Aleph Ω, Aleph Ω)], [Hyper Aleph Ω (Aleph Ω, Aleph Ω)]) Lorentz factor, you can travel about Aleph Hyper [Hyper Aleph Ω (Aleph Ω, Aleph Ω)]([Hyper Aleph Ω (Aleph Ω, Aleph Ω)], [Hyper Aleph Ω (Aleph Ω, Aleph Ω)]) through space in one-year ship-time and Aleph Hyper [Hyper

Aleph Ω (Aleph Ω, Aleph Ω)]([Hyper Aleph Ω (Aleph Ω, Aleph Ω)], [Hyper Aleph Ω (Aleph Ω, Aleph Ω)]) years into the future in one-year ship-frame.

At [Hyper-[Hyper Uncountable Infinities (Uncountable Infinities, Uncountable Infinities)] ([Hyper Uncountable Infinities (Uncountable Infinities, Uncountable Infinities)], [Hyper Uncountable Infinities (Uncountable Infinities, Uncountable Infinities)] Lorentz factor, you can travel about [Hyper-[Hyper Uncountable Infinities (Uncountable Infinities, Uncountable Infinities)] ([Hyper Uncountable Infinities (Uncountable Infinities, Uncountable Infinities)], [Hyper Uncountable Infinities (Uncountable Infinities, Uncountable Infinities)] through space in one-year ship-time and [Hyper-[Hyper Uncountable Infinities (Uncountable Infinities, Uncountable Infinities)] ([Hyper Uncountable Infinities (Uncountable Infinities, Uncountable Infinities)], [Hyper Uncountable Infinities (Uncountable Infinities, Uncountable Infinities)] years into the future in one-year ship-frame.

At [Hyper [Hyper infinities that are so extreme they are larger than themselves by extreme extents (infinities that are so extreme they are larger than themselves by extreme extents, infinities that are so extreme they are larger than themselves by extreme extents)] ([Hyper infinities that are so extreme they are larger than themselves by extreme extents (infinities that are so extreme they are larger than themselves by extreme extents, infinities that are so extreme they are larger than themselves by extreme extents)], [Hyper infinities that are so extreme they are larger than themselves by extreme extents (infinities that are so extreme they are larger than themselves by extreme extents, infinities that are so extreme they are larger than themselves by extreme extents)] Lorentz factor, you can travel about [Hyper [Hyper infinities that are so extreme they are larger than themselves by extreme extents (infinities that are so extreme they are larger than themselves by extreme extents, infinities that are so extreme they are larger than themselves by extreme extents)] ([Hyper infinities that are so extreme they are larger than themselves by extreme extents (infinities that are so extreme they are larger than themselves by extreme extents, infinities that are so extreme they are larger than themselves by extreme extents)], [Hyper infinities that are so extreme they are larger than themselves by extreme extents (infinities that are so extreme they are larger than themselves by extreme extents, infinities that are so extreme they are larger than themselves by extreme extents)] through space in one-year ship-time and [Hyper [Hyper infinities that are so extreme they are larger than themselves by extreme extents (infinities that are so extreme they are larger than themselves by extreme extents, infinities that are so extreme they are larger than themselves by extreme extents)] ([Hyper infinities that are so extreme they are larger than themselves by extreme extents (infinities that are so extreme they are larger than themselves by extreme extents, infinities that are so extreme they are larger than themselves by extreme extents)], [Hyper infinities that are so extreme they are larger than themselves by extreme extents (infinities that are so extreme they are larger than themselves by extreme extents, infinities that are so extreme they are larger than themselves by extreme extents)] years into the future in one-year ship-frame.

These infinities can be as large as you would like.

We may affix the following operator to formulas for gamma and related parameters to denote the unbounded infinite Lorentz factors that are mathematically constructible.

Exemplar Lorentz factors in light-speed travel as mathematically constructible such as the following:

> Ω → Aleph 1 → Aleph 2 → Aleph 3 → and so on to Aleph Ω → [Hyper Aleph Ω (Aleph Ω, Aleph Ω)] → Hyper [Hyper Aleph Ω (Aleph Ω, Aleph Ω)]([Hyper Aleph Ω (Aleph Ω, Aleph Ω)], [Hyper Aleph Ω (Aleph Ω, Aleph Ω)]) → [Hyper-[Hyper Uncountable Infinities (Uncountable Infinities, Uncountable Infinities)] ([Hyper Uncountable Infinities (Uncountable Infinities, Uncountable Infinities)], [Hyper Uncountable Infinities (Uncountable Infinities, Uncountable Infinities)] → [Hyper [Hyper infinities that are so extreme they are larger than themselves by extreme extents (infinities that are so extreme they are larger than themselves by extreme extents, infinities that are so extreme they are larger than themselves by extreme extents)] ([Hyper infinities that are so extreme they are larger than themselves by extreme extents (infinities that are so extreme they are larger than themselves by extreme extents, infinities that are so extreme they are larger than themselves by extreme extents)], [Hyper infinities that are so extreme they are larger than themselves by extreme extents (infinities that are so extreme they are larger than themselves by extreme extents, infinities that are so extreme they are larger than themselves by extreme extents)]

ESSAY 9) Infinite Lorentz Factor Spacecraft Breakout of Their Own World Lines

Now, according to special and general relativity, we all have our unique world lines that only we as individuals can travel.

We are also familiar with the relativistic process of time dilation.

So we may surmise that while traveling at the speed of light in infinite Lorentz factors, we can travel infinite distances through space and forward in time. The degree of infinite Lorentz factors is likely arbitrary since there are greater and lesser infinities.

We also can intuit that a spacecraft traveling inertially through ordinary 4-D space-time might run out of future time to travel forward perhaps thus leading the craft to leave ordinary 4-D space-time to enter a larger realm. We can even consider that such a spacecraft, having entered a larger realm, will eventually run out of temporal room to travel forward in said larger realm time, which might be multidimensional time as a nonlimiting option.

Eventually, a spacecraft Lorentz factor may become so infinite that the spacecraft and crew break out of their own world lines to enter another quasi-metaphysical existential or outright metaphysical existential state.

From there, a spacecraft may eventually break out of the first metaphysical state, which we will refer to as metaphysical-state-1 to enter another, broader metaphysical state, which we refer to as metaphysical-state-2.

Likewise, a spacecraft obtaining still greater Lorentz factors may break out of metaphysical-state-2 to enter a more broad metaphysical-state-3.

The number of ascending levels of metaphysical states may be undefined and infinite, as well as increasing with cosmic creation and evolution.

We may affix the following operator to formulas for gamma and related parameters to denote the possibilities of spacecraft entering these levels of metaphysical states and spacecraft arbitrary ordered derivatives and integrals of progression in the set of metaphysical-state-k's with respect to spacecraft time, acceleration, kinetic energy, and gamma.

> [w(Spacecraft metaphysical states and arbitrary ordered derivatives and integrals of progression in the set of metaphysical-state-k's with respect to tship, a, KE, and γ)]

So fasten your seat belts, folks! We are in for an eternal ride.

ESSAY 10) Extradimensional Travel at Light-Speed

Now, there are countable and uncountable infinities.

But before we go further, consider the hyper-operator notation that was designed to express huge values not otherwise expressible.

For example, note that Hyper4(a, n) is equal to a tetrated n, or a raised to the power of itself n-1 times. The latter value is symbolically written as n subscript a.

For example,

3 EXP 4 = 81, but 4 subscript 3 is approximately equal to 10 EXP (1,000,000,000,000).

Alternatively,

4 subscript 2 = 2 EXP 2 EXP 2 EXP 2 = 2 EXP [2 EXP [2 EXP 2]] = 2 EXP (2 EXP 4) = 2 EXP 16 = 65,536

For example,

Hyper5(4, 4) is equal to 4 tetrated 4 tetrated 4 tetrated 4. This value is commonly referred to as 4 pentated 4.

Hyper 6, (4,4) is 4 pentated 4 pentated 4 pentated 4 and is also referred to as 4 hexataed 4.

Hyper 7, (4,4) is 4 hexated 4 hexated 4 hexated 4 and so on.

Aleph 0 is the infinite number of integers.

Aleph 1, according to the perhaps-unprovable, and thus unfalsifiable, continuum hypothesis, is the number of real numbers that is greater than Aleph 0 by a multiplicative factor of infinity.

Aleph 2 is similarly greater than Aleph 1.

Aleph 3 is similarly greater than Aleph 2.

Aleph 4 is similarly greater than Aleph 3.

And so on.

In general: Aleph n = 2 EXP [Aleph (n-1)].

The number Ω is commonly stated as the least infinite positive integer or ordinal.

Now here is a real zinger.

So we can produce the abstraction of Hyper Aleph Ω (Aleph Ω, Aleph Ω).

So we can produce the abstraction of Hyper (High End Uncountable Infinity) ((High End Uncountable Infinity), (High End Uncountable Infinity)).

Now, the latter infinity is really, really, huge.

However, we can still go further to consider infinities that are so great that they cannot be defined.

Likely, such infinities apply only to GOD, such as the number of GOD's ontological transcendentals.

By ontological transcendentals, I mean qualities such as ontological goodness, value, purpose, and the like instead of existential properties such as immutability, being all-powerful, being all-knowing, immortality, and the like.

However, there certainly may be infinities of universes where the infinities are pre-undefinable. As such, these infinities are huge but definable.

We might refer to these pre-undefinable infinities and undefinable infinity scrapers.

So we can, by corollary, make the following hyper-operator expression:

Hyper (High End Undefinable Infinity Scraper) ((High End Undefinable Infinity Scraper), (High End Undefinable Infinity Scraper)).

There may be the above number of universes and more.

Since the invariant mass of a spacecraft might be reduced to zero, we may have reason to hope that in the depths of future eternity, we can attain spacecraft Lorentz factors in the range of values defined by the above operator or values equal to the above expression.

We may affix the following operator to formulas for gamma and related parameters to denote the open-ended plausibilities from a purely mathematical perspective to denote the above high-end undefinable infinity scraper prospects for light-speed or ever-so-slightly-faster-than-light-speed travel:

[w [γ = [Hyper (High End Undefinable Infinity Scraper) ((High End Undefinable Infinity Scraper), (High End Undefinable Infinity Scraper))]]]

Commensurate to the above awesomely large prospects of Lorentz factors, we can contemplate scenarios for which the velocity and acceleration of such a light-speed spacecraft would be super-orthogonal to its propulsive drive force.

We can also consider spacecraft forward time travel in directions that are super-orthogonal to the light-speed arrow(s) of time flow, in the background and in the spacecraft reference frames. Thus, the sin of the angles of such vectors with respect to the plane orthogonal to the spacecraft propulsive thrust are greater than 1, and perhaps in some cases, infinitely greater than 1.

So we might ordinarily affix the following operator to the lengthy formulas for gamma and related parameters to denote the super-orthogonal relations:

[z[[Sin (Angles of vectors of V and a with respect to the plane orthogonal to the spacecraft propulsive thrust and angle ship-time arrows with respect background time arrows)] → ∞↑]]

However, we affix the following operator instead for brevity:

[z[Sin angles (a,V,tship)] > 1 → ∞↑]

Note that notions of superunitary sin values require new mathematics, but perhaps we can begin the new math starting here and now with the above conjectures.

ESSAY 11) Infinite Lorentz Factor Spacecraft in Super-Relativistic Aberration

Now, the reality that light-speed infinite Lorentz factor travel may super-relativistically aberrate the background of space-time to contract to a fraction of a geometric point in front of a spacecraft seems to indicate that super-aberrated background reference frame would act as a door to other aspects of ordinary 4-D space-time.

Accordingly, other compartments or components of ordinary 4-D space-time may manifest for a light-speed inertial spacecraft.

Normally, such extradimensional extensities would be trivial and suppressed at even extreme but reasonably finite Lorentz factors, not to mention every ordinary velocities of travel on Earth by cars, planes, trains, and the like.

The possibilities of light-speed spacecraft attaining ever greater infinite Lorentz factors seem to imply super-aberrational space-time effects and the entrance of a light-speed inertial spacecraft into appended extensions and/or manifest such extensions in cases where these extensions would not exist relative to ordinary everyday velocities of travel of common transport vehicles.

Analogues may present themselves in higher dimensional space-time travel such as travel at light-speed in hyperspaces.

Now, travel at the speed of light is commonly considered associated with infinite Lorentz factors.

Accordingly, I suggest that once a spacecraft attains a suitably infinite Lorentz factor in light-speed travel, the spacecraft would run out of temporal room in which to travel forward in time in, as well as experience such an extreme relativistic background aberration such that the spacecraft leaves the universe or larger realm of travel to enter another, broader realm of travel.

However, I believe that if a spacecraft has attained light-speed in a suitably infinite Lorentz factor and then pops into another realm, and then back into the realm of origin, the effective velocity of spacecraft displacement in the realm of origin could not be greater than the speed of light in spirit.

A spacecraft having attained a suitably infinite Lorentz factor in a first broader realm of travel, I believe, may likewise pop into a higher or second alternative realm of travel.

For a spacecraft that would thus reach a velocity of light in a first larger realm of travel having popped into a second higher realm of travel and then back into the first higher realm of travel, I believe that the velocity of displacement in the first higher realm of travel cannot be greater than the velocity of light in the first higher realm of travel. Even in cases where the spacecraft had originated in light-speed travel in the realm of origin and then having popped into the first higher realm, and then into the second higher realm, then directly back into the realm of the origin or first back into the first higher realm then back into the realm of origin, I believe the distance of displacement in the realm of origin cannot exceed the velocity of light in the realm of origin.

We can likewise consider scenarios of still higher levels of realms, with the analogous velocity limits being equal to the respective velocities of light.

We can formulate such light-speed limits as follows.

$$[\Sigma(j = 1; j = \infty\uparrow):[[d[r[(x,1), (x, 2), …, (x,n)]]/dt]j]] \leq \text{(realm specific value of c)}$$

We may also consider derivatives with respect to other parameters—such as gamma, acceleration in the ship-frame, and ship kinetic energy—as follows.

$$[\Sigma(j = 1; j = \infty\uparrow):[[d[r[(x,1), (x, 2), …, (x,n)]]/d\gamma]j]]$$

$$[\Sigma(j = 1; j = \infty\uparrow):[[d[r[(x,1), (x, 2), …, (x,n)]]/d \text{ ship-frame acceleration}]j]]$$

$$[\Sigma(j = 1; j = \infty\uparrow):[[d[r[(x,1), (x, 2), …, (x,n)]]/d \text{ K.E}]j]]$$

And others.

We can likewise take higher-order derivatives of the above function with respect to any variable used as a first variable of derivation. Some of these higher-order derivatives will be equal to zero, and many others will require several orders of differentiation to return a value of zero. However, some infinite order derivatives may return nonzero values.

Integration of the above derivatives can, in a sense, provide measures of distances traveled and analogues thereof.

For example, we can take the following integrals:

$$\int [\Sigma(j = 1; j = \infty\uparrow):[[d[r[(x,1), (x, 2), ..., (x,n)]]/dt]j]] \; dtship$$

$$\int [\Sigma(j = 1; j = \infty\uparrow):[[d[r[(x,1), (x, 2), ..., (x,n)]]/dt]j]] \; d\gamma$$

$$\int [\Sigma(j = 1; j = \infty\uparrow):[[d[r[(x,1), (x, 2), ..., (x,n)]]/dt]j]] \; d \; acceleration \; ship$$

$$\int [\Sigma(j = 1; j = \infty\uparrow):[[d[r[(x,1), (x, 2), ..., (x,n)]]/dt]j]] \; dt \; KE$$

$$\int [\Sigma(j = 1; j = \infty\uparrow):[[d[r[(x,1), (x, 2), ..., (x,n)]]/d \; \gamma]j]] \; dtship$$

$$\int [\Sigma(j = 1; j = \infty\uparrow):[[d[r[(x,1), (x, 2), ..., (x,n)]]/d \; \gamma \;]j]] \; d\gamma$$

$$\int [\Sigma(j = 1; j = \infty\uparrow):[[d[r[(x,1), (x, 2), ..., (x,n)]]/d \; \gamma \;]j]] \; d \; acceleration \; ship$$

$$\int [\Sigma(j = 1; j = \infty\uparrow):[[d[r[(x,1), (x, 2), ..., (x,n)]]/d \; \gamma]j]] \; dt \; KE$$

$$\int [\Sigma(j = 1; j = \infty\uparrow):[[d[r[(x,1), (x, 2), ..., (x,n)]]/d \; acceleration]j]] \; dtship$$

$$\int [\Sigma(j = 1; j = \infty\uparrow):[[d[r[(x,1), (x, 2), ..., (x,n)]]/d \; acceleration]j]] \; d\gamma$$

$$\int [\Sigma(j = 1; j = \infty\uparrow):[[d[r[(x,1), (x, 2), ..., (x,n)]]/d \; acceleration]j]] \; d \; acceleration \; ship$$

$$\int [\Sigma(j = 1; j = \infty\uparrow):[[d[r[(x,1), (x, 2), ..., (x,n)]]/d \; acceleration]j]] \; dt \; KE$$

$$\int [\Sigma(j = 1; j = \infty\uparrow):[[d[r[(x,1), (x, 2), ..., (x,n)]]/d \; KE]j]] \; dtship$$

$$\int [\Sigma(j = 1; j = \infty\uparrow):[[d[r[(x,1), (x, 2), ..., (x,n)]]/d \; KE \;]j]] \; d\gamma$$

$$\int [\Sigma(j = 1; j = \infty\uparrow):[[d[r[(x,1), (x, 2), ..., (x,n)]]/d \; KE \;]j]] \; d \; acceleration \; ship$$

$$\int [\Sigma(j = 1; j = \infty\uparrow):[[d[r[(x,1), (x, 2), ..., (x,n)]]/d \; ke]j]] \; dt \; KE$$

We can take higher-order derivatives accordingly.

We may affix the following operator to formulas for gamma and related parameters to denote the above range of possibilities in derivatives and integrals:

$\{w\{\{\{[\Sigma(j = 1; j = \infty\uparrow):[[d[r[(x,1), (x, 2), ..., (x,n)]]/dt]j]] \leq$ (realm specific value of c)$\};\{[\Sigma(j = 1; j = \infty\uparrow):[[d[r[(x,1), (x, 2), ..., (x,n)]]/d\gamma]j]]\};\{[\Sigma(j = 1; j = \infty\uparrow):[[d[r[(x,1), (x, 2), ..., (x,n)]]/d$ ship-frame acceleration$]j]]\};\{[\Sigma(j = 1; j = \infty\uparrow):[[d[r[(x,1), (x, 2), ..., (x,n)]]/d \; K.E]j]]\};\{\int [\Sigma(j = 1; j = \infty\uparrow):[[d[r[(x,1), (x, 2), ..., (x,n)]]/dt]j]] \; dtship\}; \; \{\int [\Sigma(j = 1; j = \infty\uparrow):[[d[r[(x,1), (x, 2), ..., (x,n)]]/dt]j]] \; d\gamma\}; \; \{\int [\Sigma(j$

= 1; j = ∞↑):[[d[r[(x,1), (x, 2), ..., (x,n)]]/dt]j]] d acceleration ship}; {∫ [Σ(j = 1; j = ∞↑):[[d[r[(x,1), (x, 2), ..., (x,n)]]/dt]j]] dt KE};{∫ [Σ(j = 1; j = ∞↑):[[d[r[(x,1), (x, 2), ..., (x,n)]]/d γ]j]] dtship}; {∫ [Σ(j = 1; j = ∞↑):[[d[r[(x,1), (x, 2), ..., (x,n)]]/d γ]j]] dγ}; {∫ [Σ(j = 1; j = ∞↑):[[d[r[(x,1), (x, 2), ..., (x,n)]]/d γ]j]] d acceleration ship}; {∫ [Σ(j = 1; j = ∞↑):[[d[r[(x,1), (x, 2), ..., (x,n)]]/d γ]j]] dt KE};{∫ [Σ(j = 1; j = ∞↑):[[d[r[(x,1), (x, 2), ..., (x,n)]]/d acceleration]j]] dtship}; {∫ [Σ(j = 1; j = ∞↑):[[d[r[(x,1), (x, 2), ..., (x,n)]]/d acceleration]j]] dγ}; {∫ [Σ(j = 1; j = ∞↑):[[d[r[(x,1), (x, 2), ..., (x,n)]]/d acceleration]j]] d acceleration ship}; {∫ [Σ(j = 1; j = ∞↑):[[d[r[(x,1), (x, 2), ..., (x,n)]]/d acceleration]j]] dt KE};{∫ [Σ(j = 1; j = ∞↑):[[d[r[(x,1), (x, 2), ..., (x,n)]]/d KE]j]] dtship}; {∫ [Σ(j = 1; j = ∞↑):[[d[r[(x,1), (x, 2), ..., (x,n)]]/d KE]j]] dγ}; {∫ [Σ(j = 1; j = ∞↑):[[d[r[(x,1), (x, 2), ..., (x,n)]]/d KE]j]] d acceleration ship}; {∫ [Σ(j = 1; j = ∞↑):[[d[r[(x,1), (x, 2), ..., (x,n)]]/d ke]j]] dt KE}} as well as arbitrarily higher-order derivatives and integrals with regard to respective or other variables}}

Each of the following roster elements can be the basis of expressions of differentiated functions for power, energy, velocity, momentum, gamma, and the like by differentiation in the following manners:

d f/dt,1   d(d f/dt,1)/dt,2   d(d f/dt,2)/dt,1   d[d(d f/dt,1)/dt,2]/dt,3   d[d(d f/dt,1)/dt,3]/dt,2

d[d(d f/dt,2)/dt,1]/dt,3   d[d(d f/dt,2)/dt,3]/dt,1   d[d(d f/dt,3)/dt,1]/dt,2   d[d(d f/dt,3)/dt,2]/dt,1

d f/dT,1   d(d f/dT,1)/dT,2   d(d f/dT,2)/dT,1   d[d(d f/dT,1)/dT,2]/dT,3   d[d(d f/dT,1)/dT,3]/dT,2

d[d(d f/dT,2)/dT,1]/dT,3   d[d(d f/dT,2)/dT,3]/dT,1   d[d(d f/dT,3)/dT,1]/dT,2   d[d(d f/dT,3)/dT,2]/dT,1

T in each case is a timelike blend of various sorts of multiple time dimensions that appropriately and usefully enables derivation in one iteration.

We can go further to consider the following:

d f/dꞆ,1   d(d f/dꞆ,1)/dꞆ,2   d(d f/dꞆ,2)/dꞆ,1   d[d(d f/dꞆ,1)/dꞆ,2]/dꞆ,3   d[d(d f/dꞆ,1)/dꞆ,3]`/dꞆ,2

d[d(d f/dꞆ,2)/dꞆ,1]/dꞆ,3   d[d(d f/dꞆ,2)/dꞆ,3]/dꞆ,1   d[d(d f/dꞆ,3)/dꞆ,1]/dꞆ,2   d[d(d f/dꞆ,3)/dꞆ,2]/dꞆ,1

Ꞇ in each case is a timelike blend of timelike blends of various sorts of multiple time dimensions that appropriately and usefully enables derivation in one iteration.

By the same token, we can also integrate functions for power, energy, velocity, momentum, gamma, and the like with respect to multiple time dimensions such as following:

$\int f(dt,1) \int \int d\ f(dt,1)(dt,2) \int \int d\ f(dt,2)(dt,1) \iiint f(dt,1)(dt,2)(dt,3) \iiint f(dt,1)(dt,3)(dt,2)$

$\iiint f(dt,2)(dt,1)(dt,3) \iiint f(dt,2)(dt,3)(dt,1) \iiint f(dt,3)(dt,1)(dt,2) \iiint f(dt,3)(dt,2)(dt,1)$

$\int f(dT,1) \int\int f(dT,1) (dT,2) \int\int f(dT,2) ((dT,1)) \iiint f(dT,1) (dT,2) (dT,3) \iiint f(dT,1)$ $(dT,3) (dT,2)$

$\iiint f(dT,2) (dT,1) (dT,3) \iiint f(dT,2) (dT,3) (dT,1) \iiint f(dT,3) (dT,1) (dT,2) \iiint f(dT,3)$ $(dT,2) (dT,1)$

We can go further to consider the following:

$\int f(d\underline{T},1) \int\int f(d\underline{T},1)(d\underline{T},2) \int\int f(d\underline{T},2)(d\underline{T},1) \int\int\int f(d\underline{T},1)(d\underline{T},2)(d\underline{T},3) \int\int\int$ $f(d\underline{T},1)(d\underline{T},3)(d\underline{T},2)$

$\int\int\int f(d\underline{T},2)(d\underline{T},1)(d\underline{T},3) \int\int\int f(d\underline{T},2)(d\underline{T},3)(d\underline{T},1) \int\int\int f(d\underline{T},3)(d\underline{T},1)(d\underline{T},2) \int\int\int$ $f(d\underline{T},3)(d\underline{T},2)(d\underline{T},1)$

We can continue to yet higher-order derivatives and integrals. Suffice it to say that the general patterns continue for higher numbers of dimensions.

We can also consider higher-order blends of time and thus are not limited to the blends defined by T and $\underline{T}$. We chose these two symbols for their resemblance to the letter t.

As for hyperspatial applications, we consider the following hyperspaces:

Such hyperspaces can be *N-Time-M-Space*, where N is an integer greater than or equal to *3* and *M* is any counting number.

Alternatively, N can be any rational number equal to *3* or greater, and M is any rational number greater than or equal to *1*.

As another set of scenarios, N can be any irrational number equal to *3* or greater, and *M* is any irrational number greater than or equal to *1*.

Alternatively, N can be any rational number greater than *0*, and M is any rational number greater than *0*.

As another set of scenarios, N can be any irrational number greater than *0*, and *M* is any irrational number greater than *0*.

The *N-Space-M-Time* may optionally be as follows:

flat, positively curved, negatively curved, positively curved and torsioned at one or more scales in arbitrary patterns including, but not limited to, fractals, negatively curved and torsioned at one or more scales in arbitrary patterns including, but not limited to, fractals, positively super-curved, negatively super-curved, positively super-curved and torsioned at one or more scales in arbitrary patterns including but not limited to fractals, negatively super-curved and torsioned at one or more scales in arbitrary patterns including, but not limited to, fractals, positively super-curved and super-torsioned at one or more scales in arbitrary patterns including, but not limited to, fractals, negatively super-curved and super-torsioned at one or more scales in arbitrary patterns including, but not limited to, fractals, positively curved and positively torsioned at one or more scales in arbitrary patterns including, but not limited to, fractals, negatively curved and positively torsioned at one or more scales in arbitrary patterns including, but not limited to, fractals, positively super-curved, negatively super-curved, positively super-curved and positively torsioned at one or more scales in arbitrary patterns including, but not limited to, fractals, negatively super-curved and positively torsioned at one or more scales in arbitrary patterns including, but not limited to, fractals, positively super-curved and positively super-torsioned at one or more scales in arbitrary patterns including, but not limited to, fractals, negatively super-curved and positively super-torsioned at one or more scales in arbitrary patterns including, but not limited to fractals, positively curved and negatively torsioned at one or more scales in arbitrary patterns , but not limited to, fractals, negatively curved and negatively torsioned at one or more scales in arbitrary patterns including, but not limited to, fractals, positively super-curved, negatively super-curved, positively super-curved and negatively torsioned at one or more scales in arbitrary patterns including, but not limited to, fractals, negatively super-curved and negatively torsioned at one or more scales in arbitrary patterns including, but not limited to, fractals, positively super-curved and negatively super-torsioned at one or more scales in arbitrary patterns including, but not limited to, fractals, negatively super-curved and negatively super-torsioned at one or more scales in arbitrary patterns including, but not limited to, fractals, positively super-...-super-curved, negatively super-...-super-curved, positively super-...-super-curved and torsioned at one or more scales in arbitrary patterns including, but not limited to, fractals, negatively super-...-super-curved and torsioned at one or more scales in arbitrary patterns including, but not limited to, fractals, positively super-...-super-curved and super-...-super-torsioned at one or more scales in arbitrary patterns including, but not limited to, fractals, negatively super-...-super-curved and super-...-super-torsioned at one or more scales in arbitrary patterns including, but not limited to, fractals, positively curved and positively torsioned at one or more scales in arbitrary patterns including, but not limited to, fractals, negatively curved and positively torsioned at one or more scales in arbitrary patterns including, but not limited to, fractals, positively super-...-super-curved, negatively super-...-super-curved, positively super-...-super-curved and positively torsioned at one or more scales in arbitrary patterns including, but not limited to, fractals, negatively super-...-super-curved and positively torsioned at one or more scales in arbitrary patterns including, but not limited to, fractals, positively super-...-super-curved and positively super-...-super-torsioned at one or more scales in arbitrary patterns including, but not limited to, fractals, negatively super-...-super-curved and positively super-...-super-torsioned at one or more scales in arbitrary patterns including, but not limited to, fractals, positively curved and negatively torsioned at one or more scales in arbitrary patterns including, but not limited to, fractals, negatively curved and negatively torsioned at one or more scales in arbitrary patterns including, but not limited to, fractals, positively super-...-super-curved, negatively super-...-super-curved, positively super-...-super-curved and negatively torsioned at one or more scales in arbitrary patterns including, but not limited to, fractals, negatively super-...-super-curved and negatively torsioned at one or more scales in arbitrary patterns including, but not limited to, fractals, positively super-...-super-curved and negatively super-...-super-torsioned at one or more scales in arbitrary patterns including, but not limited to, fractals, negatively super-...-super-curved and negatively super-...-super-torsioned at one or more scales in arbitrary patterns including, but not limited to fractals

All these formulations assume distance traveled through space irrespective of additional recessionary velocities that may accrue in expanding universes or higher realms including, but not limited to, hyperspaces. Recessional velocities can exceed c with respect to a location of origin but such velocities do not accrue without space-time expansion.

So, quite amazing travel itineraries are possible with infinite distances traversable in finite ship-time, as well as infinite future time travel in finite ship-time.

The following limits may apply:

Lim light-speed travel $\rightarrow$ 1st level hidden extensities manifest for travel into.

Δ gamma $\rightarrow$ (Aleph 0) EXP (1/n), where n is finite and greater than 1.

Lim light-speed travel $\rightarrow$ 2nd level hidden extensities manifest for travel into.

Δ gamma $\rightarrow$ (Aleph 0) EXP n, where n is finite and greater than 1.

Lim light-speed travel $\rightarrow$ 3rd level hidden extensities manifest for travel into.

Δ gamma $\rightarrow$ (Aleph 1) EXP (1/n), where n is finite and greater than 1.

Lim light-speed travel $\rightarrow$ 4th level hidden extensities manifest for travel into.

Δ gamma $\rightarrow$ (Aleph 1) EXP n, where n is finite and greater than 1.

Lim light-speed travel $\rightarrow$ 5th level hidden extensities manifest for travel into.

Δ gamma $\rightarrow$ (Aleph 2) EXP (1/n), where n is finite and greater than 1.

Lim light-speed travel $\rightarrow$ 6th level hidden extensities manifest for travel into.

Δ gamma $\rightarrow$ (Aleph 2) EXP n, where n is finite and greater than 1.

Lim light-speed travel $\rightarrow$ 7th level hidden extensities manifest for travel into.

Δ gamma $\rightarrow$ (Aleph 3) EXP (1/n), where n is finite and greater than 1.

Lim light-speed travel $\rightarrow$ 8th level hidden extensities manifest for travel into.

Δ gamma $\rightarrow$ (Aleph 3) EXP n, where n is finite and greater than 1.

…

…

…

Lim light-speed travel $\rightarrow$ [(Aleph 0) − 1]th level hidden extensities manifest for travel into.

Δ gamma $\rightarrow$ [Aleph (Aleph 0)] EXP (1/n), where n is finite and greater than 1.

Lim light-speed travel $\rightarrow$ [(Aleph 0) − 1]th level hidden extensities manifest for travel into.

Δ gamma $\rightarrow$ [Aleph (Aleph 0)] EXP n, where n is finite and greater than 1.

And we continue from here without end.

The following limits may apply:

Lim light-speed travel $\rightarrow$ 1st level super-aberration manifests super-orthonormal spacecraft placement, velocity, acceleration, and ship-time arrow vectors.

Δ gamma $\rightarrow$ (Aleph 0) EXP (1/n), where n is finite and greater than 1.

Lim light-speed travel $\rightarrow$ 2nd level super-aberration manifests super-orthonormal spacecraft placement, velocity, acceleration, and ship-time arrow vectors.

Δ gamma → (Aleph 0) EXP n, where n is finite and greater than 1.

Lim light-speed travel → 3rd level super-aberration manifests super-orthonormal spacecraft placement, velocity, acceleration, and ship-time arrow vectors.

Δ gamma → (Aleph 1) EXP (1/n), where n is finite and greater than 1.

Lim light-speed travel → 4th level super-aberration manifests super-orthonormal spacecraft placement, velocity, acceleration, and ship-time arrow vectors.

Δ gamma → (Aleph 1) EXP n, where n is finite and greater than 1.

Lim light-speed travel → 5th level super-aberration manifests super-orthonormal spacecraft placement, velocity, acceleration, and ship-time arrow vectors.

Δ gamma → (Aleph 2) EXP (1/n), where n is finite and greater than 1.

Lim light-speed travel → 6th level super-aberration manifests super-orthonormal spacecraft placement, velocity, acceleration, and ship-time arrow vectors.

Δ gamma → (Aleph 2) EXP n, where n is finite and greater than 1.

Lim light-speed travel → 7th level super-aberration manifests super-orthonormal spacecraft placement, velocity, acceleration, and ship-time arrow vectors.

Δ gamma → (Aleph 3) EXP (1/n), where n is finite and greater than 1.

Lim light-speed travel $\rightarrow$ 8th level super-aberration manifests super-orthonormal spacecraft placement, velocity, acceleration, and ship-time arrow vectors.

Δ gamma $\rightarrow$ (Aleph 3) EXP n, where n is finite and greater than 1.

…

…

…

Lim light-speed travel $\rightarrow$ [(Aleph 0) − 1]th level super-aberration manifests super-orthonormal spacecraft placement, velocity, acceleration, and ship-time arrow vectors.

Δ gamma $\rightarrow$ [Aleph (Aleph 0)] EXP (1/n), where n is finite and greater than 1.

Lim light-speed travel $\rightarrow$ [(Aleph 0) − 1]th level super-aberration manifests super-orthonormal spacecraft placement, velocity, acceleration, and ship-time arrow vectors.

Δ gamma $\rightarrow$ [Aleph (Aleph 0)] EXP n, where n is finite and greater than 1.

And we continue from here without end.

The following limits may apply:

Lim [f(light-speed)] travel in hyperspace $\rightarrow$ 1st level hidden extensities manifest for travel into.

Δ gamma $\rightarrow$ (Aleph 0) EXP (1/n), where n is finite and greater than 1.

Lim [f(light-speed)] travel in hyperspace → 2nd level hidden extensities manifest for travel into.

$$\Delta \text{ gamma} \to (\text{Aleph } 0) \text{ EXP } n,$$ where n is finite and greater than 1.

Lim [f(light-speed)] travel in hyperspace → 3rd level hidden extensities manifest for travel into.

$$\Delta \text{ gamma} \to (\text{Aleph } 1) \text{ EXP } (1/n),$$ where n is finite and greater than 1.

Lim [f(light-speed)] travel in hyperspace → 4th level hidden extensities manifest for travel into.

$$\Delta \text{ gamma} \to (\text{Aleph } 1) \text{ EXP } n,$$ where n is finite and greater than 1.

Lim [f(light-speed)] travel in hyperspace → 5th level hidden extensities manifest for travel into.

$$\Delta \text{ gamma} \to (\text{Aleph } 2) \text{ EXP } (1/n),$$ where n is finite and greater than 1.

Lim [f(light-speed)] travel in hyperspace → 6th level hidden extensities manifest for travel into.

$$\Delta \text{ gamma} \to (\text{Aleph } 2) \text{ EXP } n,$$ where n is finite and greater than 1.

Lim [f(light-speed)] travel in hyperspace → 7th level hidden extensities manifest for travel into.

Δ gamma $\rightarrow$ (Aleph 3) EXP (1/n), where n is finite and greater than 1.

Lim [f(light-speed)] travel in hyperspace $\rightarrow$ 8th level hidden extensities manifest for travel into.

Δ gamma $\rightarrow$ (Aleph 3) EXP n, where n is finite and greater than 1.

…

…

…

Lim [f(light-speed)] travel in hyperspace $\rightarrow$ [(Aleph 0) − 1]th level hidden extensities manifest for travel into.

Δ gamma $\rightarrow$ [Aleph (Aleph 0)] EXP (1/n), where n is finite and greater than 1.

Lim [f(light-speed)] travel in hyperspace $\rightarrow$ [(Aleph 0) − 1]th level hidden extensities manifest for travel into.

Δ gamma $\rightarrow$ [Aleph (Aleph 0)] EXP n, where n is finite and greater than 1.

And we continue from here without end.

The following limits may apply:

Lim [f(light-speed)] travel in hyperspace $\rightarrow$ 1st level super-aberration manifests super-orthonormal spacecraft placement, velocity, acceleration, and ship-time arrow vectors.

Δ gamma → (Aleph 0) EXP (1/n), where *n* is finite and greater than 1.

Lim [f(light-speed)] travel in hyperspace → 2nd level super-aberration manifests super-orthonormal spacecraft placement, velocity, acceleration, and ship-time arrow vectors.

Δ gamma → (Aleph 0) EXP n, where *n* is finite and greater than 1.

Lim [f(light-speed)] travel in hyperspace → 3rd level super-aberration manifests super-orthonormal spacecraft placement, velocity, acceleration, and ship-time arrow vectors.

Δ gamma → (Aleph 1) EXP (1/n), where *n* is finite and greater than *1*.

Lim [f(light-speed)] travel in hyperspace → 4th level super-aberration manifests super-orthonormal spacecraft placement, velocity, acceleration, and ship-time arrow vectors.

Δ gamma → (Aleph 1) EXP n, where *n* is finite and greater than 1.

Lim [f(light-speed)] travel in hyperspace → 5th level super-aberration manifests super-orthonormal spacecraft placement, velocity, acceleration, and ship-time arrow vectors.

Δ gamma → (Aleph 2) EXP (1/n), where *n* is finite and greater than 1.

Lim [f(light-speed)] travel in hyperspace → 6th level super-aberration manifests super-orthonormal spacecraft placement, velocity, acceleration, and ship-time arrow vectors.

Δ gamma → (Aleph 2) EXP n, where *n* is finite and greater than 1.

Lim [f(light-speed)] travel in hyperspace → 7th level super-aberration manifests super-orthonormal spacecraft placement, velocity, acceleration, and ship-time arrow vectors.

Δ gamma → (Aleph 3) EXP (1/n), where *n* is finite and greater than 1.

Lim [f(light-speed)] travel in hyperspace → 8th level super-aberration manifests super-orthonormal spacecraft placement, velocity, acceleration, and ship-time arrow vectors.

Δ gamma → (Aleph 3) EXP n, where *n* is finite and greater than 1.

...

...

...

Lim [f(light-speed)] travel in hyperspace → [(Aleph 0) − 1]th level super-aberration manifests super-orthonormal spacecraft placement, velocity, acceleration, and ship-time arrow vectors.

Δ gamma → [Aleph (Aleph 0)] EXP (1/n), where *n* is finite and greater than 1.

Lim [f(light-speed)] travel in hyperspace → [(Aleph 0) − 1]th level super-aberration manifests super-orthonormal spacecraft placement, velocity, acceleration, and ship-time arrow vectors.

Δ gamma → [Aleph (Aleph 0)] EXP n, where *n* is finite and greater than 1.

And we continue from here without end.

The following limits may apply:

Lim [g(light-speed travel)] in scalar fields → 1st level hidden extensities manifest for travel into.

Δ gamma → (Aleph 0) EXP (1/n), where n is finite and greater than 1.

Lim [g(light-speed travel)] in scalar fields → 2nd level hidden extensities manifest for travel into.

Δ gamma → (Aleph 0) EXP n, where n is finite and greater than 1.

Lim [g(light-speed travel)] in scalar fields → 3rd level hidden extensities manifest for travel into.

Δ gamma → (Aleph 1) EXP (1/n), where n is finite and greater than 1.

Lim [g(light-speed travel)] in scalar fields → 4th level hidden extensities manifest for travel into.

Δ gamma → (Aleph 1) EXP n, where n is finite and greater than 1.

Lim [g(light-speed travel)] in scalar fields → 5th level hidden extensities manifest for travel into.

Δ gamma → (Aleph 2) EXP (1/n), where n is finite and greater than 1.

Lim [g(light-speed travel)] in scalar fields → 6th level hidden extensities manifest for travel into.

Δ gamma → (Aleph 2) EXP n, where n is finite and greater than 1.

Lim [g(light-speed travel)] in scalar fields → 7th level hidden extensities manifest for travel into.

Δ gamma → (Aleph 3) EXP (1/n), where n is finite and greater than 1.

Lim [g(light-speed travel)] in scalar fields → 8th level hidden extensities manifest for travel into.

Δ gamma → (Aleph 3) EXP n, where n is finite and greater than 1.

…

…

…

Lim [g(light-speed travel)] in scalar fields → [(Aleph 0) − 1]th level hidden extensities manifest for travel into.

Δ gamma → [Aleph (Aleph 0)] EXP (1/n), where n is finite and greater than 1.

Lim [g(light-speed travel)] in scalar fields → [(Aleph 0) − 1]th level hidden extensities manifest for travel into.

Δ gamma → [Aleph (Aleph 0)] EXP n, where n is finite and greater than 1.

And we continue from here without end.

The following limits may apply:

Lim [g(light-speed travel)] in scalar fields → 1st level super-aberration manifests super-orthonormal spacecraft placement, velocity, acceleration, and ship-time arrow vectors.

Δ gamma → (Aleph 0) EXP (1/n), where n is finite and greater than 1.

Lim [g(light-speed travel)] in scalar fields → 2nd level super-aberration manifests super-orthonormal spacecraft placement, velocity, acceleration, and ship-time arrow vectors.

Δ gamma → (Aleph 0) EXP n, where n is finite and greater than 1.

Lim [g(light-speed travel)] in scalar fields → 3rd level super-aberration manifests super-orthonormal spacecraft placement, velocity, acceleration, and ship-time arrow vectors.

Δ gamma → (Aleph 1) EXP (1/n), where n is finite and greater than 1.

Lim [g(light-speed travel)] in scalar fields → 4th level super-aberration manifests super-orthonormal spacecraft placement, velocity, acceleration, and ship-time arrow vectors.

Δ gamma → (Aleph 1) EXP n, where n is finite and greater than 1.

Lim [g(light-speed travel)] in scalar fields → 5th level super-aberration manifests super-orthonormal spacecraft placement, velocity, acceleration, and ship-time arrow vectors.

Δ gamma → (Aleph 2) EXP (1/n), where n is finite and greater than 1.

Lim [g(light-speed travel)] in scalar fields → 6th level super-aberration manifests super-orthonormal spacecraft placement, velocity, acceleration, and ship-time arrow vectors.

Δ gamma → (Aleph 2) EXP n, where n is finite and greater than 1.

Lim [g(light-speed travel)] in scalar fields → 7th level super-aberration manifests super-orthonormal spacecraft placement, velocity, acceleration, and ship-time arrow vectors.

Δ gamma → (Aleph 3) EXP (1/n), where n is finite and greater than 1.

Lim [g(light-speed travel)] in scalar fields → 8th level super-aberration manifests super-orthonormal spacecraft placement, velocity, acceleration, and ship-time arrow vectors.

Δ gamma → (Aleph 3) EXP n, where n is finite and greater than 1.

...

...

...

Lim [g(light-speed travel)] in scalar fields → [(Aleph 0) − 1]th level super-aberration manifests super-orthonormal spacecraft placement, velocity, acceleration, and ship-time arrow vectors.

Δ gamma → [Aleph (Aleph 0)] EXP (1/n), where n is finite and greater than 1.

Lim [g(light-speed travel)] in scalar fields → [(Aleph 0) − 1]th level super-aberration manifests super-orthonormal spacecraft placement, velocity, acceleration, and ship-time arrow vectors.

Δ gamma → [Aleph (Aleph 0)] EXP n, where *n* is finite and greater than 1.

And we continue from here without end.

Herein, we assume Lorentz factors apply in hyperspaces and in scalar fields.

Each of the following roster elements can be the basis of expressions of differentiated functions for power, energy, velocity, momentum, gamma, and the like by differentiation in the following manners:

d f/dt,1 d(d f/dt,1)/dt,2 d(d f/dt,2)/dt,1 d[d(d f/dt,1)/dt,2]/dt,3 d[d(d f/dt,1)/dt,3]/dt,2

d[d(d f/dt,2)/dt,1]/dt,3 d[d(d f/dt,2)/dt,3]/dt,1 d[d(d f/dt,3)/dt,1]/dt,2 d[d(d f/dt,3)/dt,2]/dt,1

d f/dT,1 d(d f/dT,1)/dT,2 d(d f/dT,2)/dT,1 d[d(d f/dT,1)/dT,2]/dT,3 d[d(d f/dT,1)/dT,3]/dT,2

d[d(d f/dT,2)/dT,1]/dT,3 d[d(d f/dT,2)/dT,3]/dT,1 d[d(d f/dT,3)/dT,1]/dT,2 d[d(d f/dT,3)/dT,2]/dT,1

T in each case is a timelike blend of various sorts of multiple time dimensions that appropriately and usefully enables derivation in one iteration.

We can go further to consider the following:

d f/dT,1 d(d f/dT,1)/dT,2 d(d f/dT,2)/dT,1 d[d(d f/dT,1)/dT,2]/dT,3 d[d(d f/dT,1)/dT,3]`/dT,2

d[d(d f/dT,2)/dT,1]/dT,3 d[d(d f/dT,2)/dT,3]/dT,1 d[d(d f/dT,3)/dT,1]/dT,2 d[d(d f/dT,3)/dT,2]/dT,1

T in each case is a timelike blend of timelike blends of various sorts of multiple time dimensions that appropriately and usefully enables derivation in one iteration.

By the same token, we can also integrate functions for power, energy, velocity, momentum, gamma, and the like with respect to multiple time dimensions as follows.

∫ f(dt,1) ∫ ∫d f(dt,1)(dt,2) ∫ ∫d f(dt,2)(dt,1) ∭ f(dt,1)(dt,2)(dt,3) ∭ f(dt,1)(dt,3)(dt,2)

∭ f(dt,2)(dt,1)(dt,3) ∭ f(dt,2)(dt,3)(dt,1) ∭ f(dt,3)(dt,1)(dt,2) ∭ f(dt,3)(dt,2)(dt,1)

∫ f(dT,1) ∬ f(dT,1) (dT,2) ∬ f(dT,2) ((dT,1)) ∭ f(dT,1) (dT,2) (dT,3) ∭ f(dT,1) (dT,3) (dT,2)

∭ f(dT,2) (dT,1) (dT,3) ∭ f(dT,2) (dT,3) (dT,1) ∭ f(dT,3) (dT,1) (dT,2) ∭ f(dT,3) (dT,2) (dT,1)

We can go further to consider the following:

$$\int f(d\mathcal{T},1) \quad \int \int f(d\mathcal{T},1)(d\mathcal{T},2) \quad \int \int f(d\mathcal{T},2)(d\mathcal{T},1) \quad \int \int \int f(d\mathcal{T},1)(d\mathcal{T},2)(d\mathcal{T},3) \quad \int \int \int f(d\mathcal{T},1)(d\mathcal{T},3)(d\mathcal{T},2)$$

$$\int \int \int f(d\mathcal{T},2)(d\mathcal{T},1)(d\mathcal{T},3) \quad \int \int \int f(d\mathcal{T},2)(d\mathcal{T},3)(d\mathcal{T},1) \quad \int \int \int f(d\mathcal{T},3)(d\mathcal{T},1)(d\mathcal{T},2) \quad \int \int \int f(d\mathcal{T},3)(d\mathcal{T},2)(d\mathcal{T},1)$$

We can continue to yet higher-order derivatives and integrals. Suffice it to say that the general patterns continue for higher numbers of dimensions.

We can also consider higher-order blends of time and thus are not limited to the blends defined by T and $\mathcal{T}$. We chose these two symbols for their resemblance to the letter t.

For cases where the invariant mass of the spacecraft would be reduced perhaps by evacuating or screening the Higgs field in proximity or location of the spacecraft, the spacecraft Lorentz factor would undergo an increase in Lorentz factor by a superunitary multiple of approximately

[(original non-reduced inertial mass)/(reduced inertial mass)]

The above relation would become precise in the limits of infinite invariant mass reduction, that is, where the mass of the spacecraft would be reduced to (1/said implied infinity) of the spacecraft's initial mass, all other things remaining the same.

Now, a spacecraft having achieved true light-speed presumably would have an infinite Lorentz factor. Thus, the entire universe of travel would be relativistically aberrated to be located directly in front of the spacecraft.

Accordingly, the extension of a least transfinite ordinal light-years–wide universe of travel would appear relativistically contracted to a scale of only one light-year in the spacecraft reference frame time.

As a consequence, the spacecraft would be pulled forward by general relativistic gravitational effects or perhaps quantum gravitational effects with infinite ship-frame acceleration.

For universes having an extensity of Aleph 1 light-year, there are prospects for the spacecraft to attain a Lorentz factor of Aleph 1 to thus result in the universe of travel being contracted to point directly in front of the spacecraft and extending only one light-year ahead of the spacecraft. Note that in the above computation, we assume that only the material in the path of a reasonably sized spacecraft is upswept as mass-energy fuel or species of reaction and that the spacecraft makes only one effective pass through the universe.

If such an Aleph 1 light-year–width universe has three spatial dimensions and is substantially static, then a spacecraft traveling at light-speed in the universe may obtain a Lorentz factor of roughly (Aleph 1) EXP 3. Note that in the above computation, we assume that only the material in the path of a reasonably sized spacecraft is upswept as mass-energy fuel or species of reaction and that the spacecraft makes an effective pass through the entire universe. Thus, all material in the said universe of travel is reacted against with assumed finite universal mass-energy density.

If the latter universe is expanding with time over periods much greater than Aleph 1 years, then the spacecraft may attain a Lorentz factor of roughly [k(Aleph 1)] EXP 3, where k ranges from a small superunitary real number to suitable infinite real numbers. This, of course, assumes that real positive mass-energy stocks continue being created, perhaps out of zero-point fields or analogues thereof from some forms of dark energy. Provided that k ranges from Aleph 0 to Aleph 3, then the spacecraft Lorentz factor will be limited to a range of roughly from [(Aleph 0)(Aleph 1)] EXP 3 to [(Aleph 3)(Aleph 1)] EXP 3. Note that in the above computations, we assume that only the material in the path of a reasonably sized spacecraft is upswept as mass-energy fuel or species of reaction and that the spacecraft makes effectively passes through the entire universe. Thus all material in the said universe of travel is reacted against with assumed finite universal mass-energy density.

For universes having an extensity of Aleph 2 light-years, there are prospects for the spacecraft to attain a Lorentz factor of roughly Aleph 2 to thus result in the universe of travel being contracted to point directly in front of the spacecraft and extending only one light-year ahead of the spacecraft. Note that in the above computations, we assume that only the material in the path of a reasonably sized spacecraft is upswept as mass-energy fuel or species of reaction and that the spacecraft makes only one effective pass through each universe.

If such an Aleph 2 light-year–width universe has three spatial dimensions and is substantially static, then a spacecraft traveling at light-speed in the universe may obtain a Lorentz factor of roughly (Aleph 2) EXP 3. Note that in the above computations, we assume that only the material in the path of a reasonably sized spacecraft is upswept as mass-energy fuel or species of reaction and that the spacecraft makes an effective pass through the entire universe. Thus, all material in the said universe of travel is reacted against with assumed finite universal mass-energy density.

If the latter universe is expanding with time over periods much greater than Aleph 2 years, then the spacecraft may attain a Lorentz factor of roughly [k(Aleph 2)] EXP 3, where *k* ranges from a small superunitary real number to suitable infinite real numbers. This, of course, assumes that real positive mass-energy stocks continue being created, perhaps out of zero-point fields or analogues thereof from some forms of dark energy. Provided that *k* ranges from Aleph 0 to Aleph 4, then the spacecraft Lorentz factor will be limited to a range of roughly from [(Aleph 0)(Aleph 2)] EXP 3 to [(Aleph 4)(Aleph 2)] EXP 3. Note that in the above computations, we assume that only the material in the path of a reasonably sized spacecraft is upswept as mass-energy fuel or species of reaction and that the spacecraft makes an effective pass through the entire universe. Thus, all material in the said universe of travel is reacted against with assumed finite universal mass-energy density.

For universes having an extensity of Aleph 3 light-years, there are prospects for the spacecraft to attain a Lorentz factor of roughly Aleph 3 to thus result in the universe of travel being contracted to point directly in front of the spacecraft and extending only one light-year ahead of the spacecraft. Note that in the above computations, we assume that only the material in the path of a reasonably sized spacecraft is upswept as mass-energy fuel or species of reaction and that the spacecraft makes only one effective pass through the universe.

If such an Aleph 3 light-year–width universe has three spatial dimensions and is substantially static, then a spacecraft traveling at light-speed in the universe may obtain a Lorentz factor of roughly (Aleph 3) EXP 3. Note that in the above computations, we assume that only the material in the path of a reasonably sized spacecraft is upswept as mass-energy fuel or species of reaction and that the spacecraft makes an effective pass through the entire universe. Thus, all material in said universe of travel is reacted against with assumed finite universal mass-energy density.

If the latter universe is expanding with time over periods much greater than Aleph 3 years, then the spacecraft may attain a Lorentz factor of roughly [k(Aleph 3)] EXP 3, where *k* ranges from a small superunitary real number to suitable infinite real numbers. This, of course, assumes that real positive mass-energy stocks continue being created, perhaps out of zero-point fields or analogues thereof from some forms of dark energy. Provided that *k* ranges from Aleph 0 to Aleph 5, then the spacecraft Lorentz factor will be limited to a range of roughly from [(Aleph 0)(Aleph 3)] EXP 3 to [(Aleph 5)(Aleph 3)] EXP 3. Note that in the above computations, we assume that only the material in the path of a reasonably sized spacecraft is upswept as mass-energy fuel or species of reaction and that the spacecraft makes an effective pass through the entire universe. Thus, all material in the said universe of travel is reacted against with assumed finite universal mass-energy density.

For universes having an extensity of Aleph 4 light-years, there are prospects for the spacecraft to attain a Lorentz factor of roughly Aleph 4 to thus result in the universe of travel being contracted to point directly in front of the spacecraft and extending only one light-year ahead of the spacecraft. Note that in the above computations, we assume that only the material in the path of a reasonably sized spacecraft is upswept as mass-energy fuel or species of reaction and that the spacecraft makes only one effective pass through the universe.

If such an Aleph 4 light-year–width universe has three spatial dimensions and is substantially static, then a spacecraft traveling at light-speed in the universe may obtain a Lorentz factor of roughly (Aleph 4) EXP 3. Note that in the above computations, we assume that only the material in the path of a reasonably sized spacecraft is upswept as mass-energy fuel or species of reaction and that the spacecraft makes an effective pass through the entire universe. Thus, all material in the said universe of travel is reacted against with assumed finite universal mass-energy density.

If the latter universe is expanding with time over periods much greater than Aleph 4 years, then the spacecraft may attain a Lorentz factor of roughly [k(Aleph 4)] EXP 3, where k ranges from a small superunitary real number to suitable infinite real numbers. This, of course, assumes that real positive mass-energy stocks continue being created, perhaps out of zero-point fields or analogues thereof from some forms of dark energy. Provided that k ranges from Aleph 0 to Aleph 6, then the spacecraft Lorentz factor will be limited to a range of roughly from [(Aleph 0)(Aleph 4)] EXP 3 to [(Aleph 6)(Aleph 4)] EXP 3. Note that in the above computations, we assume that only the material in the path of a reasonably sized spacecraft is upswept as mass-energy fuel or species of reaction and that the spacecraft makes an effective pass through the entire universe. Thus, all material in the said universe of travel is reacted against with assumed finite universal mass-energy density.

For universes having an extensity of Aleph G light-years, there are prospects for the spacecraft to attain a Lorentz factor of roughly Aleph G to thus result in the universe of travel being contracted to point directly in front of the spacecraft and extending only one light-year ahead of the spacecraft. Note that in the above computations, we assume that only the material in the path of a reasonably sized spacecraft is upswept as mass-energy fuel or species of reaction and that the spacecraft makes only one effective pass through the universe.

If such an Aleph G light-year–width universe has three spatial dimensions and is substantially static, then a spacecraft traveling at light-speed in the universe may obtain a Lorentz factor of roughly (Aleph G) EXP 3. Note that in the above computations, we assume that only the material in the path of a reasonably sized spacecraft is upswept as mass-energy fuel or species of reaction and that the spacecraft makes an effective pass through the entire universe. Thus, all material in the said universe of travel is reacted against with assumed finite universal mass-energy density.

If the latter universe is expanding with time over periods much greater than Aleph G years, then the spacecraft may attain a Lorentz factor of roughly [k(Aleph G)] EXP 3, where k ranges from a small superunitary real number to suitable infinite real numbers. This, of course, assumes that real positive mass-energy stocks continue being created, perhaps out of zero-point fields or analogues thereof from some forms of dark energy. Provided k ranges from Aleph 0 to Aleph (G + 2), then the spacecraft Lorentz factor will be limited to a range of roughly from [(Aleph 0)(Aleph G)] EXP 3 to [[Aleph (G + 2)](Aleph G)] EXP 3. Note that in the above computations, we assume that only the material in the path of a reasonably sized spacecraft is upswept as mass-energy fuel or species of reaction and that the spacecraft makes an effective pass through the entire universe. Thus, all material in the said universe of travel is reacted against with assumed finite universal mass-energy density.

Now we consider spacecraft travel in a multiverse of Aleph H universes in width of three dimensions, where H is an integer greater than G + 2, and where said multiverse contains [(Aleph H) EXP 3] universes.

For universes in said 3-D-space multiverse having an extensity of (Aleph 1) light-years, there are prospects for the spacecraft to attain a Lorentz factor of roughly [(Aleph H) EXP 3] (Aleph 1) to thus result in the multiverse of travel being contracted to point directly in front of the spacecraft and Lorentz contracted by a factor of roughly [(Aleph H) EXP 3] (Aleph 1) relative to the spacecraft. Note that in the above computations, we assume that only the material in the path of a reasonably sized spacecraft is upswept as mass-energy fuel or species of reaction and that the spacecraft makes only one effective pass through each universe.

If such Aleph 1 light-year–width universes have three spatial dimensions and are substantially static, then a spacecraft traveling at light-speed in the multiverse of consideration may obtain a Lorentz factor of roughly [(Aleph H) EXP 3] [(Aleph 1) EXP 3]. Note that in the above computations, we assume that only the material in the path of a reasonably sized spacecraft is upswept as mass-energy fuel or species of reaction and that the spacecraft makes an effective pass entirely through each universe. Thus, all material in said universes of travel is reacted against with assumed finite universal mass-energy density.

If the latter universes are expanding with time over periods much greater than Aleph 1 years, then the spacecraft may attain a Lorentz factor of roughly [(Aleph H) EXP 3] [[k(Aleph 1)] EXP 3], where k ranges from a small superunitary real number to suitable infinite real numbers. This, of course, assumes that real positive mass-energy stocks continue being created, perhaps out of zero-point fields or analogues thereof from some forms of dark energy. Provided that k ranges from Aleph 0 to Aleph 3, then the spacecraft Lorentz factor will be limited to a range of roughly from [(Aleph H) EXP 3] [[(Aleph 0)(Aleph 1)] EXP 3] to [(Aleph H) EXP 3] [[(Aleph 3)(Aleph 1)] EXP 3]. Note that in the above computations, we assume that only the material in the path of a reasonably sized spacecraft is upswept as mass-energy fuel or species of reaction and that the spacecraft makes an effective entire pass through each universe. Thus, all material in said universes of travel is reacted against with assumed finite universal mass-energy density.

For universes in said multiverse having an extensity of Aleph 2 light-years, there are prospects for the spacecraft to attain a Lorentz factor of roughly [(Aleph H) EXP 3] (Aleph 2) to thus result in the multiverse of travel being contracted to a point directly in front of the spacecraft and by a factor of [(Aleph H) EXP 3] (Aleph 2). Note that in the above computations, we assume that only the material in the path of a reasonably sized spacecraft is upswept as mass-energy fuel or species of reaction and that the spacecraft makes only one effective pass through each universe.

If such Aleph 2 light-year width universes have three spatial dimensions and are substantially static, then a spacecraft traveling at light-speed in the multiverse considered may obtain a Lorentz factor of roughly [(Aleph H) EXP 3] [(Aleph 2) EXP 3]. Note that in the above computations, we assume that only the material in the path of a reasonably sized spacecraft is upswept as mass-energy fuel or species of reaction and that the spacecraft makes an effective pass entirely through each universe. Thus, all material in said universes of travel is reacted against with assumed finite universal mass-energy density.

If the latter universes are expanding with time over periods much greater than Aleph 2 years, then the spacecraft may attain a Lorentz factor of roughly [(Aleph H) EXP 3] [[k(Aleph 2)] EXP 3], where k ranges from a small superunitary real number to suitable infinite real numbers. This, of course, assumes that real positive mass-energy stocks continue being created, perhaps out of zero-point fields or analogues thereof from some forms of dark energy. Provided that k ranges from Aleph 0 to Aleph 4, then the spacecraft Lorentz factor will be limited to a range of roughly from [(Aleph H) EXP 3] [[(Aleph 0)(Aleph 2)] EXP 3] to [(Aleph H) EXP 3] [[(Aleph 4)(Aleph 2)] EXP 3]. Note that in the above computations, we assume that only the material in the path of a reasonably sized spacecraft is upswept as mass-energy fuel or species of reaction and that the spacecraft makes an effective pass entirely through each universe. Thus, all material in said universes of travel is reacted against with assumed finite universal mass-energy density.

For universes in said multiverse having an extensity of Aleph 3 light-years, there are prospects for the spacecraft to attain a Lorentz factor of roughly [(Aleph H) EXP 3] (Aleph 3) to thus result in the multiverse of travel being contracted to point directly in front of the spacecraft and contracted by a factor of [(Aleph H) EXP 3](Aleph 3). Note that in the above computations, we assume that only the material in the path of a reasonably sized spacecraft is upswept as mass-energy fuel or species of reaction and that the spacecraft makes only one effective pass through each universe.

If such Aleph 3 light-year–width universes have three spatial dimensions and are substantially static, then a spacecraft traveling at light-speed in the multiverse of consideration may obtain a Lorentz factor of roughly [(Aleph H) EXP 3] [(Aleph 3) EXP 3]. Note that in the above computations, we assume that only the material in the path of a reasonably sized spacecraft is upswept as mass-energy fuel or species of reaction and that the spacecraft makes an effective pass entirely through each universe. Thus, all material in said universes of travel is reacted against with assumed finite universal mass-energy density.

If the latter universes are expanding with time over periods much greater than Aleph 3 years, then the spacecraft may attain a Lorentz factor of roughly [(Aleph H) EXP 3] [[k(Aleph 3)] EXP 3], where *k* ranges from a small superunitary real number to suitable infinite real numbers. This, of course, assumes that real positive mass-energy stocks continue being created, perhaps out of zero-point fields or analogues thereof from some forms of dark energy. Provided that *k* ranges from Aleph 0 to Aleph 5, then the spacecraft Lorentz factor will be limited to a range of roughly from [(Aleph H) EXP 3] [[(Aleph 0)(Aleph 3)] EXP 3] to [(Aleph H) EXP 3] [[(Aleph 5)(Aleph 3)] EXP 3]. Note that in the above computations, we assume that only the material in the path of a reasonably sized spacecraft is upswept as mass-energy fuel or species of reaction and that the spacecraft makes an effective pass entirely through each universe. Thus, all material in said universes of travel is reacted against with assumed finite universal mass-energy density.

For universes having an extensity of Aleph 4 light-years, there are prospects for the spacecraft to attain a Lorentz factor of roughly [(Aleph H) EXP 3] (Aleph 4) to thus result in the multiverse of travel being contracted to point directly in front of the spacecraft and being contracted relative to the spacecraft by roughly a factor of [(Aleph H) EXP 3](Aleph 4). Note that in the above computations, we assume that only the material in the path of a reasonably sized spacecraft is upswept as mass-energy fuel or species of reaction and that the spacecraft makes only one effective pass through each universe.

If such Aleph 4 light-year–width universes have three spatial dimensions and are substantially static, then a spacecraft traveling at light-speed in the multiverse may obtain a Lorentz factor of roughly [(Aleph H) EXP 3] [(Aleph 4) EXP 3]. Note that in the above computations, we assume that only the material in the path of a reasonably sized spacecraft is upswept as mass-energy fuel or species of reaction and that the spacecraft makes an effective pass entirely through each universe. Thus, all material in said universes of travel is reacted against with assumed finite universal mass-energy density.

If the latter universes are expanding with time over periods much greater than Aleph 4 years, then the spacecraft may attain a Lorentz factor of roughly [(Aleph H) EXP 3] [[k(Aleph 4)] EXP 3], where k ranges from a small superunitary real number to suitable infinite real numbers. This, of course, assumes that real positive mass-energy stocks continue being created, perhaps out of zero-point fields or analogues thereof from some forms of dark energy. Provided that k ranges roughly from Aleph 0 to Aleph 6, then the spacecraft Lorentz factor will be limited to a range from [(Aleph H) EXP 3] [[(Aleph 0)(Aleph 4)] EXP 3] to [(Aleph H) EXP 3] [[(Aleph 6)(Aleph 4)] EXP 3]. Note that in the above computations, we assume that only the material in the path of a reasonably sized spacecraft is upswept as mass-energy fuel or species of reaction and that the spacecraft makes an effective pass entirely through each universe. Thus, all material in said universes of travel is reacted against with assumed finite universal mass-energy density.

For universes having an extensity of Aleph G light-years, there are prospects for the spacecraft to attain a Lorentz factor of roughly [(Aleph H) EXP 3] (Aleph G) to thus result in the multiverse of travel being contracted to point directly in front of the spacecraft with a Lorentz contraction of roughly [(Aleph H) EXP 3] (Aleph G). Note that in the above computations, we assume that only the material in the path of a reasonably sized spacecraft is upswept as mass-energy fuel or species of reaction and that the spacecraft makes only one effective pass through each universe.

If such Aleph G light-year–width universes have three spatial dimensions and are substantially static, then a spacecraft traveling at light-speed in the multiverse may obtain a Lorentz factor of roughly [(Aleph H) EXP 3] [(Aleph G) EXP 3]. Note that in the above computations, we assume that only the material in the path of a reasonably sized spacecraft is upswept as mass-energy fuel or species of reaction and that the spacecraft makes an effective pass entirely through each universe. Thus, all material in said universes of travel is reacted against with assumed finite universal mass-energy density.

If the latter universes are expanding with time over periods much greater than Aleph G years, then the spacecraft may attain a Lorentz factor of roughly [(Aleph H) EXP 3] [[k(Aleph G)] EXP 3], where k ranges from a small superunitary real number to suitable infinite real numbers. This, of course, assumes that real positive mass-energy stocks continue being created, perhaps out of zero-point fields or analogues thereof from some forms of dark energy. Provided that k ranges from Aleph 0 to Aleph (G + 2), then the spacecraft Lorentz factor will be limited to a range of roughly from [(Aleph H) EXP 3] [[(Aleph 0)(Aleph G)] EXP 3] to [(Aleph H) EXP 3] [[[Aleph (G + 2)](Aleph G)] EXP 3]. Note that in the above computations, we assume that only the material in the path of a reasonably sized spacecraft is upswept as mass-energy fuel or species of reaction and that the spacecraft makes an effective pass entirely through each universe. Thus, all material in said universes of travel is reacted against with assumed finite universal mass-energy density.

Now we consider spacecraft travel in a multiverse of Aleph H universes in a width of four dimensions, where H is an integer greater than G + 2 and where said multiverse contains [(Aleph H) EXP 4] universes.

For universes in said 4-D-space multiverse having an extensity of (Aleph 1) light-years, there are prospects for the spacecraft to attain a Lorentz factor of roughly [(Aleph H) EXP 4] (Aleph 1) to thus result in the multiverse of travel being contracted to point directly in front of the spacecraft and Lorentz contracted by a factor of roughly [(Aleph H) EXP 4] (Aleph 1) relative to the spacecraft. Note that in the above computations, we assume that only the material in the path of a reasonably sized spacecraft is upswept as mass-energy fuel or species of reaction and that the spacecraft makes only one effective pass through each universe.

If such Aleph 1 light-year-width universes have three spatial dimensions and are substantially static, then a spacecraft traveling at light-speed in the multiverse of consideration may obtain a Lorentz factor of roughly [(Aleph H) EXP 4] [(Aleph 1) EXP 3]. Note that in the above computations, we assume that only the material in the path of a reasonably sized spacecraft is upswept as mass-energy fuel or species of reaction and that the spacecraft makes an effective pass entirely through each universe. Thus, all material in said universes of travel is reacted against with assumed finite universal mass-energy density.

If the latter universes are expanding with time over periods much greater than Aleph 1 years, then the spacecraft may attain a Lorentz factor of roughly [(Aleph H) EXP 4] [[k(Aleph 1)] EXP 3], where k ranges from a small superunitary real number to suitable infinite real numbers. This, of course, assumes that real positive mass-energy stocks continue being created, perhaps out of zero-point fields or analogues thereof from some forms of dark energy. Provided that k ranges from Aleph 0 to Aleph 3, then the spacecraft Lorentz factor will be limited to a range of roughly from [(Aleph H) EXP 4] [[(Aleph 0)(Aleph 1)] EXP 3] to [(Aleph H) EXP 4] [[(Aleph 3)(Aleph 1)] EXP 3]. Note that in the above computations, we assume that only the material in the path of a reasonably sized spacecraft is upswept as mass-energy fuel or species of reaction and that the spacecraft makes an effective entire pass through each universe. Thus, all material in

said universes of travel is reacted against with assumed finite universal mass-energy density.

For universes in said multiverse having an extensity of Aleph 2 light-years, there are prospects for the spacecraft to attain a Lorentz factor of roughly [(Aleph H) EXP 4] (Aleph 2) to thus result in the multiverse of travel being contracted to a point directly in front of the spacecraft and by a factor of [(Aleph H) EXP 4] (Aleph 2). Note that in the above computations, we assume that only the material in the path of a reasonably sized spacecraft is upswept as mass-energy fuel or species of reaction and that the spacecraft makes only one effective pass through each universe.

If such Aleph 2 light-year–width universes have three spatial dimensions and are substantially static, then a spacecraft traveling at light-speed in the multiverse considered may obtain a Lorentz factor of roughly [(Aleph H) EXP 4] [(Aleph 2) EXP 3]. Note that in the above computations, we assume that only the material in the path of a reasonably sized spacecraft is upswept as mass-energy fuel or species of reaction and that the spacecraft makes an effective pass entirely through each universe. Thus, all material in said universes of travel is reacted against with assumed finite universal mass-energy density.

If the latter universes are expanding with time over periods much greater than Aleph 2 years, then the spacecraft may attain a Lorentz factor of roughly [(Aleph H) EXP 4] [[k(Aleph 2)] EXP 3], where k ranges from a small superunitary real number to suitable infinite real numbers. This, of course, assumes that real positive mass-energy stocks continue being created, perhaps out of zero-point fields or analogues thereof from some forms of dark energy. Provided that k ranges from Aleph 0 to Aleph 4, then the spacecraft Lorentz factor will be limited to a range of roughly from [(Aleph H) EXP 4] [[(Aleph 0)(Aleph 2)] EXP 3] to [(Aleph H) EXP 4] [[(Aleph 4)(Aleph 2)] EXP 3]. Note that in the above computations, we assume that only the material in the path of a reasonably sized spacecraft is upswept as mass-energy fuel or species of reaction and that the spacecraft makes an effective pass entirely through each universe. Thus, all material in said universes of travel is reacted against with assumed finite universal mass-energy density.

For universes in said multiverse having an extensity of Aleph 3 light-years, there are prospects for the spacecraft to attain a Lorentz factor of roughly [(Aleph H) EXP 4] (Aleph 3) to thus result in the multiverse of travel being contracted to point directly in front of the spacecraft and contracted by a factor of [(Aleph H) EXP 4](Aleph 3). Note that in the above computations, we assume that only the material in the path of a reasonably sized spacecraft is upswept as mass-energy fuel or species of reaction and that the spacecraft makes only one effective pass through each universe.

If such Aleph 3 light-year–width universes have three spatial dimensions and are substantially static, then a spacecraft traveling at light-speed in the multiverse of consideration may obtain a Lorentz factor of roughly [(Aleph H) EXP 4] [(Aleph 3) EXP 3]. Note that in the above computations, we assume that only the material in the path of a reasonably sized spacecraft is upswept as mass-energy fuel or species of reaction and that the spacecraft makes an effective pass entirely through each universe. Thus, all material in said universes of travel is reacted against with assumed finite universal mass-energy density.

If the latter universes are expanding with time over periods much greater than Aleph 3 years, then the spacecraft may attain a Lorentz factor of roughly [(Aleph H) EXP 4] [[k(Aleph 3)] EXP 3], where k ranges from a small superunitary real number to suitable infinite real numbers. This, of course, assumes that real positive mass-energy stocks continue being created, perhaps out of zero-point fields or analogues thereof from some forms of dark energy. Provided that k ranges from Aleph 0 to Aleph 5, then the spacecraft Lorentz factor will be limited to a range of roughly from [(Aleph H) EXP 4] [[(Aleph 0)(Aleph 3)] EXP 3] to [(Aleph H) EXP 4] [[(Aleph 5)(Aleph 3)] EXP 3]. Note that in the above computations, we assume that only the material in the path of a reasonably sized spacecraft is upswept as mass-energy fuel or species of reaction and that the spacecraft makes an effective pass entirely through each universe. Thus, all material in said universes of travel is reacted against with assumed finite universal mass-energy density.

For universes having an extensity of Aleph 4 light-years, there are prospects for the spacecraft to attain a Lorentz factor of roughly [(Aleph H) EXP 4] (Aleph 4) to thus result in the multiverse of travel being contracted to point directly in front of the spacecraft and being contracted relative to the spacecraft by roughly a factor of [(Aleph H) EXP 4](Aleph 4). Note that in the above computations, we assume that only the material in the path of a reasonably sized spacecraft is upswept as mass-energy fuel or species of reaction and that the spacecraft makes only one effective pass through each universe.

If such Aleph 4 light-year–width universes have three spatial dimensions and are substantially static, then a spacecraft traveling at light-speed in the multiverse may obtain a Lorentz factor of roughly [(Aleph H) EXP 4] [(Aleph 4) EXP 3]. Note that in the above computations, we assume that only the material in the path of a reasonably sized spacecraft is upswept as mass-energy fuel or species of reaction and that the spacecraft makes an effective pass entirely through each universe. Thus, all material in said universes of travel is reacted against with assumed finite universal mass-energy density.

If the latter universes are expanding with time over periods much greater than Aleph 4 years, then the spacecraft may attain a Lorentz factor of roughly [(Aleph H) EXP 4] [[k(Aleph 4)] EXP 3], where *k* ranges from a small superunitary real number to suitable infinite real numbers. This, of course, assumes that real positive mass-energy stocks continue being created, perhaps out of zero-point fields or analogues thereof from some forms of dark energy. Provided that *k* ranges roughly from Aleph 0 to Aleph 6, then the spacecraft Lorentz factor will be limited to a range from [(Aleph H) EXP 4] [[(Aleph 0)(Aleph 4)] EXP 3] to [(Aleph H) EXP 4] [[(Aleph 6)(Aleph 4)] EXP 3]. Note that in the above computations, we assume that only the material in the path of a reasonably sized spacecraft is upswept as mass-energy fuel or species of reaction and that the spacecraft makes an effective pass entirely through each universe. Thus, all material in said universes of travel is reacted against with assumed finite universal mass-energy density.

For universes having an extensity of Aleph G light-years, there are prospects for the spacecraft to attain a Lorentz factor of roughly [(Aleph H) EXP 4] (Aleph G) to thus result in the multiverse of travel being contracted to point directly in front of the spacecraft with a Lorentz contraction of roughly [(Aleph H) EXP 4] (Aleph G). Note that in the above computations, we assume that only the material in the path of a reasonably sized spacecraft is upswept as mass-energy fuel or species of reaction and that the spacecraft makes only one effective pass through each universe.

If such Aleph G light-year–width universes have three spatial dimensions and are substantially static, then a spacecraft traveling at light-speed in the multiverse may obtain a Lorentz factor of roughly [(Aleph H) EXP 4] [(Aleph G) EXP 3]. Note that in the above computations, we assume that only the material in the path of a reasonably sized spacecraft is upswept as mass-energy fuel or species of reaction and that the spacecraft makes an effective pass entirely through each universe. Thus, all material in said universes of travel is reacted against with assumed finite universal mass-energy density.

If the latter universes are expanding with time over periods much greater than Aleph G years, then the spacecraft may attain a Lorentz factor of roughly [(Aleph H) EXP 4] [[k(Aleph G)] EXP 3], where *k* ranges from a small superunitary real number to suitable infinite real numbers. This, of course, assumes that real positive mass-energy stocks continue being created, perhaps out of zero-point fields or analogues thereof from some forms of dark energy. Provided that *k* ranges from Aleph 0 to Aleph (G + 2), then the spacecraft Lorentz factor will be limited to a range of roughly from [(Aleph H) EXP 4] [[(Aleph 0)(Aleph G)] EXP 3] to [(Aleph H) EXP 4] [[[Aleph (G + 2)](Aleph G)] EXP 3]. Note that in the above computations, we assume that only the material in the path of a reasonably sized spacecraft is upswept as mass-energy fuel or species of reaction and that the spacecraft makes an effective pass entirely through each universe. Thus, all material in said universes of travel is reacted against with assumed finite universal mass-energy density.

Now we consider spacecraft travel in a multiverse of Aleph H universes in width of five dimensions, where H is an integer greater than G + 2 and where said multiverse contains [(Aleph H) EXP 5] universes.

For universes in said 5-D–space multiverse, having an extensity of (Aleph 1) light-years, there are prospects for the spacecraft to attain a Lorentz factor of roughly [(Aleph H) EXP 5] (Aleph 1) to thus result in the multiverse of travel being contracted to point directly in front of the spacecraft and Lorentz contracted by a factor of roughly [(Aleph H) EXP 5] (Aleph 1) relative to the spacecraft. Note that in the above computations, we assume that only the material in the path of a reasonably sized spacecraft is upswept as mass-energy fuel or species of reaction and that the spacecraft makes only one effective pass through each universe.

If such Aleph 1 light-year-width universes have three spatial dimensions and are substantially static, then a spacecraft traveling at light-speed in the multiverse of consideration may obtain a Lorentz factor of roughly [(Aleph H) EXP 5] [(Aleph 1) EXP 3]. Note that in the above computations, we assume that only the material in the path of a reasonably sized spacecraft is upswept as mass-energy fuel or species of reaction and that the spacecraft makes an effective pass entirely through each universe. Thus, all material in said universes of travel is reacted against with assumed finite universal mass-energy density.

If the latter universes are expanding with time over periods much greater than Aleph 1 years, then the spacecraft may attain a Lorentz factor of roughly [(Aleph H) EXP 5] [[k(Aleph 1)] EXP 3], where k ranges from a small superunitary real number to suitable infinite real numbers. This, of course, assumes that real positive mass-energy stocks continue being created, perhaps out of zero-point fields or analogues thereof from some forms of dark energy. Provided that k ranges from Aleph 0 to Aleph 3, then the spacecraft Lorentz factor will be limited to a range roughly from [(Aleph H) EXP 5] [[(Aleph 0)(Aleph 1)] EXP 3] to [(Aleph H) EXP 5] [[(Aleph 3)(Aleph 1)] EXP 3]. Note that in the above computations, we assume that only the material in the path of a reasonably sized spacecraft is upswept as mass-energy fuel or species of reaction and that the spacecraft makes an effective entire pass through each universe. Thus, all material in said universes of travel is reacted against with assumed finite universal mass-energy density.

For universes in said multiverse having an extensity of Aleph 2 light-years, there are prospects for the spacecraft to attain a Lorentz factor of roughly [(Aleph H) EXP 5] (Aleph 2) to thus result in the multiverse of travel being contracted to a point directly in front of the spacecraft and by a factor of [(Aleph H) EXP 5] (Aleph 2). Note that in the above computations, we assume that only the material in the path of a reasonably sized spacecraft is upswept as mass-energy fuel or species of reaction and that the spacecraft makes only one effective pass through each universe.

If such Aleph 2 light-year–width universes have three spatial dimensions and are substantially static, then a spacecraft traveling at light-speed in the multiverse considered may obtain a Lorentz factor of roughly [(Aleph H) EXP 5] [(Aleph 2) EXP 3]. Note that in the above computations, we assume that only the material in the path of a reasonably sized spacecraft is upswept as mass-energy fuel or species of reaction and that the spacecraft makes an effective pass entirely through each universe. Thus, all material in said universes of travel is reacted against with assumed finite universal mass-energy density.

If the latter universes are expanding with time over periods much greater than Aleph 2 years, then the spacecraft may attain a Lorentz factor of roughly [(Aleph H) EXP 5] [[k(Aleph 2)] EXP 3], where k ranges from a small superunitary real number to suitable infinite real numbers. This, of course, assumes that real positive mass-energy stocks continue being created, perhaps out of zero-point fields or analogues thereof from some forms of dark energy. Provided that k ranges from Aleph 0 to Aleph 4, then the spacecraft Lorentz factor will be limited to a range of roughly from [(Aleph H) EXP 5] [[(Aleph 0)(Aleph 2)] EXP 3] to [(Aleph H) EXP 5] [[(Aleph 4)(Aleph 2)] EXP 3]. Note that in the above computations, we assume that only the material in the path of a reasonably sized spacecraft is upswept as mass-energy fuel or species of reaction and that the spacecraft makes an effective pass entirely through each universe. Thus, all material in said universes of travel is reacted against with assumed finite universal mass-energy density.

For universes in said multiverse having an extensity of Aleph 3 light-years, there are prospects for the spacecraft to attain a Lorentz factor of roughly [(Aleph H) EXP 5] (Aleph 3) to thus result in the multiverse of travel being contracted to point directly in front of the spacecraft and contracted by a factor of [(Aleph H) EXP 4](Aleph 3). Note that in the above computations, we assume that only the material in the path of a reasonably sized spacecraft is upswept as mass-energy fuel or species of reaction and that the spacecraft makes only one effective pass through each universe.

If such Aleph 3 light-year–width universes have three spatial dimensions and are substantially static, then a spacecraft traveling at light-speed in the multiverse of consideration may obtain a Lorentz factor of roughly [(Aleph H) EXP 5] [(Aleph 3) EXP 3]. Note that in the above computations, we assume that only the material in the path of a reasonably sized spacecraft is upswept as mass-energy fuel or species of reaction and that the spacecraft makes an effective pass entirely through each universe. Thus, all material in said universes of travel is reacted against with assumed finite universal mass-energy density.

If the latter universes are expanding with time over periods much greater than Aleph 3 years, then the spacecraft may attain a Lorentz factor of roughly [(Aleph H) EXP 5] [[k(Aleph 3)] EXP 3], where *k* ranges from a small superunitary real number to suitable infinite real numbers. This, of course, assumes that real positive mass-energy stocks continue being created, perhaps out of zero-point fields or analogues thereof from some forms of dark energy. Provided that *k* ranges from Aleph 0 to Aleph 5, then the spacecraft Lorentz factor will be limited to a range of roughly from [(Aleph H) EXP 5] [[(Aleph 0)(Aleph 3)] EXP 3] to [(Aleph H) EXP 5] [[(Aleph 5)(Aleph 3)] EXP 3]. Note that in the above computations, we assume that only the material in the path of a reasonably sized spacecraft is upswept as mass-energy fuel or species of reaction and that the spacecraft makes an effective pass entirely through each universe. Thus, all material in said universes of travel is reacted against with assumed finite universal mass-energy density.

For universes having an extensity of Aleph 4 light-years, there are prospects for the spacecraft to attain a Lorentz factor of roughly [(Aleph H) EXP 5] (Aleph 4) to thus result in the multiverse of travel being contracted to point directly in front of the spacecraft and being contracted relative to the spacecraft by roughly a factor of [(Aleph H) EXP 4](Aleph 4). Note that in the above computations, we assume that only the material in the path of a reasonably sized spacecraft is upswept as mass-energy fuel or species of reaction and that the spacecraft makes only one effective pass through each universe.

If such Aleph 4 light-year–width universes have three spatial dimensions and are substantially static, then a spacecraft traveling at light-speed in the multiverse may obtain a Lorentz factor of roughly [(Aleph H) EXP 5] [(Aleph 4) EXP 3]. Note that in the above computations, we assume that only the material in the path of a reasonably sized spacecraft is upswept as mass-energy fuel or species of reaction and that the spacecraft makes an effective pass entirely through each universe. Thus, all material in said universes of travel is reacted against with assumed finite universal mass-energy density.

If the latter universes are expanding with time over periods much greater than Aleph 4 years, then the spacecraft may attain a Lorentz factor of roughly [(Aleph H) EXP 5] [[k(Aleph 4)] EXP 3], where *k* ranges from a small superunitary real number to suitable infinite real numbers. This, of course, assumes that real positive mass-energy stocks continue being created, perhaps out of zero-point fields or analogues thereof from some forms of dark energy. Provided that *k* ranges roughly from Aleph 0 to Aleph 6, then the spacecraft Lorentz factor will be limited to a range from [(Aleph H) EXP 5] [[(Aleph 0)(Aleph 4)] EXP 3] to [(Aleph H) EXP 5] [[(Aleph 6)(Aleph 4)] EXP 3]. Note that in the above computations, we assume that only the material in the path of a reasonably sized spacecraft is upswept as mass-energy fuel or species of reaction and that the spacecraft makes an effective pass entirely through each universe. Thus, all material in

said universes of travel is reacted against with assumed finite universal mass-energy density.

For universes having an extensity of Aleph G light-years, there are prospects for the spacecraft to attain a Lorentz factor of roughly [(Aleph H) EXP 5] (Aleph G) to thus result in the multiverse of travel being contracted to point directly in front of the spacecraft with a Lorentz contraction of roughly [(Aleph H) EXP 5] (Aleph G). Note that in the above computations, we assume that only the material in the path of a reasonably sized spacecraft is upswept as mass-energy fuel or species of reaction and that the spacecraft makes only one effective pass through each universe.

If such Aleph G light-year–width universes have three spatial dimensions and are substantially static, then a spacecraft traveling at light-speed in the multiverse may obtain a Lorentz factor of roughly [(Aleph H) EXP 5] [(Aleph G) EXP 3]. Note that in the above computations, we assume that only the material in the path of a reasonably sized spacecraft is upswept as mass-energy fuel or species of reaction and that the spacecraft makes an effective pass entirely through each universe. Thus, all material in said universes of travel is reacted against with assumed finite universal mass-energy density.

If the latter universes are expanding with time over periods much greater than Aleph G years, then the spacecraft may attain a Lorentz factor of roughly [(Aleph H) EXP 5] [[k(Aleph G)] EXP 3], where *k* ranges from a small superunitary real number to suitable infinite real numbers. This, of course, assumes that real positive mass-energy stocks continue being created, perhaps out of zero-point fields or analogues thereof from some forms of dark energy. Provided that *k* ranges from Aleph 0 to Aleph (G + 2), then the spacecraft Lorentz factor will be limited to a range of roughly from [(Aleph H) EXP 5] [[(Aleph 0)(Aleph G)] EXP 3] to [(Aleph H) EXP 5] [[[Aleph (G + 2)](Aleph G)] EXP 3]. Note that in the above computations, we assume that only the material in the path of a reasonably sized spacecraft is upswept as mass-energy fuel or species of reaction and that the spacecraft makes an effective pass entirely through each universe. Thus, all material in said universes of travel is reacted against with assumed finite universal mass-energy density.

Now we consider spacecraft travel in a multiverse of Aleph H universes in width of six dimensions, where *H* is an integer greater than G + 2, and where said multiverse contains [(Aleph H) EXP 6] universes.

For universes in said 6-D–space multiverse having an extensity of (Aleph 1) light-years, there are prospects for the spacecraft to attain a Lorentz factor of roughly [(Aleph H) EXP 6] (Aleph 1) to thus result in the multiverse of travel being contracted to point directly in front of the spacecraft and Lorentz contracted by a factor of roughly [(Aleph

H) EXP 6] (Aleph 1) relative to the spacecraft. Note that in the above computations, we assume that only the material in the path of a reasonably sized spacecraft is upswept as mass-energy fuel or species of reaction and that the spacecraft makes only one effective pass through each universe.

If such Aleph 1 light-year-width universes have three spatial dimensions and are substantially static, then a spacecraft traveling at light-speed in the multiverse of consideration may obtain a Lorentz factor of roughly [(Aleph H) EXP 6] [(Aleph 1) EXP 3]. Note that in the above computations, we assume that only the material in the path of a reasonably sized spacecraft is upswept as mass-energy fuel or species of reaction and that the spacecraft makes an effective pass entirely through each universe. Thus, all material in said universes of travel is reacted against with assumed finite universal mass-energy density.

If the latter universes are expanding with time over periods much greater than Aleph 1 years, then the spacecraft may attain a Lorentz factor of roughly [(Aleph H) EXP 6] [[k(Aleph 1)] EXP 3], where *k* ranges from a small superunitary real number to suitable infinite real numbers. This, of course, assumes that real positive mass-energy stocks continue being created, perhaps out of zero-point fields or analogues thereof from some forms of dark energy. Provided that *k* ranges from Aleph 0 to Aleph 3, then the spacecraft Lorentz factor will be limited to a range roughly from [(Aleph H) EXP 6] [[(Aleph 0)(Aleph 1)] EXP 3] to [(Aleph H) EXP 6] [[(Aleph 3)(Aleph 1)] EXP 3]. Note that in the above computations, we assume that only the material in the path of a reasonably sized spacecraft is upswept as mass-energy fuel or species of reaction and that the spacecraft makes an effective entire pass through each universe. Thus, all material in said universes of travel is reacted against with assumed finite universal mass-energy density.

For universes in the said multiverse, having an extensity of Aleph 2 light-years, there are prospects for the spacecraft to attain a Lorentz factor of roughly [(Aleph H) EXP 6] (Aleph 2) to thus result in the multiverse of travel being contracted to a point directly in front of the spacecraft and by a factor of [(Aleph H) EXP 6] (Aleph 2). Note that in the above computations, we assume that only the material in the path of a reasonably sized spacecraft is upswept as mass-energy fuel or species of reaction and that the spacecraft makes only one effective pass through each universe.

If such Aleph 2 light-year–width universes have three spatial dimensions and are substantially static, then a spacecraft traveling at light-speed in the multiverse considered may obtain a Lorentz factor of roughly [(Aleph H) EXP 6] [(Aleph 2) EXP 3]. Note that in the above computations, we assume that only the material in the path of a reasonably sized spacecraft is upswept as mass-energy fuel or species of reaction and that the spacecraft makes an effective pass entirely through each universe. Thus, all material in said universes of travel is reacted against with assumed finite universal mass-energy density.

If the latter universes are expanding with time over periods much greater than Aleph 2 years, then the spacecraft may attain a Lorentz factor of roughly [(Aleph H) EXP 6] [[k(Aleph 2)] EXP 3], where k ranges from a small superunitary real number to suitable infinite real numbers. This, of course, assumes that real positive mass-energy stocks continue being created, perhaps out of zero-point fields or analogues thereof from some forms of dark energy. Provided that k ranges from Aleph 0 to Aleph 4, then the spacecraft Lorentz factor will be limited to a range of roughly from [(Aleph H) EXP 6] [[(Aleph 0)(Aleph 2)] EXP 3] to [(Aleph H) EXP 6] [[(Aleph 4)(Aleph 2)] EXP 3]. Note that in the above computations, we assume that only the material in the path of a reasonably sized spacecraft is upswept as mass-energy fuel or species of reaction and that the spacecraft makes an effective pass entirely through each universe. Thus, all material in said universes of travel is reacted against with assumed finite universal mass-energy density.

For universes in said multiverse having an extensity of Aleph 3 light-years, there are prospects for the spacecraft to attain a Lorentz factor of roughly [(Aleph H) EXP 6] (Aleph 3) to thus result in the multiverse of travel being contracted to point directly in front of the spacecraft and contracted by a factor of [(Aleph H) EXP 4](Aleph 3). Note that in the above computations, we assume that only the material in the path of a reasonably sized spacecraft is upswept as mass-energy fuel or species of reaction and that the spacecraft makes only one effective pass through each universe.

If such Aleph 3 light-year–width universes have three spatial dimensions and are substantially static, then a spacecraft traveling at light-speed in the multiverse of consideration may obtain a Lorentz factor of roughly [(Aleph H) EXP 6] [(Aleph 3) EXP 3]. Note that in the above computations, we assume that only the material in the path of a reasonably sized spacecraft is upswept as mass-energy fuel or species of reaction and that the spacecraft makes an effective pass entirely through each universe. Thus, all material in said universes of travel is reacted against with assumed finite universal mass-energy density.

If the latter universes are expanding with time over periods much greater than Aleph 3 years, then the spacecraft may attain a Lorentz factor of roughly [(Aleph H) EXP 6] [[k(Aleph 3)] EXP 3], where k ranges from a small superunitary real number to suitable infinite real numbers. This, of course, assumes that real positive mass-energy stocks continue being created, perhaps out of zero-point fields or analogues thereof from some forms of dark energy. Provided that k ranges from Aleph 0 to Aleph 5, then the spacecraft Lorentz factor will be limited to a range of roughly from [(Aleph H) EXP 6] [[(Aleph 0)(Aleph 3)] EXP 3] to [(Aleph H) EXP 6] [[(Aleph 5)(Aleph 3)] EXP 3]. Note that in the above computations, we assume that only the material in the path of a reasonably sized spacecraft is upswept as mass-energy fuel or species of reaction and that the spacecraft makes an effective pass entirely through each universe. Thus, all material in said universes of travel is reacted against with assumed finite universal mass-energy density.

For universes having an extensity of Aleph 4 light-years, there are prospects for the spacecraft to attain a Lorentz factor of roughly [(Aleph H) EXP 6] (Aleph 4) to thus result in the multiverse of travel being contracted to point directly in front of the spacecraft and being contracted relative to the spacecraft by roughly a factor of [(Aleph H) EXP 4](Aleph 4). Note that in the above computations, we assume that only the material in the path of a reasonably sized spacecraft is upswept as mass-energy fuel or species of reaction and that the spacecraft makes only one effective pass through each universe.

If such Aleph 4 light-year–width universes have three spatial dimensions and are substantially static, then a spacecraft traveling at light-speed in the multiverse may obtain a Lorentz factor of roughly [(Aleph H) EXP 6] [(Aleph 4) EXP 3]. Note that in the above computations, we assume that only the material in the path of a reasonably sized spacecraft is upswept as mass-energy fuel or species of reaction and that the spacecraft makes an effective pass entirely through each universe. Thus, all material in said universes of travel is reacted against with assumed finite universal mass-energy density.

If the latter universes are expanding with time over periods much greater than Aleph 4 years, then the spacecraft may attain a Lorentz factor of roughly [(Aleph H) EXP 6] [[k(Aleph 4)] EXP 3], where k ranges from a small superunitary real number to suitable infinite real numbers. This, of course, assumes that real positive mass-energy stocks continue being created, perhaps out of zero-point fields or analogues thereof from some forms of dark energy. Provided that k ranges roughly from Aleph 0 to Aleph 6, then the spacecraft Lorentz factor will be limited to a range from [(Aleph H) EXP 6] [[(Aleph 0)(Aleph 4)] EXP 3] to [(Aleph H) EXP 6] [[(Aleph 6)(Aleph 4)] EXP 3]. Note that in the above computations, we assume that only the material in the path of a reasonably sized spacecraft is upswept as mass-energy fuel or species of reaction and that the spacecraft makes an effective pass entirely through each universe. Thus, all material in said universes of travel is reacted against with assumed finite universal mass-energy density.

For universes having an extensity of Aleph G light-years, there are prospects for the spacecraft to attain a Lorentz factor of roughly [(Aleph H) EXP 6] (Aleph G) to thus result in the multiverse of travel being contracted to point directly in front of the spacecraft with a Lorentz contraction of roughly [(Aleph H) EXP 6] (Aleph G). Note that in the above computations, we assume that only the material in the path of a reasonably sized spacecraft is upswept as mass-energy fuel or species of reaction and that the spacecraft makes only one effective pass through each universe.

If such Aleph G light-year–width universes have three spatial dimensions and are substantially static, then a spacecraft traveling at light-speed in the multiverse may obtain a Lorentz factor of roughly [(Aleph H) EXP 6] [(Aleph G) EXP 3]. Note that in the above computations, we assume that only the material in the path of a reasonably sized spacecraft is upswept as mass-energy fuel or species of reaction and that the spacecraft makes an effective pass entirely through each universe. Thus, all material in said universes of travel is reacted against with assumed finite universal mass-energy density.

If the latter universes are expanding with time over periods much greater than Aleph G years, then the spacecraft may attain a Lorentz factor of roughly [(Aleph H) EXP 6] [[k(Aleph G)] EXP 3], where k ranges from a small superunitary real number to suitable infinite real numbers. This, of course, assumes that real positive mass-energy stocks continue being created, perhaps out of zero-point fields or analogues thereof from some forms of dark energy. Provided that k ranges from Aleph 0 to Aleph (G + 2), then the spacecraft Lorentz factor will be limited to a range of roughly from [(Aleph H) EXP 6] [[(Aleph 0)(Aleph G)] EXP 3] to [(Aleph H) EXP 6] [[[Aleph (G + 2)](Aleph G)] EXP 3]. Note that in the above computations, we assume that only the material in the path of a reasonably sized spacecraft is upswept as mass-energy fuel or species of reaction and that the spacecraft makes an effective pass entirely through each universe. Thus, all material in said universes of travel is reacted against with assumed finite universal mass-energy density.

Now we consider spacecraft travel in a multiverse of Aleph H universes in width of N dimensions, where H is an integer greater than G + 2, and where said multiverse contains [(Aleph H) EXP N] universes.

For universes in said N-D–space multiverse having an extensity of (Aleph 1) light-years, there are prospects for the spacecraft to attain a Lorentz factor of roughly [(Aleph H) EXP N] (Aleph 1) to thus result in the multiverse of travel being contracted to point directly in front of the spacecraft and Lorentz contracted by a factor of roughly [(Aleph H) EXP N] (Aleph 1) relative to the spacecraft. Note that in the above computations, we assume that only the material in the path of a reasonably sized spacecraft is upswept as mass-energy fuel or species of reaction and that the spacecraft makes only one effective pass through each universe.

If such Aleph 1 light-year–width universes have three spatial dimensions and are substantially static, then a spacecraft traveling at light-speed in the multiverse of consideration may obtain a Lorentz factor of roughly [(Aleph H) EXP N] [(Aleph 1) EXP

3]. Note that in the above computations, we assume that only the material in the path of a reasonably sized spacecraft is upswept as mass-energy fuel or species of reaction and that the spacecraft makes an effective pass entirely through each universe. Thus, all material in said universes of travel is reacted against with assumed finite universal mass-energy density.

If the latter universes are expanding with time over periods much greater than Aleph 1 years, then the spacecraft may attain a Lorentz factor of roughly [(Aleph H) EXP N] [[k(Aleph 1)] EXP 3], where k ranges from a small superunitary real number to suitable infinite real numbers. This, of course, assumes that real positive mass-energy stocks continue being created, perhaps out of zero-point fields or analogues thereof from some forms of dark energy. Provided that k ranges from Aleph 0 to Aleph 3, then the spacecraft Lorentz factor will be limited to a range of roughly from [(Aleph H) EXP N] [[(Aleph 0)(Aleph 1)] EXP 3] to [(Aleph H) EXP N] [[(Aleph 3)(Aleph 1)] EXP 3]. Note that in the above computations, we assume that only the material in the path of a reasonably sized spacecraft is upswept as mass-energy fuel or species of reaction and that the spacecraft makes an effective entire pass through each universe. Thus, all material in said universes of travel is reacted against with assumed finite universal mass-energy density.

For universes in said multiverse having an extensity of Aleph 2 light-years, there are prospects for the spacecraft to attain a Lorentz factor of roughly [(Aleph H) EXP N] (Aleph 2) to thus result in the multiverse of travel being contracted to a point directly in front of the spacecraft and by a factor of [(Aleph H) EXP N] (Aleph 2). Note that in the above computations, we assume that only the material in the path of a reasonably sized spacecraft is upswept as mass-energy fuel or species of reaction and that the spacecraft makes only one effective pass through each universe.

If such Aleph 2 light-year–width universes have three spatial dimensions and are substantially static, then a spacecraft traveling at light-speed in the multiverse considered may obtain a Lorentz factor of roughly [(Aleph H) EXP N] [(Aleph 2) EXP 3]. Note that in the above computations, we assume that only the material in the path of a reasonably sized spacecraft is upswept as mass-energy fuel or species of reaction and that the spacecraft makes an effective pass entirely through each universe. Thus, all material in said universes of travel is reacted against with assumed finite universal mass-energy density.

If the latter universes are expanding with time over periods much greater than Aleph 2 years, then the spacecraft may attain a Lorentz factor of roughly [(Aleph H) EXP N] [[k(Aleph 2)] EXP 3], where k ranges from a small superunitary real number to suitable infinite real numbers. This, of course, assumes that real positive mass-energy stocks continue being created, perhaps out of zero-point fields or analogues thereof from some forms of dark energy. Provided that *k* ranges from Aleph 0 to Aleph 4, then the spacecraft Lorentz factor will be limited to a range of roughly from [(Aleph H) EXP N] [[(Aleph 0)(Aleph 2)] EXP 3] to [(Aleph H) EXP N] [[(Aleph 4)(Aleph 2)] EXP 3]. Note that in the above computations, we assume that only the material in the path of a reasonably sized spacecraft is upswept as mass-energy fuel or species of reaction and that the spacecraft makes an effective pass entirely through each universe. Thus, all material in said universes of travel is reacted against with assumed finite universal mass-energy density.

For universes in said multiverse having an extensity of Aleph 3 light-years, there are prospects for the spacecraft to attain a Lorentz factor of roughly [(Aleph H) EXP N] (Aleph 3) to thus result in the multiverse of travel being contracted to point directly in front of the spacecraft and contracted by a factor of [(Aleph H) EXP 4](Aleph 3). Note that in the above computations, we assume that only the material in the path of a reasonably sized spacecraft is upswept as mass-energy fuel or species of reaction and that the spacecraft makes only one effective pass through each universe.

If such Aleph 3 light-year–width universes have three spatial dimensions and are substantially static, then a spacecraft traveling at light-speed in the multiverse of consideration may obtain a Lorentz factor of roughly [(Aleph H) EXP N] [(Aleph 3) EXP 3]. Note that in the above computations, we assume that only the material in the path of a reasonably sized spacecraft is upswept as mass-energy fuel or species of reaction and that the spacecraft makes an effective pass entirely through each universe. Thus, all material in said universes of travel is reacted against with assumed finite universal mass-energy density.

If the latter universes are expanding with time over periods much greater than Aleph 3 years, then the spacecraft may attain a Lorentz factor of roughly [(Aleph H) EXP N] [[k(Aleph 3)] EXP 3], where *k* ranges from a small superunitary real number to suitable infinite real numbers. This, of course, assumes that real positive mass-energy stocks continue being created, perhaps out of zero-point fields or analogues thereof from some forms of dark energy. Provided that *k* ranges from Aleph 0 to Aleph 5, then the spacecraft Lorentz factor will be limited to a range of roughly from [(Aleph H) EXP N] [[(Aleph 0)(Aleph 3)] EXP 3] to [(Aleph H) EXP N] [[(Aleph 5)(Aleph 3)] EXP 3]. Note that in the above computations, we assume that only the material in the path of a reasonably sized spacecraft is upswept as mass-energy fuel or species of reaction and that the spacecraft makes an effective pass entirely through each universe. Thus, all material in said universes of travel is reacted against with assumed finite universal mass-energy density.

For universes having an extensity of Aleph 4 light-years, there are prospects for the spacecraft to attain a Lorentz factor of roughly [(Aleph H) EXP N] (Aleph 4) to thus result in the multiverse of travel being contracted to point directly in front of the spacecraft and being contracted relative to the spacecraft by roughly a factor of [(Aleph H) EXP 4](Aleph 4). Note that in the above computations, we assume that only the material in the path of a reasonably sized spacecraft is upswept as mass-energy fuel or species of reaction and that the spacecraft makes only one effective pass through each universe.

If such Aleph 4 light-year–width universes have three spatial dimensions and are substantially static, then a spacecraft traveling at light-speed in the multiverse may obtain a Lorentz factor of roughly [(Aleph H) EXP N] [(Aleph 4) EXP 3]. Note that in the above computations, we assume that only the material in the path of a reasonably sized spacecraft is upswept as mass-energy fuel or species of reaction and that the spacecraft makes an effective pass entirely through each universe. Thus, all material in said universes of travel is reacted against with assumed finite universal mass-energy density.

If the latter universes are expanding with time over periods much greater than Aleph 4 years, then the spacecraft may attain a Lorentz factor of roughly [(Aleph H) EXP N] [[k(Aleph 4)] EXP 3], where k ranges from a small superunitary real number to suitable infinite real numbers. This, of course, assumes that real positive mass-energy stocks continue being created, perhaps out of zero-point fields or analogues thereof from some forms of dark energy. Provided that k ranges roughly from Aleph 0 to Aleph 6, then the spacecraft Lorentz factor will be limited to a range from [(Aleph H) EXP N] [[(Aleph 0)(Aleph 4)] EXP 3] to [(Aleph H) EXP N] [[(Aleph 6)(Aleph 4)] EXP 3]. Note that in the above computations, we assume that only the material in the path of a reasonably sized spacecraft is upswept as mass-energy fuel or species of reaction and that the spacecraft makes an effective pass entirely through each universe. Thus, all material in said universes of travel is reacted against with assumed finite universal mass-energy density.

For universes having an extensity of Aleph G light-years, there are prospects for the spacecraft to attain a Lorentz factor of roughly [(Aleph H) EXP N] (Aleph G) to thus result in the multiverse of travel being contracted to point directly in front of the spacecraft with a Lorentz contraction of roughly [(Aleph H) EXP N] (Aleph G). Note that in the above computations, we assume that only the material in the path of a reasonably sized spacecraft is upswept as mass-energy fuel or species of reaction and that the spacecraft makes only one effective pass through each universe.

If such Aleph G light-year–width universes have three spatial dimensions and are substantially static, then a spacecraft traveling at light-speed in the multiverse may obtain a Lorentz factor of roughly [(Aleph H) EXP N] [(Aleph G) EXP 3]. Note that in the above computations, we assume that only the material in the path of a reasonably sized spacecraft is upswept as mass-energy fuel or species of reaction and that the spacecraft makes an effective pass entirely through each universe. Thus, all material in said universes of travel is reacted against with assumed finite universal mass-energy density.

If the latter universes are expanding with time over periods much greater than Aleph G years, then the spacecraft may attain a Lorentz factor of roughly [(Aleph H) EXP N] [[k(Aleph G)] EXP 3], where k ranges from a small superunitary real number to suitable infinite real numbers. This, of course, assumes that real positive mass-energy stocks continue being created, perhaps out of zero-point fields or analogues thereof from some forms of dark energy. Provided that k ranges from Aleph 0 to Aleph (G + 2), then the spacecraft Lorentz factor will be limited to a range of roughly from [(Aleph H) EXP N] [[(Aleph 0)(Aleph G)] EXP 3] to [(Aleph H) EXP N] [[[Aleph (G + 2)](Aleph G)] EXP 3]. Note that in the above computations, we assume that only the material in the path of a reasonably sized spacecraft is upswept as mass-energy fuel or species of reaction and that the spacecraft makes an effective pass entirely through each universe. Thus, all material in said universes of travel is reacted against with assumed finite universal mass-energy density.

Now we consider spacecraft travel in a multiverse of Aleph H universes in width of four dimensions, where H is an integer greater than G + 2, and where said multiverse contains [(Aleph H) EXP 4] universes.

For universes in said 4-D–space multiverse having an extensity of (Aleph 1) light-years, there are prospects for the spacecraft to attain a Lorentz factor of roughly [(Aleph H) EXP 4] (Aleph 1) to thus result in the multiverse of travel being contracted to point directly in front of the spacecraft and Lorentz contracted by a factor of roughly [(Aleph H) EXP 4] (Aleph 1) relative to the spacecraft. Note that in the above computations, we assume that only the material in the path of a reasonably sized spacecraft is upswept as mass-energy fuel or species of reaction and that the spacecraft makes only one effective pass through each universe.

If such Aleph 1 light-year–width universes have four spatial dimensions and are substantially static, then a spacecraft traveling at light-speed in the multiverse of consideration may obtain a Lorentz factor of roughly [(Aleph H) EXP 4] [(Aleph 1) EXP 4]. Note that in the above computations, we assume that only the material in the path of a reasonably sized spacecraft is upswept as mass-energy fuel or species of reaction and that the spacecraft makes an effective pass entirely through each universe. Thus, all material in said universes of travel is reacted against with assumed finite universal mass-energy density.

If the latter universes are expanding with time over periods much greater than Aleph 1 years, then the spacecraft may attain a Lorentz factor of roughly [(Aleph H) EXP 4] [[k(Aleph 1)] EXP 4], where k ranges from a small superunitary real number to suitable infinite real numbers. This, of course, assumes that real positive mass-energy stocks continue being created, perhaps out of zero-point fields or analogues thereof from some forms of dark energy. Provided that k ranges from Aleph 0 to Aleph 3, then the spacecraft Lorentz factor will be limited to a range of roughly from [(Aleph H) EXP 4] [[(Aleph 0)(Aleph 1)] EXP 4] to [(Aleph H) EXP 4] [[(Aleph 3)(Aleph 1)] EXP 4]. Note that in the above computations, we assume that only the material in the path of a reasonably sized spacecraft is upswept as mass-energy fuel or species of reaction and that the spacecraft makes an effective entire pass through each universe. Thus, all material in said universes of travel is reacted against with assumed finite universal mass-energy density.

For universes in said multiverse having an extensity of Aleph 2 light-years, there are prospects for the spacecraft to attain a Lorentz factor of roughly [(Aleph H) EXP 4] (Aleph 2) to thus result in the multiverse of travel being contracted to a point directly in front of the spacecraft and by a factor of [(Aleph H) EXP 4] (Aleph 2). Note that in the above computations, we assume that only the material in the path of a reasonably sized spacecraft is upswept as mass-energy fuel or species of reaction and that the spacecraft makes only one effective pass through each universe.

If such Aleph 2 light-year–width universes have four spatial dimensions and are substantially static, then a spacecraft traveling at light-speed in the multiverse considered may obtain a Lorentz factor of roughly [(Aleph H) EXP 4] [(Aleph 2) EXP 4]. Note that in the above computations, we assume that only the material in the path of a reasonably sized spacecraft is upswept as mass-energy fuel or species of reaction and that the spacecraft makes an effective pass entirely through each universe. Thus, all material in said universes of travel is reacted against with assumed finite universal mass-energy density.

If the latter universes are expanding with time over periods much greater than Aleph 2 years, then the spacecraft may attain a Lorentz factor of roughly [(Aleph H) EXP 4] [[k(Aleph 2)] EXP 4], where k ranges from a small superunitary real number to suitable infinite real numbers. This, of course, assumes that real positive mass-energy stocks continue being created, perhaps out of zero-point fields or analogues thereof from some forms of dark energy. Provided that k ranges from Aleph 0 to Aleph 4, then the spacecraft Lorentz factor will be limited to a range of roughly from [(Aleph H) EXP 4] [[(Aleph 0)(Aleph 2)] EXP 4] to [(Aleph H) EXP 4] [[(Aleph 4)(Aleph 2)] EXP 4]. Note that in the above computations, we assume that only the material in the path of a reasonably sized spacecraft is upswept as mass-energy fuel or species of reaction and that the spacecraft makes an effective pass entirely through each universe. Thus, all material in

said universes of travel is reacted against with assumed finite universal mass-energy density.

For universes in said multiverse having an extensity of Aleph 3 light-years, there are prospects for the spacecraft to attain a Lorentz factor of roughly [(Aleph H) EXP 4] (Aleph 3) to thus result in the multiverse of travel being contracted to point directly in front of the spacecraft and contracted by a factor of [(Aleph H) EXP 4](Aleph 3). Note that in the above computations, we assume that only the material in the path of a reasonably sized spacecraft is upswept as mass-energy fuel or species of reaction and that the spacecraft makes only one effective pass through each universe.

If such Aleph 3 light-year–width universes have four spatial dimensions and are substantially static, then a spacecraft traveling at light-speed in the multiverse of consideration may obtain a Lorentz factor of roughly [(Aleph H) EXP 4] [(Aleph 3) EXP 4]. Note that in the above computations, we assume that only the material in the path of a reasonably sized spacecraft is upswept as mass-energy fuel or species of reaction and that the spacecraft makes an effective pass entirely through each universe. Thus, all material in said universes of travel is reacted against with assumed finite universal mass-energy density.

If the latter universes are expanding with time over periods much greater than Aleph 3 years, then the spacecraft may attain a Lorentz factor of roughly [(Aleph H) EXP 4] [[k(Aleph 3)] EXP 4], where k ranges from a small superunitary real number to suitable infinite real numbers. This, of course, assumes that real positive mass-energy stocks continue being created, perhaps out of zero-point fields or analogues thereof from some forms of dark energy. Provided that k ranges from Aleph 0 to Aleph 5, then the spacecraft Lorentz factor will be limited to a range of roughly from [(Aleph H) EXP 4] [[(Aleph 0)(Aleph 3)] EXP 4] to [(Aleph H) EXP 4] [[(Aleph 5)(Aleph 3)] EXP 4]. Note that in the above computations, we assume that only the material in the path of a reasonably sized spacecraft is upswept as mass-energy fuel or species of reaction and that the spacecraft makes an effective pass entirely through each universe. Thus, all material in said universes of travel is reacted against with assumed finite universal mass-energy density.

For universes having an extensity of Aleph 4 light-years, there are prospects for the spacecraft to attain a Lorentz factor of roughly [(Aleph H) EXP 4] (Aleph 4) to thus result in the multiverse of travel being contracted to point directly in front of the spacecraft and being contracted relative to the spacecraft by roughly a factor of [(Aleph H) EXP 4](Aleph 4). Note that in the above computations, we assume that only the material in the path of a reasonably sized spacecraft is upswept as mass-energy fuel or species of reaction and that the spacecraft makes only one effective pass through each universe.

If such Aleph 4 light-year–width universes have four spatial dimensions and are substantially static, then a spacecraft traveling at light-speed in the multiverse may obtain a Lorentz factor of roughly [(Aleph H) EXP 4] [(Aleph 4) EXP 4]. Note that in the above computations, we assume that only the material in the path of a reasonably sized spacecraft is upswept as mass-energy fuel or species of reaction and that the spacecraft makes an effective pass entirely through each universe. Thus, all material in said universes of travel is reacted against with assumed finite universal mass-energy density.

If the latter universes are expanding with time over periods much greater than Aleph 4 years, then the spacecraft may attain a Lorentz factor of roughly [(Aleph H) EXP 4] [[k(Aleph 4)] EXP 4], where k ranges from a small superunitary real number to suitable infinite real numbers. This, of course, assumes that real positive mass-energy stocks continue being created, perhaps out of zero-point fields or analogues thereof from some forms of dark energy. Provided that k ranges roughly from Aleph 0 to Aleph 6, then the spacecraft Lorentz factor will be limited to a range from [(Aleph H) EXP 4] [[(Aleph 0)(Aleph 4)] EXP 4] to [(Aleph H) EXP 4] [[(Aleph 6)(Aleph 4)] EXP 4]. Note that in the above computations, we assume that only the material in the path of a reasonably sized spacecraft is upswept as mass-energy fuel or species of reaction and that the spacecraft makes an effective pass entirely through each universe. Thus, all material in said universes of travel is reacted against with assumed finite universal mass-energy density.

For universes having an extensity of Aleph G light-years, there are prospects for the spacecraft to attain a Lorentz factor of roughly [(Aleph H) EXP 4] (Aleph G) to thus result in the multiverse of travel being contracted to point directly in front of the spacecraft with a Lorentz contraction of roughly [(Aleph H) EXP 4] (Aleph G). Note that in the above computations, we assume that only the material in the path of a reasonably sized spacecraft is upswept as mass-energy fuel or species of reaction and that the spacecraft makes only one effective pass through each universe.

If such Aleph G light-year–width universes have four spatial dimensions and are substantially static, then a spacecraft traveling at light-speed in the multiverse may obtain a Lorentz factor of roughly [(Aleph H) EXP 4] [(Aleph G) EXP 4]. Note that in the above computations, we assume that only the material in the path of a reasonably sized spacecraft is upswept as mass-energy fuel or species of reaction and that the spacecraft makes an effective pass entirely through each universe. Thus, all material in said universes of travel is reacted against with assumed finite universal mass-energy density.

If the latter universes are expanding with time over periods much greater than Aleph G years, then the spacecraft may attain a Lorentz factor of roughly [(Aleph H) EXP 4] [[k(Aleph G)] EXP 4], where *k* ranges from a small superunitary real number to suitable infinite real numbers. This, of course, assumes that real positive mass-energy stocks continue being created, perhaps out of zero-point fields or analogues thereof from some forms of dark energy. Provided that *k* ranges from Aleph 0 to Aleph (G + 2), then the spacecraft Lorentz factor will be limited to a range of roughly from [(Aleph H) EXP 4] [[(Aleph 0)(Aleph G)] EXP 4] to [(Aleph H) EXP 4] [[[Aleph (G + 2)](Aleph G)] EXP 4]. Note that in the above computations, we assume that only the material in the path of a reasonably sized spacecraft is upswept as mass-energy fuel or species of reaction and that the spacecraft makes an effective pass entirely through each universe. Thus, all material in said universes of travel is reacted against with assumed finite universal mass-energy density.

Now we consider spacecraft travel in a multiverse of Aleph H universes in width of five dimensions, where *H* is an integer greater than G + 2, and where said multiverse contains [(Aleph H) EXP 5] universes.

For universes in said 5-D–space multiverse having an extensity of (Aleph 1) light-years, there are prospects for the spacecraft to attain a Lorentz factor of roughly [(Aleph H) EXP 5] (Aleph 1) to thus result in the multiverse of travel being contracted to point directly in front of the spacecraft and Lorentz contracted by a factor of roughly [(Aleph H) EXP 5] (Aleph 1) relative to the spacecraft. Note that in the above computations, we assume that only the material in the path of a reasonably sized spacecraft is upswept as mass-energy fuel or species of reaction and that the spacecraft makes only one effective pass through each universe.

If such Aleph 1 light-year–width universes have four spatial dimensions and are substantially static, then a spacecraft traveling at light-speed in the multiverse of consideration may obtain a Lorentz factor of roughly [(Aleph H) EXP 5] [(Aleph 1) EXP 4]. Note that in the above computations, we assume that only the material in the path of a reasonably sized spacecraft is upswept as mass-energy fuel or species of reaction and that the spacecraft makes an effective pass entirely through each universe. Thus, all material in said universes of travel is reacted against with assumed finite universal mass-energy density.

If the latter universes are expanding with time over periods much greater than Aleph 1 years, then the spacecraft may attain a Lorentz factor of roughly [(Aleph H) EXP 5] [[k(Aleph 1)] EXP 4], where *k* ranges from a small superunitary real number to suitable infinite real numbers. This, of course, assumes that real positive mass-energy stocks continue being created, perhaps out of zero-point fields or analogues thereof from some forms of dark energy. Provided that *k* ranges from Aleph 0 to Aleph 3, then the spacecraft Lorentz factor will be limited to a range of roughly from [(Aleph H) EXP 5] [[(Aleph 0)(Aleph 1)] EXP 4] to [(Aleph H) EXP 5] [[(Aleph 3)(Aleph 1)] EXP 4]. Note that in the above computations, we assume that only the material in the path of a reasonably sized spacecraft is upswept as mass-energy fuel or species of reaction and that the spacecraft makes an effective entire pass through each universe. Thus, all material in said universes of travel is reacted against with assumed finite universal mass-energy density.

For universes in said multiverse having an extensity of Aleph 2 light-years, there are prospects for the spacecraft to attain a Lorentz factor of roughly [(Aleph H) EXP 5] (Aleph 2) to thus result in the multiverse of travel being contracted to a point directly in front of the spacecraft and by a factor of [(Aleph H) EXP 5] (Aleph 2). Note that in the above computations, we assume that only the material in the path of a reasonably sized spacecraft is upswept as mass-energy fuel or species of reaction and that the spacecraft makes only one effective pass through each universe.

If such Aleph 2 light-year–width universes have four spatial dimensions and are substantially static, then a spacecraft traveling at light-speed in the multiverse considered may obtain a Lorentz factor of roughly [(Aleph H) EXP 5] [(Aleph 2) EXP 4]. Note that in the above computations, we assume that only the material in the path of a reasonably sized spacecraft is upswept as mass-energy fuel or species of reaction and that the spacecraft makes an effective pass entirely through each universe. Thus, all material in said universes of travel is reacted against with assumed finite universal mass-energy density.

If the latter universes are expanding with time over periods much greater than Aleph 2 years, then the spacecraft may attain a Lorentz factor of roughly [(Aleph H) EXP 5] [[k(Aleph 2)] EXP 4], where k ranges from a small superunitary real number to suitable infinite real numbers. This, of course, assumes that real positive mass-energy stocks continue being created, perhaps out of zero-point fields or analogues thereof from some forms of dark energy. Provided that k ranges from Aleph 0 to Aleph 4, then the spacecraft Lorentz factor will be limited to a range of roughly from [(Aleph H) EXP 5] [[(Aleph 0)(Aleph 2)] EXP 4] to [(Aleph H) EXP 5] [[(Aleph 4)(Aleph 2)] EXP 4]. Note that in the above computations, we assume that only the material in the path of a reasonably sized spacecraft is upswept as mass-energy fuel or species of reaction and that the spacecraft makes an effective pass entirely through each universe. Thus, all material in said universes of travel is reacted against with assumed finite universal mass-energy density.

For universes in said multiverse having an extensity of Aleph 3 light-years, there are prospects for the spacecraft to attain a Lorentz factor of roughly [(Aleph H) EXP 5] (Aleph 3) to thus result in the multiverse of travel being contracted to point directly in front of the spacecraft and contracted by a factor of [(Aleph H) EXP 4](Aleph 3). Note that in the above computations, we assume that only the material in the path of a

reasonably sized spacecraft is upswept as mass-energy fuel or species of reaction and that the spacecraft makes only one effective pass through each universe.

If such Aleph 3 light-year–width universes have four spatial dimensions and are substantially static, then a spacecraft traveling at light-speed in the multiverse of consideration may obtain a Lorentz factor of roughly [(Aleph H) EXP 5] [(Aleph 3) EXP 4]. Note that in the above computations, we assume that only the material in the path of a reasonably sized spacecraft is upswept as mass-energy fuel or species of reaction and that the spacecraft makes an effective pass entirely through each universe. Thus, all material in said universes of travel is reacted against with assumed finite universal mass-energy density.

If the latter universes are expanding with time over periods much greater than Aleph 3 years, then the spacecraft may attain a Lorentz factor of roughly [(Aleph H) EXP 5] [[k(Aleph 3)] EXP 4], where k ranges from a small superunitary real number to suitable infinite real numbers. This, of course, assumes that real positive mass-energy stocks continue being created, perhaps out of zero-point fields or analogues thereof from some forms of dark energy. Provided that k ranges from Aleph 0 to Aleph 5, then the spacecraft Lorentz factor will be limited to a range of roughly from [(Aleph H) EXP 5] [[(Aleph 0)(Aleph 3)] EXP 4] to [(Aleph H) EXP 5] [[(Aleph 5)(Aleph 3)] EXP 4]. Note that in the above computations, we assume that only the material in the path of a reasonably sized spacecraft is upswept as mass-energy fuel or species of reaction and that the spacecraft makes an effective pass entirely through each universe. Thus, all material in said universes of travel is reacted against with assumed finite universal mass-energy density.

For universes having an extensity of Aleph 4 light-years, there are prospects for the spacecraft to attain a Lorentz factor of roughly [(Aleph H) EXP 5] (Aleph 4) to thus result in the multiverse of travel being contracted to point directly in front of the spacecraft and being contracted relative to the spacecraft by roughly a factor of [(Aleph H) EXP 4](Aleph 4). Note that in the above computations, we assume that only the material in the path of a reasonably sized spacecraft is upswept as mass-energy fuel or species of reaction and that the spacecraft makes only one effective pass through each universe.

If such Aleph 4 light-year–width universes have four spatial dimensions and are substantially static, then a spacecraft traveling at light-speed in the multiverse may obtain a Lorentz factor of roughly [(Aleph H) EXP 5] [(Aleph 4) EXP 4]. Note that in the above computations, we assume that only the material in the path of a reasonably sized spacecraft is upswept as mass-energy fuel or species of reaction and that the spacecraft makes an effective pass entirely through each universe. Thus, all material in said universes of travel is reacted against with assumed finite universal mass-energy density.

If the latter universes are expanding with time over periods much greater than Aleph 4 years, then the spacecraft may attain a Lorentz factor of roughly [(Aleph H) EXP 5] [[k(Aleph 4)] EXP 4], where k ranges from a small superunitary real number to suitable infinite real numbers. This, of course, assumes that real positive mass-energy stocks continue being created, perhaps out of zero-point fields or analogues thereof from some forms of dark energy. Provided that k ranges roughly from Aleph 0 to Aleph 6, then the spacecraft Lorentz factor will be limited to a range from [(Aleph H) EXP 5] [[(Aleph 0)(Aleph 4)] EXP 4] to [(Aleph H) EXP 5] [[(Aleph 6)(Aleph 4)] EXP 4]. Note that in the above computations, we assume that only the material in the path of a reasonably sized spacecraft is upswept as mass-energy fuel or species of reaction and that the spacecraft makes an effective pass entirely through each universe. Thus, all material in said universes of travel is reacted against with assumed finite universal mass-energy density.

For universes having an extensity of Aleph G light-years, there are prospects for the spacecraft to attain a Lorentz factor of roughly [(Aleph H) EXP 5] (Aleph G) to thus result in the multiverse of travel being contracted to point directly in front of the spacecraft with a Lorentz contraction of roughly [(Aleph H) EXP 5] (Aleph G). Note that in the above computations, we assume that only the material in the path of a reasonably sized spacecraft is upswept as mass-energy fuel or species of reaction and that the spacecraft makes only one effective pass through each universe.

If such Aleph G light-year–width universes have four spatial dimensions and are substantially static, then a spacecraft traveling at light-speed in the multiverse may obtain a Lorentz factor of roughly [(Aleph H) EXP 5] [(Aleph G) EXP 4]. Note that in the above computations, we assume that only the material in the path of a reasonably sized spacecraft is upswept as mass-energy fuel or species of reaction and that the spacecraft makes an effective pass entirely through each universe. Thus, all material in said universes of travel is reacted against with assumed finite universal mass-energy density.

If the latter universes are expanding with time over periods much greater than Aleph G years, then the spacecraft may attain a Lorentz factor of roughly [(Aleph H) EXP 5] [[k(Aleph G)] EXP 4], where k ranges from a small superunitary real number to suitable infinite real numbers. This, of course, assumes that real positive mass-energy stocks continue being created, perhaps out of zero-point fields or analogues thereof from some forms of dark energy. Provided that k ranges from Aleph 0 to Aleph (G + 2), then the spacecraft Lorentz factor will be limited to a range of roughly from [(Aleph H) EXP 5] [[(Aleph 0)(Aleph G)] EXP 4] to [(Aleph H) EXP 5] [[[Aleph (G + 2)](Aleph G)] EXP 4]. Note that in the above computations, we assume that only the material in the path of a reasonably sized spacecraft is upswept as mass-energy fuel or species of reaction and that the spacecraft makes an effective pass entirely through each universe. Thus, all material in said universes of travel is reacted against with assumed finite universal mass-energy density.

Now we consider spacecraft travel in a multiverse of Aleph H universes in width of six dimensions, where H is an integer greater than G + 2, and where said multiverse contains [(Aleph H) EXP 6] universes.

For universes in said 6-D–space multiverse having an extensity of (Aleph 1) light-years, there are prospects for the spacecraft to attain a Lorentz factor of roughly [(Aleph H) EXP 6] (Aleph 1) to thus result in the multiverse of travel being contracted to point directly in front of the spacecraft and Lorentz contracted by a factor of roughly [(Aleph H) EXP 6] (Aleph 1) relative to the spacecraft. Note that in the above computations, we assume that only the material in the path of a reasonably sized spacecraft is upswept as mass-energy fuel or species of reaction and that the spacecraft makes only one effective pass through each universe.

If such Aleph 1 light-year–width universes have four spatial dimensions and are substantially static, then a spacecraft traveling at light-speed in the multiverse of consideration may obtain a Lorentz factor of roughly [(Aleph H) EXP 6] [(Aleph 1) EXP 4]. Note that in the above computations, we assume that only the material in the path of a reasonably sized spacecraft is upswept as mass-energy fuel or species of reaction and that the spacecraft makes an effective pass entirely through each universe. Thus, all material in said universes of travel is reacted against with assumed finite universal mass-energy density.

If the latter universes are expanding with time over periods much greater than Aleph 1 years, then the spacecraft may attain a Lorentz factor of roughly [(Aleph H) EXP 6] [[k(Aleph 1)] EXP 4], where k ranges from a small superunitary real number to suitable infinite real numbers. This, of course, assumes that real positive mass-energy stocks continue being created, perhaps out of zero-point fields or analogues thereof from some forms of dark energy. Provided that k ranges from Aleph 0 to Aleph 3, then the spacecraft Lorentz factor will be limited to a range of roughly from [(Aleph H) EXP 6] [[(Aleph 0)(Aleph 1)] EXP 4] to [(Aleph H) EXP 6] [[(Aleph 3)(Aleph 1)] EXP 4]. Note that in the above computations, we assume that only the material in the path of a reasonably sized spacecraft is upswept as mass-energy fuel or species of reaction and that the spacecraft makes an effective entire pass through each universe. Thus, all material in said universes of travel is reacted against with assumed finite universal mass-energy density.

For universes in said multiverse having an extensity of Aleph 2 light-years, there are prospects for the spacecraft to attain a Lorentz factor of roughly [(Aleph H) EXP 6] (Aleph 2) to thus result in the multiverse of travel being contracted to a point directly in front of the spacecraft and by a factor of [(Aleph H) EXP 6] (Aleph 2). Note that in the above computations, we assume that only the material in the path of a reasonably sized spacecraft is upswept as mass-energy fuel or species of reaction and that the spacecraft makes only one effective pass through each universe.

If such Aleph 2 light-year–width universes have four spatial dimensions and are substantially static, then a spacecraft traveling at light-speed in the multiverse considered may obtain a Lorentz factor of roughly [(Aleph H) EXP 6] [(Aleph 2) EXP 4]. Note that in the above computations, we assume that only the material in the path of a reasonably sized spacecraft is upswept as mass-energy fuel or species of reaction and that the spacecraft makes an effective pass entirely through each universe. Thus, all material in said universes of travel is reacted against with assumed finite universal mass-energy density.

If the latter universes are expanding with time over periods much greater than Aleph 2 years, then the spacecraft may attain a Lorentz factor of roughly [(Aleph H) EXP 6] [[k(Aleph 2)] EXP 4], where k ranges from a small superunitary real number to suitable infinite real numbers. This, of course, assumes that real positive mass-energy stocks continue being created, perhaps out of zero-point fields or analogues thereof from some forms of dark energy. Provided that k ranges from Aleph 0 to Aleph 4, then the spacecraft Lorentz factor will be limited to a range of roughly from [(Aleph H) EXP 6] [[(Aleph 0)(Aleph 2)] EXP 4] to [(Aleph H) EXP 6] [[(Aleph 4)(Aleph 2)] EXP 4]. Note that in the above computations, we assume that only the material in the path of a reasonably sized spacecraft is upswept as mass-energy fuel or species of reaction and that the spacecraft makes an effective pass entirely through each universe. Thus, all material in said universes of travel is reacted against with assumed finite universal mass-energy density.

For universes in said multiverse having an extensity of Aleph 3 light-years, there are prospects for the spacecraft to attain a Lorentz factor of roughly [(Aleph H) EXP 6] (Aleph 3) to thus result in the multiverse of travel being contracted to point directly in front of the spacecraft and contracted by a factor of [(Aleph H) EXP 4](Aleph 3). Note that in the above computations, we assume that only the material in the path of a reasonably sized spacecraft is upswept as mass-energy fuel or species of reaction and that the spacecraft makes only one effective pass through each universe.

If such Aleph 3 light-year–width universes have four spatial dimensions and are substantially static, then a spacecraft traveling at light-speed in the multiverse of consideration may obtain a Lorentz factor of roughly [(Aleph H) EXP 6] [(Aleph 3) EXP 4]. Note that in the above computations, we assume that only the material in the path of a reasonably sized spacecraft is upswept as mass-energy fuel or species of reaction and that the spacecraft makes an effective pass entirely through each universe. Thus, all material in said universes of travel is reacted against with assumed finite universal mass-energy density.

If the latter universes are expanding with time over periods much greater than Aleph 3 years, then the spacecraft may attain a Lorentz factor of roughly [(Aleph H) EXP 6] [[k(Aleph 3)] EXP 4], where k ranges from a small superunitary real number to suitable infinite real numbers. This, of course, assumes that real positive mass-energy stocks continue being created, perhaps out of zero-point fields or analogues thereof from some forms of dark energy. Provided that k ranges from Aleph 0 to Aleph 5, then the spacecraft Lorentz factor will be limited to a range of roughly from [(Aleph H) EXP 6] [[(Aleph 0)(Aleph 3)] EXP 4] to [(Aleph H) EXP 6] [[(Aleph 5)(Aleph 3)] EXP 4]. Note that in the above computations, we assume that only the material in the path of a reasonably sized spacecraft is upswept as mass-energy fuel or species of reaction and that the spacecraft makes an effective pass entirely through each universe. Thus, all material in said universes of travel is reacted against with assumed finite universal mass-energy density.

For universes having an extensity of Aleph 4 light-years, there are prospects for the spacecraft to attain a Lorentz factor of roughly [(Aleph H) EXP 6] (Aleph 4) to thus result in the multiverse of travel being contracted to point directly in front of the spacecraft and being contracted relative to the spacecraft by roughly a factor of [(Aleph H) EXP 4](Aleph 4). Note that in the above computations, we assume that only the material in the path of a reasonably sized spacecraft is upswept as mass-energy fuel or species of reaction and that the spacecraft makes only one effective pass through each universe.

If such Aleph 4 light-year–width universes have four spatial dimensions and are substantially static, then a spacecraft traveling at light-speed in the multiverse may obtain a Lorentz factor of roughly [(Aleph H) EXP 6] [(Aleph 4) EXP 4]. Note that in the above computations, we assume that only the material in the path of a reasonably sized spacecraft is upswept as mass-energy fuel or species of reaction and that the spacecraft makes an effective pass entirely through each universe. Thus, all material in said universes of travel is reacted against with assumed finite universal mass-energy density.

If the latter universes are expanding with time over periods much greater than Aleph 4 years, then the spacecraft may attain a Lorentz factor of roughly [(Aleph H) EXP 6] [[k(Aleph 4)] EXP 4], where k ranges from a small superunitary real number to suitable infinite real numbers. This, of course, assumes that real positive mass-energy stocks continue being created, perhaps out of zero-point fields or analogues thereof from some forms of dark energy. Provided that k ranges roughly from Aleph 0 to Aleph 6, then the spacecraft Lorentz factor will be limited to a range from [(Aleph H) EXP 6] [[(Aleph 0)(Aleph 4)] EXP 4] to [(Aleph H) EXP 6] [[(Aleph 6)(Aleph 4)] EXP 4]. Note that in the above computations, we assume that only the material in the path of a reasonably sized spacecraft is upswept as mass-energy fuel or species of reaction and that the spacecraft makes an effective pass entirely through each universe. Thus, all material in said universes of travel is reacted against with assumed finite universal mass-energy density.

For universes having an extensity of Aleph G light-years, there are prospects for the spacecraft to attain a Lorentz factor of roughly [(Aleph H) EXP 6] (Aleph G) to thus result in the multiverse of travel being contracted to point directly in front of the spacecraft with a Lorentz contraction of roughly [(Aleph H) EXP 6] (Aleph G). Note that in the above computations, we assume that only the material in the path of a reasonably sized spacecraft is upswept as mass-energy fuel or species of reaction and that the spacecraft makes only one effective pass through each universe.

If such Aleph G light-year–width universes have four spatial dimensions and are substantially static, then a spacecraft traveling at light-speed in the multiverse may obtain a Lorentz factor of roughly [(Aleph H) EXP 6] [(Aleph G) EXP 4]. Note that in the above computations, we assume that only the material in the path of a reasonably sized spacecraft is upswept as mass-energy fuel or species of reaction and that the spacecraft makes an effective pass entirely through each universe. Thus, all material in said universes of travel is reacted against with assumed finite universal mass-energy density.

If the latter universes are expanding with time over periods much greater than Aleph G years, then the spacecraft may attain a Lorentz factor of roughly [(Aleph H) EXP 6] [[k(Aleph G)] EXP 4], where k ranges from a small superunitary real number to suitable infinite real numbers. This, of course, assumes that real positive mass-energy stocks continue being created, perhaps out of zero-point fields or analogues thereof from some forms of dark energy. Provided that k ranges from Aleph 0 to Aleph (G + 2), then the spacecraft Lorentz factor will be limited to a range of roughly from [(Aleph H) EXP 6] [[(Aleph 0)(Aleph G)] EXP 4] to [(Aleph H) EXP 6] [[[Aleph (G + 2)](Aleph G)] EXP 4]. Note that in the above computations, we assume that only the material in the path of a reasonably sized spacecraft is upswept as mass-energy fuel or species of reaction and that the spacecraft makes an effective pass entirely through each universe. Thus, all material in said universes of travel is reacted against with assumed finite universal mass-energy density.

Now we consider spacecraft travel in a multiverse of Aleph H universes in width of N dimensions, where H is an integer greater than G + 2, and where said multiverse contains [(Aleph H) EXP N] universes.

For universes in said N-D–space multiverse having an extensity of (Aleph 1) light-years, there are prospects for the spacecraft to attain a Lorentz factor of roughly [(Aleph H) EXP N] (Aleph 1) to thus result in the multiverse of travel being contracted to point directly in front of the spacecraft and Lorentz contracted by a factor of roughly [(Aleph H) EXP N] (Aleph 1) relative to the spacecraft. Note that in the above computations, we assume that only the material in the path of a reasonably sized spacecraft is upswept as mass-energy fuel or species of reaction and that the spacecraft makes only one effective pass through each universe.

If such Aleph 1 light-year–width universes have four spatial dimensions and are substantially static, then a spacecraft traveling at light-speed in the multiverse of consideration may obtain a Lorentz factor of roughly [(Aleph H) EXP N] [(Aleph 1) EXP 4]. Note that in the above computations, we assume that only the material in the path of a reasonably sized spacecraft is upswept as mass-energy fuel or species of reaction and that the spacecraft makes an effective pass entirely through each universe. Thus, all material in said universes of travel is reacted against with assumed finite universal mass-energy density.

If the latter universes are expanding with time over periods much greater than Aleph 1 years, then the spacecraft may attain a Lorentz factor of roughly [(Aleph H) EXP N] [[k(Aleph 1)] EXP 4], where k ranges from a small superunitary real number to suitable infinite real numbers. This, of course, assumes that real positive mass-energy stocks continue being created, perhaps out of zero-point fields or analogues thereof from some forms of dark energy. Provided that k ranges from Aleph 0 to Aleph 3, then the spacecraft Lorentz factor will be limited to a range of roughly from [(Aleph H) EXP N] [[(Aleph 0)(Aleph 1)] EXP 4] to [(Aleph H) EXP N] [[(Aleph 3)(Aleph 1)] EXP 4]. Note that in the above computations, we assume that only the material in the path of a reasonably sized spacecraft is upswept as mass-energy fuel or species of reaction and that the spacecraft makes an effective entire pass through each universe. Thus, all material in said universes of travel is reacted against with assumed finite universal mass-energy density.

For universes in said multiverse having an extensity of Aleph 2 light-years, there are prospects for the spacecraft to attain a Lorentz factor of roughly [(Aleph H) EXP N] (Aleph 2) to thus result in the multiverse of travel being contracted to a point directly in front of the spacecraft and by a factor of [(Aleph H) EXP N] (Aleph 2). Note that in the above computations, we assume that only the material in the path of a reasonably sized spacecraft is upswept as mass-energy fuel or species of reaction and that the spacecraft makes only one effective pass through each universe.

If such Aleph 2 light-year–width universes have four spatial dimensions and are substantially static, then a spacecraft traveling at light-speed in the multiverse considered may obtain a Lorentz factor of roughly [(Aleph H) EXP N] [(Aleph 2) EXP 4]. Note that in the above computations, we assume that only the material in the path of a reasonably sized spacecraft is upswept as mass-energy fuel or species of reaction and that the spacecraft makes an effective pass entirely through each universe. Thus, all material in said universes of travel is reacted against with assumed finite universal mass-energy density.

If the latter universes are expanding with time over periods much greater than Aleph 2 years, then the spacecraft may attain a Lorentz factor of roughly [(Aleph H) EXP N] [[k(Aleph 2)] EXP 4], where k ranges from a small superunitary real number to suitable infinite real numbers. This, of course, assumes that real positive mass-energy stocks continue being created, perhaps out of zero-point fields or analogues thereof from some forms of dark energy. Provided that k ranges from Aleph 0 to Aleph 4, then the spacecraft Lorentz factor will be limited to a range of roughly from [(Aleph H) EXP N] [[(Aleph 0)(Aleph 2)] EXP 4] to [(Aleph H) EXP N] [[(Aleph 4)(Aleph 2)] EXP 4]. Note that in the above computations, we assume that only the material in the path of a reasonably sized spacecraft is upswept as mass-energy fuel or species of reaction and that the spacecraft makes an effective pass entirely through each universe. Thus, all material in said universes of travel is reacted against with assumed finite universal mass-energy density.

For universes in said multiverse having an extensity of Aleph 3 light-years, there are prospects for the spacecraft to attain a Lorentz factor of roughly [(Aleph H) EXP N] (Aleph 3) to thus result in the multiverse of travel being contracted to point directly in front of the spacecraft and contracted by a factor of [(Aleph H) EXP 4](Aleph 3). Note that in the above computations, we assume that only the material in the path of a reasonably sized spacecraft is upswept as mass-energy fuel or species of reaction and that the spacecraft makes only one effective pass through each universe.

If such Aleph 3 light-year–width universes have four spatial dimensions and are substantially static, then a spacecraft traveling at light-speed in the multiverse of consideration may obtain a Lorentz factor of roughly [(Aleph H) EXP N] [(Aleph 3) EXP 4]. Note that in the above computations, we assume that only the material in the path of a reasonably sized spacecraft is upswept as mass-energy fuel or species of reaction and that the spacecraft makes an effective pass entirely through each universe. Thus, all material in said universes of travel is reacted against with assumed finite universal mass-energy density.

If the latter universes are expanding with time over periods much greater than Aleph 3 years, then the spacecraft may attain a Lorentz factor of roughly [(Aleph H) EXP N] [[k(Aleph 3)] EXP 4], where k ranges from a small superunitary real number to suitable infinite real numbers. This, of course, assumes that real positive mass-energy stocks continue being created, perhaps out of zero-point fields or analogues thereof from some forms of dark energy. Provided that k ranges from Aleph 0 to Aleph 5, then the spacecraft Lorentz factor will be limited to a range of roughly from [(Aleph H) EXP N] [[(Aleph 0)(Aleph 3)] EXP 4] to [(Aleph H) EXP N] [[(Aleph 5)(Aleph 3)] EXP 4]. Note that in the above computations, we assume that only the material in the path of a reasonably sized spacecraft is upswept as mass-energy fuel or species of reaction and that the spacecraft makes an effective pass entirely through each universe. Thus, all material in said universes of travel is reacted against with assumed finite universal mass-energy density.

For universes having an extensity of Aleph 4 light-years, there are prospects for the spacecraft to attain a Lorentz factor of roughly [(Aleph H) EXP N] (Aleph 4) to thus result in the multiverse of travel being contracted to point directly in front of the spacecraft and being contracted relative to the spacecraft by roughly a factor of [(Aleph H) EXP 4](Aleph 4). Note that in the above computations, we assume that only the material in the path of a reasonably sized spacecraft is upswept as mass-energy fuel or species of reaction and that the spacecraft makes only one effective pass through each universe.

If such Aleph 4 light-year–width universes have four spatial dimensions and are substantially static, then a spacecraft traveling at light-speed in the multiverse may obtain a Lorentz factor of roughly [(Aleph H) EXP N] [(Aleph 4) EXP 4]. Note that in the above computations, we assume that only the material in the path of a reasonably sized spacecraft is upswept as mass-energy fuel or species of reaction and that the spacecraft makes an effective pass entirely through each universe. Thus, all material in said universes of travel is reacted against with assumed finite universal mass-energy density.

If the latter universes are expanding with time over periods much greater than Aleph 4 years, then the spacecraft may attain a Lorentz factor of roughly [(Aleph H) EXP N] [[k(Aleph 4)] EXP 4], where k ranges from a small superunitary real number to suitable infinite real numbers. This, of course, assumes that real positive mass-energy stocks continue being created, perhaps out of zero-point fields or analogues thereof from some forms of dark energy. Provided that k ranges roughly from Aleph 0 to Aleph 6, then the spacecraft Lorentz factor will be limited to a range from [(Aleph H) EXP N] [[(Aleph 0)(Aleph 4)] EXP 4] to [(Aleph H) EXP N] [[(Aleph 6)(Aleph 4)] EXP 4]. Note that in the above computations, we assume that only the material in the path of a reasonably sized spacecraft is upswept as mass-energy fuel or species of reaction and that the spacecraft makes an effective pass entirely through each universe. Thus, all material in said universes of travel is reacted against with assumed finite universal mass-energy density.

For universes having an extensity of Aleph G light-years, there are prospects for the spacecraft to attain a Lorentz factor of roughly [(Aleph H) EXP N] (Aleph G) to thus result in the multiverse of travel being contracted to point directly in front of the spacecraft with a Lorentz contraction of roughly [(Aleph H) EXP N] (Aleph G). Note that in the above computations, we assume that only the material in the path of a reasonably sized spacecraft is upswept as mass-energy fuel or species of reaction and that the spacecraft makes only one effective pass through each universe.

If such Aleph G light-year–width universes have four spatial dimensions and are substantially static, then a spacecraft traveling at light-speed in the multiverse may obtain a Lorentz factor of roughly [(Aleph H) EXP N] [(Aleph G) EXP 4]. Note that in the above computations, we assume that only the material in the path of a reasonably sized spacecraft is upswept as mass-energy fuel or species of reaction and that the spacecraft makes an effective pass entirely through each universe. Thus, all material in said universes of travel is reacted against with assumed finite universal mass-energy density.

If the latter universes are expanding with time over periods much greater than Aleph G years, then the spacecraft may attain a Lorentz factor of roughly [(Aleph H) EXP N] [[k(Aleph G)] EXP 4], where *k* ranges from a small superunitary real number to suitable infinite real numbers. This, of course, assumes that real positive mass-energy stocks continue being created, perhaps out of zero-point fields or analogues thereof from some forms of dark energy. Provided that *k* ranges from Aleph 0 to Aleph (G + 2), then the spacecraft Lorentz factor will be limited to a range roughly from [(Aleph H) EXP N] [[(Aleph 0)(Aleph G)] EXP 4] to [(Aleph H) EXP N] [[[Aleph (G + 2)](Aleph G)] EXP 4]. Note that in the above computations, we assume that only the material in the path of a reasonably sized spacecraft is upswept as mass-energy fuel or species of reaction and that the spacecraft makes an effective pass entirely through each universe. Thus, all material in said universes of travel is reacted against with assumed finite universal mass-energy density.

Now we consider spacecraft travel in a multiverse of Aleph H universes in width of five dimensions, where *H* is an integer greater than G + 2, and where said multiverse contains [(Aleph H) EXP 5] universes.

For universes in said 5-D–space multiverse having an extensity of (Aleph 1) light-years, there are prospects for the spacecraft to attain a Lorentz factor of roughly [(Aleph H) EXP 5] (Aleph 1) to thus result in the multiverse of travel being contracted to point directly in front of the spacecraft and Lorentz contracted by a factor of roughly [(Aleph H) EXP 5] (Aleph 1) relative to the spacecraft. Note that in the above computations, we assume that only the material in the path of a reasonably sized spacecraft is upswept as mass-energy fuel or species of reaction and that the spacecraft makes only one effective pass through each universe.

If such Aleph 1 light-year–width universes have five spatial dimensions and are substantially static, then a spacecraft traveling at light-speed in the multiverse of consideration may obtain a Lorentz factor of roughly [(Aleph H) EXP 5] [(Aleph 1) EXP 5]. Note that in the above computations, we assume that only the material in the path of a reasonably sized spacecraft is upswept as mass-energy fuel or species of reaction and that the spacecraft makes an effective pass entirely through each universe. Thus, all material in said universes of travel is reacted against with assumed finite universal mass-energy density.

If the latter universes are expanding with time over periods much greater than Aleph 1 years, then the spacecraft may attain a Lorentz factor of roughly [(Aleph H) EXP 5] [[k(Aleph 1)] EXP 5], where k ranges from a small superunitary real number to suitable infinite real numbers. This, of course, assumes that real positive mass-energy stocks continue being created, perhaps out of zero-point fields or analogues thereof from some forms of dark energy. Provided k ranges from Aleph 0 to Aleph 3, then the spacecraft Lorentz factor will be limited to a range of roughly from [(Aleph H) EXP 5] [[(Aleph 0)(Aleph 1)] EXP 5] to [(Aleph H) EXP 5] [[(Aleph 3)(Aleph 1)] EXP 5]. Note that in the above computations, we assume that only the material in the path of a reasonably sized spacecraft is upswept as mass-energy fuel or species of reaction and that the spacecraft makes an effective entire pass through each universe. Thus, all material in said universes of travel is reacted against with assumed finite universal mass-energy density.

For universes in said multiverse having an extensity of Aleph 2 light-years, there are prospects for the spacecraft to attain a Lorentz factor of roughly [(Aleph H) EXP 5] (Aleph 2) to thus result in the multiverse of travel being contracted to a point directly in front of the spacecraft and by a factor of [(Aleph H) EXP 5] (Aleph 2). Note that in the above computations, we assume that only the material in the path of a reasonably sized spacecraft is upswept as mass-energy fuel or species of reaction and that the spacecraft makes only one effective pass through each universe.

If such Aleph 2 light-year–width universes have five spatial dimensions and are substantially static, then a spacecraft traveling at light-speed in the multiverse considered may obtain a Lorentz factor of roughly [(Aleph H) EXP 5] [(Aleph 2) EXP 5]. Note that in the above computations, we assume that only the material in the path of a reasonably sized spacecraft is upswept as mass-energy fuel or species of reaction and that the spacecraft makes an effective pass entirely through each universe. Thus, all material in said universes of travel is reacted against with assumed finite universal mass-energy density.

If the latter universes are expanding with time over periods much greater than Aleph 2 years, then the spacecraft may attain a Lorentz factor of roughly [(Aleph H) EXP 5] [[k(Aleph 2)] EXP 5], where k ranges from a small superunitary real number to suitable infinite real numbers. This, of course, assumes that real positive mass-energy stocks continue being created, perhaps out of zero-point fields or analogues thereof from some forms of dark energy. Provided that k ranges from Aleph 0 to Aleph 4, then the spacecraft Lorentz factor will be limited to a range of roughly from [(Aleph H) EXP 5] [[(Aleph 0)(Aleph 2)] EXP 5] to [(Aleph H) EXP 5] [[(Aleph 4)(Aleph 2)] EXP 5]. Note that in the above computations, we assume that only the material in the path of a reasonably sized spacecraft is upswept as mass-energy fuel or species of reaction and that the spacecraft makes an effective pass entirely through each universe. Thus, all material in said universes of travel is reacted against with assumed finite universal mass-energy density.

For universes in said multiverse having an extensity of Aleph 3 light-years, there are prospects for the spacecraft to attain a Lorentz factor of roughly [(Aleph H) EXP 5] (Aleph 3) to thus result in the multiverse of travel being contracted to point directly in front of the spacecraft and contracted by a factor of [(Aleph H) EXP 5](Aleph 3). Note that in the above computations, we assume that only the material in the path of a reasonably sized spacecraft is upswept as mass-energy fuel or species of reaction and that the spacecraft makes only one effective pass through each universe.

If such Aleph 3 light-year–width universes have five spatial dimensions and are substantially static, then a spacecraft traveling at light-speed in the multiverse of consideration may obtain a Lorentz factor of roughly [(Aleph H) EXP 5] [(Aleph 3) EXP 5]. Note that in the above computations, we assume that only the material in the path of a reasonably sized spacecraft is upswept as mass-energy fuel or species of reaction and that the spacecraft makes an effective pass entirely through each universe. Thus, all material in said universes of travel is reacted against with assumed finite universal mass-energy density.

If the latter universes are expanding with time over periods much greater than Aleph 3 years, then the spacecraft may attain a Lorentz factor of roughly [(Aleph H) EXP 5] [[k(Aleph 3)] EXP 5], where k ranges from a small superunitary real number to suitable infinite real numbers. This, of course, assumes that real positive mass-energy stocks continue being created, perhaps out of zero-point fields or analogues thereof from some forms of dark energy. Provided that k ranges from Aleph 0 to Aleph 5, then the spacecraft Lorentz factor will be limited to a range of roughly from [(Aleph H) EXP 5] [[(Aleph 0)(Aleph 3)] EXP 5] to [(Aleph H) EXP 5] [[(Aleph 5)(Aleph 3)] EXP 5]. Note that in the above computations, we assume that only the material in the path of a reasonably sized spacecraft is upswept as mass-energy fuel or species of reaction and that the spacecraft makes an effective pass entirely through each universe. Thus, all material in

said universes of travel is reacted against with assumed finite universal mass-energy density.

For universes having an extensity of Aleph 4 light-years, there are prospects for the spacecraft to attain a Lorentz factor of roughly [(Aleph H) EXP 5] (Aleph 4) to thus result in the multiverse of travel being contracted to point directly in front of the spacecraft and being contracted relative to the spacecraft by roughly a factor of [(Aleph H) EXP 5](Aleph 4). Note that in the above computations, we assume that only the material in the path of a reasonably sized spacecraft is upswept as mass-energy fuel or species of reaction and that the spacecraft makes only one effective pass through each universe.

If such Aleph 4 light-year–width universes have five spatial dimensions and are substantially static, then a spacecraft traveling at light-speed in the multiverse may obtain a Lorentz factor of roughly [(Aleph H) EXP 5] [(Aleph 4) EXP 5]. Note that in the above computations, we assume that only the material in the path of a reasonably sized spacecraft is upswept as mass-energy fuel or species of reaction and that the spacecraft makes an effective pass entirely through each universe. Thus, all material in said universes of travel is reacted against with assumed finite universal mass-energy density.

If the latter universes are expanding with time over periods much greater than Aleph 4 years, then the spacecraft may attain a Lorentz factor of roughly [(Aleph H) EXP 5] [[k(Aleph 4)] EXP 5], where k ranges from a small superunitary real number to suitable infinite real numbers. This, of course, assumes that real positive mass-energy stocks continue being created, perhaps out of zero-point fields or analogues thereof from some forms of dark energy. Provided that r ranges roughly from Aleph 0 to Aleph 6, then the spacecraft Lorentz factor will be limited to a range from [(Aleph H) EXP 5] [[(Aleph 0)(Aleph 4)] EXP 5] to [(Aleph H) EXP 5] [[(Aleph 6)(Aleph 4)] EXP 5]. Note that in the above computations, we assume that only the material in the path of a reasonably sized spacecraft is upswept as mass-energy fuel or species of reaction and that the spacecraft makes an effective pass entirely through each universe. Thus, all material in said universes of travel is reacted against with assumed finite universal mass-energy density.

For universes having an extensity of Aleph G light-years, there are prospects for the spacecraft to attain a Lorentz factor of roughly [(Aleph H) EXP 5] (Aleph G) to thus result in the multiverse of travel being contracted to point directly in front of the spacecraft with a Lorentz contraction of roughly [(Aleph H) EXP 5] (Aleph G). Note that in the above computations, we assume that only the material in the path of a reasonably sized spacecraft is upswept as mass-energy fuel or species of reaction and that the spacecraft makes only one effective pass through each universe.

If such Aleph G light-year–width universes have five spatial dimensions and are substantially static, then a spacecraft traveling at light-speed in the multiverse may obtain a Lorentz factor of roughly [(Aleph H) EXP 5] [(Aleph G) EXP 5]. Note that in the above computations, we assume that only the material in the path of a reasonably sized spacecraft is upswept as mass-energy fuel or species of reaction and that the spacecraft makes an effective pass entirely through each universe. Thus, all material in said universes of travel is reacted against with assumed finite universal mass-energy density.

If the latter universes are expanding with time over periods much greater than Aleph G years, then the spacecraft may attain a Lorentz factor of roughly [(Aleph H) EXP 5] [[k(Aleph G)] EXP 5], where k ranges from a small superunitary real number to suitable infinite real numbers. This, of course, assumes that real positive mass-energy stocks continue being created, perhaps out of zero-point fields or analogues thereof from some forms of dark energy. Provided that k ranges from Aleph 0 to Aleph (G + 2), then the spacecraft Lorentz factor will be limited to a range of roughly from [(Aleph H) EXP 5] [[(Aleph 0)(Aleph G)] EXP 5] to [(Aleph H) EXP 5] [[[Aleph (G + 2)](Aleph G)] EXP 5]. Note that in the above computations, we assume that only the material in the path of a reasonably sized spacecraft is upswept as mass-energy fuel or species of reaction and that the spacecraft makes an effective pass entirely through each universe. Thus, all material in said universes of travel is reacted against with assumed finite universal mass-energy density.

Now we consider spacecraft travel in a multiverse of Aleph H universes in width of six dimensions, where H is an integer greater than G + 2, and where said multiverse contains [(Aleph H) EXP 6] universes.

For universes in said 6-D–space multiverse having an extensity of (Aleph 1) light-years, there are prospects for the spacecraft to attain a Lorentz factor of roughly [(Aleph H) EXP 6] (Aleph 1) to thus result in the multiverse of travel being contracted to point directly in front of the spacecraft and Lorentz contracted by a factor of roughly [(Aleph H) EXP 6] (Aleph 1) relative to the spacecraft. Note that in the above computations, we assume that only the material in the path of a reasonably sized spacecraft is upswept as mass-energy fuel or species of reaction and that the spacecraft makes only one effective pass through each universe.

If such Aleph 1 light-year–width universes have five spatial dimensions and are substantially static, then a spacecraft traveling at light-speed in the multiverse of consideration may obtain a Lorentz factor of roughly [(Aleph H) EXP 6] [(Aleph 1) EXP 5]. Note that in the above computations, we assume that only the material in the path of a reasonably sized spacecraft is upswept as mass-energy fuel or species of reaction and that the spacecraft makes an effective pass entirely through each universe. Thus, all material in said universes of travel is reacted against with assumed finite universal mass-energy density.

If the latter universes are expanding with time over periods much greater than Aleph 1 years, then the spacecraft may attain a Lorentz factor of roughly [(Aleph H) EXP 6] [[k(Aleph 1)] EXP 5], where *k* ranges from a small superunitary real number to suitable infinite real numbers. This, of course, assumes that real positive mass-energy stocks continue being created, perhaps out of zero-point fields or analogues thereof from some forms of dark energy. Provided that *k* ranges from Aleph 0 to Aleph 3, then the spacecraft Lorentz factor will be limited to a range of roughly from [(Aleph H) EXP 6] [[(Aleph 0)(Aleph 1)] EXP 5] to [(Aleph H) EXP 6] [[(Aleph 3)(Aleph 1)] EXP 5]. Note that in the above computations, we assume that only the material in the path of a reasonably sized spacecraft is upswept as mass-energy fuel or species of reaction and that the spacecraft makes an effective entire pass through each universe. Thus, all material in said universes of travel is reacted against with assumed finite universal mass-energy density.

For universes in said multiverse having an extensity of Aleph 2 light-years, there are prospects for the spacecraft to attain a Lorentz factor of roughly [(Aleph H) EXP 6] (Aleph 2) to thus result in the multiverse of travel being contracted to a point directly in front of the spacecraft and by a factor of [(Aleph H) EXP 6] (Aleph 2). Note that in the above computations, we assume that only the material in the path of a reasonably sized spacecraft is upswept as mass-energy fuel or species of reaction and that the spacecraft makes only one effective pass through each universe.

If such Aleph 2 light-year–width universes have five spatial dimensions and are substantially static, then a spacecraft traveling at light-speed in the multiverse considered may obtain a Lorentz factor of roughly [(Aleph H) EXP 6] [(Aleph 2) EXP 5]. Note that in the above computations, we assume that only the material in the path of a reasonably sized spacecraft is upswept as mass-energy fuel or species of reaction and that the spacecraft makes an effective pass entirely through each universe. Thus, all material in said universes of travel is reacted against with assumed finite universal mass-energy density.

If the latter universes are expanding with time over periods much greater than Aleph 2 years, then the spacecraft may attain a Lorentz factor of roughly [(Aleph H) EXP 6] [[k(Aleph 2)] EXP 5], where *k* ranges from a small superunitary real number to suitable infinite real numbers. This, of course, assumes that real positive mass-energy stocks continue being created, perhaps out of zero-point fields or analogues thereof from some forms of dark energy. Provided that *k* ranges from Aleph 0 to Aleph 4, then the spacecraft Lorentz factor will be limited to a range of roughly from [(Aleph H) EXP 6] [[(Aleph 0)(Aleph 2)] EXP 5] to [(Aleph H) EXP 6] [[(Aleph 4)(Aleph 2)] EXP 5]. Note that in the above computations, we assume that only the material in the path of a reasonably sized spacecraft is upswept as mass-energy fuel or species of reaction and that the spacecraft makes an effective pass entirely through each universe. Thus, all material in said universes of travel is reacted against with assumed finite universal mass-energy density.

For universes in said multiverse having an extensity of Aleph 3 light-years, there are prospects for the spacecraft to attain a Lorentz factor of roughly [(Aleph H) EXP 6] (Aleph 3) to thus result in the multiverse of travel being contracted to point directly in front of the spacecraft and contracted by a factor of [(Aleph H) EXP 5](Aleph 3). Note that in the above computations, we assume that only the material in the path of a reasonably sized spacecraft is upswept as mass-energy fuel or species of reaction and that the spacecraft makes only one effective pass through each universe.

If such Aleph 3 light-year–width universes have five spatial dimensions and are substantially static, then a spacecraft traveling at light-speed in the multiverse of consideration may obtain a Lorentz factor of roughly [(Aleph H) EXP 6] [(Aleph 3) EXP 5]. Note that in the above computations, we assume that only the material in the path of a reasonably sized spacecraft is upswept as mass-energy fuel or species of reaction and that the spacecraft makes an effective pass entirely through each universe. Thus, all material in said universes of travel is reacted against with assumed finite universal mass-energy density.

If the latter universes are expanding with time over periods much greater than Aleph 3 years, then the spacecraft may attain a Lorentz factor of roughly [(Aleph H) EXP 6] [[k(Aleph 3)] EXP 5], where k ranges from a small superunitary real number to suitable infinite real numbers. This, of course, assumes that real positive mass-energy stocks continue being created, perhaps out of zero-point fields or analogues thereof from some forms of dark energy. Provided that k ranges from Aleph 0 to Aleph 5, then the spacecraft Lorentz factor will be limited to a range of roughly from [(Aleph H) EXP 6] [[(Aleph 0)(Aleph 3)] EXP 5] to [(Aleph H) EXP 6] [[(Aleph 5)(Aleph 3)] EXP 5]. Note that in the above computations, we assume that only the material in the path of a reasonably sized spacecraft is upswept as mass-energy fuel or species of reaction and that the spacecraft makes an effective pass entirely through each universe. Thus, all material in said universes of travel is reacted against with assumed finite universal mass-energy density.

For universes having an extensity of Aleph 4 light-years, there are prospects for the spacecraft to attain a Lorentz factor of roughly [(Aleph H) EXP 6] (Aleph 4) to thus result in the multiverse of travel being contracted to point directly in front of the spacecraft and being contracted relative to the spacecraft by roughly a factor of [(Aleph H) EXP 5](Aleph 4). Note that in the above computations, we assume that only the material in the path of a reasonably sized spacecraft is upswept as mass-energy fuel or species of reaction and that the spacecraft makes only one effective pass through each universe.

If such Aleph 4 light-year–width universes have five spatial dimensions and are substantially static, then a spacecraft traveling at light-speed in the multiverse may obtain a Lorentz factor of roughly [(Aleph H) EXP 6] [(Aleph 4) EXP 5]. Note that in the above computations, we assume that only the material in the path of a reasonably sized spacecraft is upswept as mass-energy fuel or species of reaction and that the spacecraft makes an effective pass entirely through each universe. Thus, all material in said universes of travel is reacted against with assumed finite universal mass-energy density.

If the latter universes are expanding with time over periods much greater than Aleph 4 years, then the spacecraft may attain a Lorentz factor of roughly [(Aleph H) EXP 6] [[k(Aleph 4)] EXP 5], where k ranges from a small superunitary real number to suitable infinite real numbers. This, of course, assumes that real positive mass-energy stocks continue being created, perhaps out of zero-point fields or analogues thereof from some forms of dark energy. Provided that k ranges roughly from Aleph 0 to Aleph 6, then the spacecraft Lorentz factor will be limited to a range from [(Aleph H) EXP 6] [[(Aleph 0)(Aleph 4)] EXP 5] to [(Aleph H) EXP 6] [[(Aleph 6)(Aleph 4)] EXP 5]. Note that in the above computations, we assume that only the material in the path of a reasonably sized spacecraft is upswept as mass-energy fuel or species of reaction and that the spacecraft makes an effective pass entirely through each universe. Thus, all material in said universes of travel is reacted against with assumed finite universal mass-energy density.

For universes having an extensity of Aleph G light-years, there are prospects for the spacecraft to attain a Lorentz factor of roughly [(Aleph H) EXP 6] (Aleph G) to thus result in the multiverse of travel being contracted to point directly in front of the spacecraft with a Lorentz contraction of roughly [(Aleph H) EXP 6] (Aleph G). Note that in the above computations, we assume that only the material in the path of a reasonably sized spacecraft is upswept as mass-energy fuel or species of reaction and that the spacecraft makes only one effective pass through each universe.

If such Aleph G light-year–width universes have five spatial dimensions and are substantially static, then a spacecraft traveling at light-speed in the multiverse may obtain a Lorentz factor of roughly [(Aleph H) EXP 6] [(Aleph G) EXP 5]. Note that in the above computations, we assume that only the material in the path of a reasonably sized spacecraft is upswept as mass-energy fuel or species of reaction and that the spacecraft makes an effective pass entirely through each universe. Thus, all material in said universes of travel is reacted against with assumed finite universal mass-energy density.

If the latter universes are expanding with time over periods much greater than Aleph G years, then the spacecraft may attain a Lorentz factor of roughly [(Aleph H) EXP 6] [[k(Aleph G)] EXP 5], where *k* ranges from a small superunitary real number to suitable infinite real numbers. This, of course, assumes that real positive mass-energy stocks continue being created, perhaps out of zero-point fields or analogues thereof from some forms of dark energy. Provided that *k* ranges from Aleph 0 to Aleph (G + 2), then the spacecraft Lorentz factor will be limited to a range of roughly from [(Aleph H) EXP 6] [[(Aleph 0)(Aleph G)] EXP 5] to [(Aleph H) EXP 6] [[[Aleph (G + 2)](Aleph G)] EXP 5]. Note that in the above computations, we assume that only the material in the path of a reasonably sized spacecraft is upswept as mass-energy fuel or species of reaction and that the spacecraft makes an effective pass entirely through each universe. Thus, all material in said universes of travel is reacted against with assumed finite universal mass-energy density.

Now, we consider spacecraft travel in a multiverse of Aleph H universes in width of N dimensions where *H* is an integer greater than G + 2, and where said multiverse contains [(Aleph H) EXP N] universes.

For universes in said N-D–space multiverse having an extensity of (Aleph 1) light-years, there are prospects for the spacecraft to attain a Lorentz factor of roughly [(Aleph H) EXP N] (Aleph 1) to thus result in the multiverse of travel being contracted to point directly in front of the spacecraft and Lorentz contracted by a factor of roughly [(Aleph H) EXP N] (Aleph 1) relative to the spacecraft. Note that in the above computations, we assume that only the material in the path of a reasonably sized spacecraft is upswept as mass-energy fuel or species of reaction and that the spacecraft makes only one effective pass through each universe.

If such Aleph 1 light-year-width universes have five spatial dimensions and are substantially static, then a spacecraft traveling at light-speed in the multiverse of consideration may obtain a Lorentz factor of roughly [(Aleph H) EXP N] [(Aleph 1) EXP 5]. Note that in the above computations, we assume that only the material in the path of a reasonably sized spacecraft is upswept as mass-energy fuel or species of reaction and that the spacecraft makes an effective pass entirely through each universe. Thus, all material in said universes of travel is reacted against with assumed finite universal mass-energy density.

If the latter universes are expanding with time over periods much greater than Aleph 1 years, then the spacecraft may attain a Lorentz factor of roughly [(Aleph H) EXP N] [[k(Aleph 1)] EXP 5], where *k* ranges from a small superunitary real number to suitable infinite real numbers. This, of course, assumes that real positive mass-energy stocks continue being created, perhaps out of zero-point fields or analogues thereof from some forms of dark energy. Provided that *k* ranges from Aleph 0 to Aleph 3, then the spacecraft Lorentz factor will be limited to a range of roughly from [(Aleph H) EXP N] [[(Aleph 0)(Aleph 1)] EXP 5] to [(Aleph H) EXP N] [[(Aleph 3)(Aleph 1)] EXP 5]. Note that in the above computations, we assume that only the material in the path of a reasonably sized spacecraft is upswept as mass-energy fuel or species of reaction and that the spacecraft makes an effective entire pass through each universe. Thus, all material in said universes of travel is reacted against with assumed finite universal mass-energy density.

For universes in said multiverse having an extensity of Aleph 2 light-years, there are prospects for the spacecraft to attain a Lorentz factor of roughly [(Aleph H) EXP N] (Aleph 2) to thus result in the multiverse of travel being contracted to a point directly in front of the spacecraft and by a factor of [(Aleph H) EXP N] (Aleph 2). Note that in the above computations, we assume that only the material in the path of a reasonably sized spacecraft is upswept as mass-energy fuel or species of reaction and that the spacecraft makes only one effective pass through each universe.

If such Aleph 2 light-year–width universes have five spatial dimensions and are substantially static, then a spacecraft traveling at light-speed in the multiverse considered may obtain a Lorentz factor of roughly [(Aleph H) EXP N] [(Aleph 2) EXP 5]. Note that in the above computations, we assume that only the material in the path of a reasonably sized spacecraft is upswept as mass-energy fuel or species of reaction and that the spacecraft makes an effective pass entirely through each universe. Thus, all material in said universes of travel is reacted against with assumed finite universal mass-energy density.

If the latter universes are expanding with time over periods much greater than Aleph 2 years, then the spacecraft may attain a Lorentz factor of roughly [(Aleph H) EXP N] [[k(Aleph 2)] EXP 5], where k ranges from a small superunitary real number to suitable infinite real numbers. This, of course, assumes that real positive mass-energy stocks continue being created, perhaps out of zero-point fields or analogues thereof from some forms of dark energy. Provided that k ranges from Aleph 0 to Aleph 4, then the spacecraft Lorentz factor will be limited to a range of roughly from [(Aleph H) EXP N] [[(Aleph 0)(Aleph 2)] EXP 5] to [(Aleph H) EXP N] [[(Aleph 4)(Aleph 2)] EXP 5]. Note that in the above computations, we assume that only the material in the path of a reasonably sized spacecraft is upswept as mass-energy fuel or species of reaction and that the spacecraft makes an effective pass entirely through each universe. Thus, all material in said universes of travel is reacted against with assumed finite universal mass-energy density.

For universes in said multiverse having an extensity of Aleph 3 light-years, there are prospects for the spacecraft to attain a Lorentz factor of roughly [(Aleph H) EXP N] (Aleph 3) to thus result in the multiverse of travel being contracted to point directly in front of the spacecraft and contracted by a factor of [(Aleph H) EXP 5](Aleph 3). Note that in the above computations, we assume that only the material in the path of a reasonably sized spacecraft is upswept as mass-energy fuel or species of reaction and that the spacecraft makes only one effective pass through each universe.

If such Aleph 3 light-year–width universes have five spatial dimensions and are substantially static, then a spacecraft traveling at light-speed in the multiverse of consideration may obtain a Lorentz factor of roughly [(Aleph H) EXP N] [(Aleph 3) EXP 5]. Note that in the above computations, we assume that only the material in the path of a reasonably sized spacecraft is upswept as mass-energy fuel or species of reaction and that the spacecraft makes an effective pass entirely through each universe. Thus, all material in said universes of travel is reacted against with assumed finite universal mass-energy density.

If the latter universes are expanding with time over periods much greater than Aleph 3 years, then the spacecraft may attain a Lorentz factor of roughly [(Aleph H) EXP N] [[k(Aleph 3)] EXP 5], where k ranges from a small superunitary real number to suitable infinite real numbers. This, of course, assumes that real positive mass-energy stocks continue being created, perhaps out of zero-point fields or analogues thereof from some forms of dark energy. Provided that k ranges from Aleph 0 to Aleph 5, then the spacecraft Lorentz factor will be limited to a range of roughly from [(Aleph H) EXP N] [[(Aleph 0)(Aleph 3)] EXP 5] to [(Aleph H) EXP N] [[(Aleph 5)(Aleph 3)] EXP 5]. Note that in the above computations, we assume that only the material in the path of a reasonably sized spacecraft is upswept as mass-energy fuel or species of reaction and that the spacecraft makes an effective pass entirely through each universe. Thus, all material in said universes of travel is reacted against with assumed finite universal mass-energy density.

For universes having an extensity of Aleph 4 light-years, there are prospects for the spacecraft to attain a Lorentz factor of roughly [(Aleph H) EXP N] (Aleph 4) to thus result in the multiverse of travel being contracted to point directly in front of the spacecraft and being contracted relative to the spacecraft by roughly a factor of [(Aleph H) EXP 5](Aleph 4). Note that in the above computations, we assume that only the material in the path of a reasonably sized spacecraft is upswept as mass-energy fuel or species of reaction and that the spacecraft makes only one effective pass through each universe.

If such Aleph 4 light-year–width universes have five spatial dimensions and are substantially static, then a spacecraft traveling at light-speed in the multiverse may obtain a Lorentz factor of roughly [(Aleph H) EXP N] [(Aleph 4) EXP 5]. Note that in the above computations, we assume that only the material in the path of a reasonably sized spacecraft is upswept as mass-energy fuel or species of reaction and that the spacecraft makes an effective pass entirely through each universe. Thus, all material in said universes of travel is reacted against with assumed finite universal mass-energy density.

If the latter universes are expanding with time over periods much greater than Aleph 4 years, then the spacecraft may attain a Lorentz factor of roughly [(Aleph H) EXP N] [[k(Aleph 4)] EXP 5], where k ranges from a small superunitary real number to suitable infinite real numbers. This, of course, assumes that real positive mass-energy stocks continue being created, perhaps out of zero-point fields or analogues thereof from some forms of dark energy. Provided that k ranges roughly from Aleph 0 to Aleph 6, then the spacecraft Lorentz factor will be limited to a range from [(Aleph H) EXP N] [[(Aleph 0)(Aleph 4)] EXP 5] to [(Aleph H) EXP N] [[(Aleph 6)(Aleph 4)] EXP 5]. Note that in the above computations, we assume that only the material in the path of a reasonably sized spacecraft is upswept as mass-energy fuel or species of reaction and that the spacecraft makes an effective pass entirely through each universe. Thus, all material in said universes of travel is reacted against with assumed finite universal mass-energy density.

For universes having an extensity of Aleph G light-years, there are prospects for the spacecraft to attain a Lorentz factor of roughly [(Aleph H) EXP N] (Aleph G) to thus result in the multiverse of travel being contracted to point directly in front of the spacecraft with a Lorentz contraction of roughly [(Aleph H) EXP N] (Aleph G). Note that in the above computations, we assume that only the material in the path of a reasonably sized spacecraft is upswept as mass-energy fuel or species of reaction and that the spacecraft makes only one effective pass through each universe.

If such Aleph G light-year–width universes have five spatial dimensions and are substantially static, then a spacecraft traveling at light-speed in the multiverse may obtain a Lorentz factor of roughly [(Aleph H) EXP N] [(Aleph G) EXP 5]. Note that in the above computations, we assume that only the material in the path of a reasonably sized spacecraft is upswept as mass-energy fuel or species of reaction and that the spacecraft makes an effective pass entirely through each universe. Thus, all material in said universes of travel is reacted against with assumed finite universal mass-energy density.

If the latter universes are expanding with time over periods much greater than Aleph G years, then the spacecraft may attain a Lorentz factor of roughly [(Aleph H) EXP N] [[k(Aleph G)] EXP 5], where k ranges from a small superunitary real number to suitable infinite real numbers. This, of course, assumes that real positive mass-energy stocks continue being created, perhaps out of zero-point fields or analogues thereof from some forms of dark energy. Provided that k ranges from Aleph 0 to Aleph (G + 2), then the spacecraft Lorentz factor will be limited to a range of roughly from [(Aleph H) EXP N]

Now we consider spacecraft travel in a multiverse of Aleph H universes in width of N dimensions, where *H* is an integer greater than G + 2, and where said multiverse contains [(Aleph H) EXP N] universes.

For universes in said N-D-space multiverse having an extensity of (Aleph 1) light-years, there are prospects for the spacecraft to attain a Lorentz factor of roughly [(Aleph H) EXP N] (Aleph 1) to thus result in the multiverse of travel being contracted to point directly in front of the spacecraft and Lorentz contracted by a factor of roughly [(Aleph H) EXP N] (Aleph 1) relative to the spacecraft. Note that in the above computations, we assume that only the material in the path of a reasonably sized spacecraft is upswept as mass-energy fuel or species of reaction and that the spacecraft makes only one effective pass through each universe.

If such Aleph 1 light-year–width universes have six spatial dimensions and are substantially static, then a spacecraft traveling at light-speed in the multiverse of consideration may obtain a Lorentz factor of roughly [(Aleph H) EXP N] [(Aleph 1) EXP 6]. Note that in the above computations, we assume that only the material in the path of a reasonably sized spacecraft is upswept as mass-energy fuel or species of reaction and that the spacecraft makes an effective pass entirely through each universe. Thus, all material in said universes of travel is reacted against with assumed finite universal mass-energy density.

If the latter universes are expanding with time over periods much greater than Aleph 1 years, then the spacecraft may attain a Lorentz factor of roughly [(Aleph H) EXP N] [[k(Aleph 1)] EXP 6], where *k* ranges from a small superunitary real number to suitable infinite real numbers. This, of course, assumes that real positive mass-energy stocks continue being created, perhaps out of zero-point fields or analogues thereof from some forms of dark energy. Provided that *k* ranges from Aleph 0 to Aleph 3, then the spacecraft Lorentz factor will be limited to a range of roughly from [(Aleph H) EXP N] [[(Aleph 0)(Aleph 1)] EXP 6] to [(Aleph H) EXP N] [[(Aleph 3)(Aleph 1)] EXP 6]. Note that in the above computations, we assume that only the material in the path of a reasonably sized spacecraft is upswept as mass-energy fuel or species of reaction and that the spacecraft makes an effective entire pass through each universe. Thus, all material in said universes of travel is reacted against with assumed finite universal mass-energy density.

For universes in said multiverse having an extensity of Aleph 2 light-years, there are prospects for the spacecraft to attain a Lorentz factor of roughly [(Aleph H) EXP N] (Aleph 2) to thus result in the multiverse of travel being contracted to a point directly in front of the spacecraft and by a factor of [(Aleph H) EXP N] (Aleph 2). Note that in the above computations, we assume that only the material in the path of a reasonably sized spacecraft is upswept as mass-energy fuel or species of reaction and that the spacecraft makes only one effective pass through each universe.

If such Aleph 2 light-year–width universes have six spatial dimensions and are substantially static, then a spacecraft traveling at light-speed in the multiverse considered may obtain a Lorentz factor of roughly [(Aleph H) EXP N] [(Aleph 2) EXP 6]. Note that in the above computations, we assume that only the material in the path of a reasonably sized spacecraft is upswept as mass-energy fuel or species of reaction and that the spacecraft makes an effective pass entirely through each universe. Thus, all material in said universes of travel is reacted against with assumed finite universal mass-energy density.

If the latter universes are expanding with time over periods much greater than Aleph 2 years, then the spacecraft may attain a Lorentz factor of roughly [(Aleph H) EXP N] [[k(Aleph 2)] EXP 6], where k ranges from a small superunitary real number to suitable infinite real numbers. This, of course, assumes that real positive mass-energy stocks continue being created, perhaps out of zero-point fields or analogues thereof from some forms of dark energy. Provided that k ranges from Aleph 0 to Aleph 4, then the spacecraft Lorentz factor will be limited to a range of roughly from [(Aleph H) EXP N] [[(Aleph 0)(Aleph 2)] EXP 6] to [(Aleph H) EXP N] [[(Aleph 4)(Aleph 2)] EXP 6]. Note that in the above computations, we assume that only the material in the path of a reasonably sized spacecraft is upswept as mass-energy fuel or species of reaction and that the spacecraft makes an effective pass entirely through each universe. Thus, all material in said universes of travel is reacted against with assumed finite universal mass-energy density.

For universes in said multiverse having an extensity of Aleph 3 light-years, there are prospects for the spacecraft to attain a Lorentz factor of roughly [(Aleph H) EXP N] (Aleph 3) to thus result in the multiverse of travel being contracted to point directly in front of the spacecraft and contracted by a factor of [(Aleph H) EXP 6](Aleph 3). Note that in the above computations, we assume that only the material in the path of a reasonably sized spacecraft is upswept as mass-energy fuel or species of reaction and that the spacecraft makes only one effective pass through each universe.

If such Aleph 3 light-year–width universes have six spatial dimensions and are substantially static, then a spacecraft traveling at light-speed in the multiverse of consideration may obtain a Lorentz factor of roughly [(Aleph H) EXP N] [(Aleph 3) EXP 6]. Note that in the above computations, we assume that only the material in the path of a reasonably sized spacecraft is upswept as mass-energy fuel or species of reaction and that the spacecraft makes an effective pass entirely through each universe. Thus, all material in said universes of travel is reacted against with assumed finite universal mass-energy density.

If the latter universes are expanding with time over periods much greater than Aleph 3 years, then the spacecraft may attain a Lorentz factor of roughly [(Aleph H) EXP N] [[k(Aleph 3)] EXP 6], where k ranges from a small superunitary real number to suitable infinite real numbers. This, of course, assumes that real positive mass-energy stocks continue being created, perhaps out of zero-point fields or analogues thereof from some forms of dark energy. Provided that k ranges from Aleph 0 to Aleph 5, then the spacecraft Lorentz factor will be limited to a range of roughly from [(Aleph H) EXP N] [[(Aleph 0)(Aleph 3)] EXP 6] to [(Aleph H) EXP N] [[(Aleph 5)(Aleph 3)] EXP 6]. Note that in the above computations, we assume that only the material in the path of a reasonably sized spacecraft is upswept as mass-energy fuel or species of reaction and that the spacecraft makes an effective pass entirely through each universe. Thus, all material in said universes of travel is reacted against with assumed finite universal mass-energy density.

For universes having an extensity of Aleph 4 light-years, there are prospects for the spacecraft to attain a Lorentz factor of roughly [(Aleph H) EXP N] (Aleph 4) to thus result in the multiverse of travel being contracted to point directly in front of the spacecraft and being contracted relative to the spacecraft by roughly a factor of [(Aleph H) EXP 6](Aleph 4). Note that in the above computations, we assume that only the material in the path of a reasonably sized spacecraft is upswept as mass-energy fuel or species of reaction and that the spacecraft makes only one effective pass through each universe.

If such Aleph 4 light-year–width universes have six spatial dimensions and are substantially static, then a spacecraft traveling at light-speed in the multiverse may obtain a Lorentz factor of roughly [(Aleph H) EXP N] [(Aleph 4) EXP 6]. Note that in the above computations, we assume that only the material in the path of a reasonably sized spacecraft is upswept as mass-energy fuel or species of reaction and that the spacecraft makes an effective pass entirely through each universe. Thus, all material in said universes of travel is reacted against with assumed finite universal mass-energy density.

If the latter universes are expanding with time over periods much greater than Aleph 4 years, then the spacecraft may attain a Lorentz factor of roughly [(Aleph H) EXP N] [[k(Aleph 4)] EXP 6], where k ranges from a small superunitary real number to suitable infinite real numbers. This, of course, assumes that real positive mass-energy stocks continue being created, perhaps out of zero-point fields or analogues thereof from some forms of dark energy. Provided that k ranges roughly from Aleph 0 to Aleph 6, then the spacecraft Lorentz factor will be limited to a range from [(Aleph H) EXP N] [[(Aleph 0)(Aleph 4)] EXP 6] to [(Aleph H) EXP N] [[(Aleph 6)(Aleph 4)] EXP 6]. Note that in the above computations, we assume that only the material in the path of a reasonably sized spacecraft is upswept as mass-energy fuel or species of reaction and that the spacecraft makes an effective pass entirely through each universe. Thus, all material in said universes of travel is reacted against with assumed finite universal mass-energy density.

For universes having an extensity of Aleph G light-years, there are prospects for the spacecraft to attain a Lorentz factor of roughly [(Aleph H) EXP N] (Aleph G) to thus result in the multiverse of travel being contracted to point directly in front of the spacecraft with a Lorentz contraction of roughly [(Aleph H) EXP N] (Aleph G). Note that in the above computations, we assume that only the material in the path of a reasonably sized spacecraft is upswept as mass-energy fuel or species of reaction and that the spacecraft makes only one effective pass through each universe.

If such Aleph G light-year–width universes have six spatial dimensions and are substantially static, then a spacecraft traveling at light-speed in the multiverse may obtain a Lorentz factor of roughly [(Aleph H) EXP N] [(Aleph G) EXP 6]. Note that in the above computations, we assume that only the material in the path of a reasonably sized spacecraft is upswept as mass-energy fuel or species of reaction and that the spacecraft makes an effective pass entirely through each universe. Thus, all material in said universes of travel is reacted against with assumed finite universal mass-energy density.

If the latter universes are expanding with time over periods much greater than Aleph G years, then the spacecraft may attain a Lorentz factor of roughly [(Aleph H) EXP N] [[k(Aleph G)] EXP 6], where k ranges from a small superunitary real number to suitable infinite real numbers. This, of course, assumes that real positive mass-energy stocks continue being created, perhaps out of zero-point fields or analogues thereof from some forms of dark energy. Provided that k ranges from Aleph 0 to Aleph (G + 2), then the spacecraft Lorentz factor will be limited to a range of roughly from [(Aleph H) EXP N] [[(Aleph 0)(Aleph G)] EXP 6] to [(Aleph H) EXP N] [[[Aleph (G + 2)](Aleph G)] EXP 6]. Note that in the above computations, we assume that only the material in the path of a reasonably sized spacecraft is upswept as mass-energy fuel or species of reaction and that the spacecraft makes an effective pass entirely through each universe. Thus, all material in said universes of travel is reacted against with assumed finite universal mass-energy density.

Now, we consider spacecraft travel in a multiverse of Aleph H universes in width of N dimensions, where H is an integer greater than G + 2, and where said multiverse contains [(Aleph H) EXP N] universes.

For universes in said N-D–space multiverse having an extensity of (Aleph 1) light-years, there are prospects for the spacecraft to attain a Lorentz factor of roughly [(Aleph H) EXP N] (Aleph 1) to thus result in the multiverse of travel being contracted to point directly in front of the spacecraft and Lorentz contracted by a factor of roughly [(Aleph H) EXP N] (Aleph 1) relative to the spacecraft. Note that in the above computations, we assume that only the material in the path of a reasonably sized spacecraft is upswept as mass-energy fuel or species of reaction and that the spacecraft makes only one effective pass through each universe.

If such Aleph 1 light-year–width universes have N spatial dimensions and are substantially static, then a spacecraft traveling at light-speed in the multiverse of consideration may obtain a Lorentz factor of roughly [(Aleph H) EXP N] [(Aleph 1) EXP N]. Note that in the above computations, we assume that only the material in the path of a reasonably sized spacecraft is upswept as mass-energy fuel or species of reaction and that the spacecraft makes an effective pass entirely through each universe. Thus, all material in said universes of travel is reacted against with assumed finite universal mass-energy density.

If the latter universes are expanding with time over periods much greater than Aleph 1 years, then the spacecraft may attain a Lorentz factor of roughly [(Aleph H) EXP N] [[k(Aleph 1)] EXP N], where k ranges from a small superunitary real number to suitable infinite real numbers. This, of course, assumes that real positive mass-energy stocks continue being created, perhaps out of zero-point fields or analogues thereof from some forms of dark energy. Provided that k ranges from Aleph 0 to Aleph 3, then the spacecraft Lorentz factor will be limited to a range of roughly from [(Aleph H) EXP N] [[(Aleph 0)(Aleph 1)] EXP N] to [(Aleph H) EXP N] [[(Aleph 3)(Aleph 1)] EXP N]. Note that in the above computations, we assume that only the material in the path of a reasonably sized spacecraft is upswept as mass-energy fuel or species of reaction and that the spacecraft makes an effective entire pass through each universe. Thus, all material in said universes of travel is reacted against with assumed finite universal mass-energy density.

For universes in said multiverse having an extensity of Aleph 2 light-years, there are prospects for the spacecraft to attain a Lorentz factor of roughly [(Aleph H) EXP N] (Aleph 2) to thus result in the multiverse of travel being contracted to a point directly in front of the spacecraft and by a factor of [(Aleph H) EXP N] (Aleph 2). Note that in the above computations, we assume that only the material in the path of a reasonably sized spacecraft is upswept as mass-energy fuel or species of reaction and that the spacecraft makes only one effective pass through each universe.

If such Aleph 2 light-year–width universes have N spatial dimensions and are substantially static, then a spacecraft traveling at light-speed in the multiverse considered may obtain a Lorentz factor of roughly [(Aleph H) EXP N] [(Aleph 2) EXP N]. Note that in the above computations, we assume that only the material in the path of a reasonably sized spacecraft is upswept as mass-energy fuel or species of reaction and that the spacecraft makes an effective pass entirely through each universe. Thus, all material in said universes of travel is reacted against with assumed finite universal mass-energy density.

If the latter universes are expanding with time over periods much greater than Aleph 2 years, then the spacecraft may attain a Lorentz factor of roughly [(Aleph H) EXP N] [[k(Aleph 2)] EXP N], where *k* ranges from a small superunitary real number to suitable infinite real numbers. This, of course, assumes that real positive mass-energy stocks continue being created, perhaps out of zero-point fields or analogues thereof from some forms of dark energy. Provided that *k* ranges from Aleph 0 to Aleph 4, then the spacecraft Lorentz factor will be limited to a range of roughly from [(Aleph H) EXP N] [[(Aleph 0)(Aleph 2)] EXP N] to [(Aleph H) EXP N] [[(Aleph 4)(Aleph 2)] EXP N]. Note that in the above computations, we assume that only the material in the path of a reasonably sized spacecraft is upswept as mass-energy fuel or species of reaction and that the spacecraft makes an effective pass entirely through each universe. Thus, all material in said universes of travel is reacted against with assumed finite universal mass-energy density.

For universes in said multiverse having an extensity of Aleph 3 light-years, there are prospects for the spacecraft to attain a Lorentz factor of roughly [(Aleph H) EXP N] (Aleph 3) to thus result in the multiverse of travel being contracted to point directly in front of the spacecraft and contracted by a factor of [(Aleph H) EXP N](Aleph 3). Note that in the above computations, we assume that only the material in the path of a reasonably sized spacecraft is upswept as mass-energy fuel or species of reaction and that the spacecraft makes only one effective pass through each universe.

If such Aleph 3 light-year–width universes have *N* spatial dimensions and are substantially static, then a spacecraft traveling at light-speed in the multiverse of consideration may obtain a Lorentz factor of roughly [(Aleph H) EXP N] [(Aleph 3) EXP N]. Note that in the above computations, we assume that only the material in the path of a reasonably sized spacecraft is upswept as mass-energy fuel or species of reaction and that the spacecraft makes an effective pass entirely through each universe. Thus, all material in said universes of travel is reacted against with assumed finite universal mass-energy density.

If the latter universes are expanding with time over periods much greater than Aleph 3 years, then the spacecraft may attain a Lorentz factor of roughly [(Aleph H) EXP N] [[k(Aleph 3)] EXP N], where *k* ranges from a small superunitary real number to suitable infinite real numbers. This, of course, assumes that real positive mass-energy stocks continue being created, perhaps out of zero-point fields or analogues thereof from some forms of dark energy. Provided that *k* ranges from Aleph 0 to Aleph 5, then the spacecraft Lorentz factor will be limited to a range of roughly from [(Aleph H) EXP N]

[[(Aleph 0)(Aleph 3)] EXP N] to [(Aleph H) EXP N] [[(Aleph 5)(Aleph 3)] EXP N]. Note that in the above computations, we assume that only the material in the path of a reasonably sized spacecraft is upswept as mass-energy fuel or species of reaction and that the spacecraft makes an effective pass entirely through each universe. Thus, all material in said universes of travel is reacted against with assumed finite universal mass-energy density.

For universes having an extensity of Aleph 4 light-years, there are prospects for the spacecraft to attain a Lorentz factor of roughly [(Aleph H) EXP N] (Aleph 4) to thus result in the multiverse of travel being contracted to point directly in front of the spacecraft and being contracted relative to the spacecraft by roughly a factor of [(Aleph H) EXP N](Aleph 4). Note that in the above computations, we assume that only the material in the path of a reasonably sized spacecraft is upswept as mass-energy fuel or species of reaction and that the spacecraft makes only one effective pass through each universe.

If such Aleph 4 light-year–width universes have *N* spatial dimensions and are substantially static, then a spacecraft traveling at light-speed in the multiverse may obtain a Lorentz factor of roughly [(Aleph H) EXP N] [(Aleph 4) EXP N]. Note that in the above computations, we assume that only the material in the path of a reasonably sized spacecraft is upswept as mass-energy fuel or species of reaction and that the spacecraft makes an effective pass entirely through each universe. Thus, all material in said universes of travel is reacted against with assumed finite universal mass-energy density.

If the latter universes are expanding with time over periods much greater than Aleph 4 years, then the spacecraft may attain a Lorentz factor of roughly [(Aleph H) EXP N] [[k(Aleph 4)] EXP N], where *k* ranges from a small superunitary real number to suitable infinite real numbers. This, of course, assumes that real positive mass-energy stocks continue being created, perhaps out of zero-point fields or analogues thereof from some forms of dark energy. Provided that *k* ranges roughly from Aleph 0 to Aleph 6, then the spacecraft Lorentz factor will be limited to a range from [(Aleph H) EXP N] [[(Aleph 0)(Aleph 4)] EXP N] to [(Aleph H) EXP N] [[(Aleph 6)(Aleph 4)] EXP N]. Note that in the above computations, we assume that only the material in the path of a reasonably sized spacecraft is upswept as mass-energy fuel or species of reaction and that the spacecraft makes an effective pass entirely through each universe. Thus, all material in said universes of travel is reacted against with assumed finite universal mass-energy density.

For universes having an extensity of Aleph G light-years, there are prospects for the spacecraft to attain a Lorentz factor of roughly [(Aleph H) EXP N] (Aleph G) to thus result in the multiverse of travel being contracted to point directly in front of the spacecraft with a Lorentz contraction of roughly [(Aleph H) EXP N] (Aleph G). Note that in the above computations, we assume that only the material in the path of a reasonably sized spacecraft is upswept as mass-energy fuel or species of reaction and that the spacecraft makes only one effective pass through each universe.

If such Aleph G light-year–width universes have *N* spatial dimensions and are substantially static, then a spacecraft traveling at light-speed in the multiverse may obtain a Lorentz factor of roughly [(Aleph H) EXP N] [(Aleph G) EXP N]. Note that in the above computations, we assume that only the material in the path of a reasonably sized spacecraft is upswept as mass-energy fuel or species of reaction and that the spacecraft makes an effective pass entirely through each universe. Thus, all material in said universes of travel is reacted against with assumed finite universal mass-energy density.

If the latter universes are expanding with time over periods much greater than Aleph G years, then the spacecraft may attain a Lorentz factor of roughly [(Aleph H) EXP N] [[k(Aleph G)] EXP N], where *k* ranges from a small superunitary real number to suitable infinite real numbers. This, of course, assumes that real positive mass-energy stocks continue being created, perhaps out of zero-point fields or analogues thereof from some forms of dark energy. Provided that *k* ranges from Aleph 0 to Aleph (G + 2), then the spacecraft Lorentz factor will be limited to a range of roughly from [(Aleph H) EXP N] [[(Aleph 0)(Aleph G)] EXP N] to [(Aleph H) EXP N] [[[Aleph (G + 2)](Aleph G)] EXP N]. Note that in the above computations, we assume that only the material in the path of a reasonably sized spacecraft is upswept as mass-energy fuel or species of reaction and that the spacecraft makes an effective pass entirely through each universe. Thus, all material in said universes of travel is reacted against with assumed finite universal mass-energy density.

Note that in the above multiversal scenarios, we assume that the universes are packed within the multiverses considered at near maximum possible packing factors.

Now, we consider spacecraft travel in a 3-D–space forest of width of Aleph J multiverses each of Aleph H universes in width of three dimensions, where *H* is an integer greater than G + 2, and where said multiverses each contain [(Aleph J) EXP 3][(Aleph H) EXP 3] universes.

For universes in said 3-D–space multiverses having an extensity of (Aleph 1) light-years, there are prospects for the spacecraft to attain a Lorentz factor of roughly [(Aleph J) EXP 3] [(Aleph J) EXP 3][(Aleph H) EXP 3] (Aleph 1) to thus result in the multiverses of travel being contracted to point directly in front of the spacecraft and Lorentz contracted by a factor of roughly [(Aleph J) EXP 3][(Aleph H) EXP 3] (Aleph 1) relative to the spacecraft. Note that in the above computations, we assume that only the material in the path of a reasonably sized spacecraft is upswept as mass-energy fuel or species of reaction and that the spacecraft makes only one effective pass through each universe.

If such Aleph 1 light-year–width universes have three spatial dimensions and are substantially static, then a spacecraft traveling at light-speed in the forests of consideration may obtain a Lorentz factor of roughly [(Aleph J) EXP 3] [(Aleph J) EXP 3][(Aleph H) EXP 3] [(Aleph 1) EXP 3]. Note that in the above computations, we assume that only the material in the path of a reasonably sized spacecraft is upswept as mass-energy fuel or species of reaction and that the spacecraft makes an effective pass entirely through each universe. Thus, all material in said universes of travel is reacted against with assumed finite universal mass-energy density.

If the latter universes are expanding with time over periods much greater than Aleph 1 years, then the spacecraft may attain a Lorentz factor of roughly [(Aleph J) EXP 3] [(Aleph J) EXP 3][(Aleph H) EXP 3] [[k(Aleph 1)] EXP 3], where *k* ranges from a small superunitary real number to suitable infinite real numbers. This, of course, assumes that real positive mass-energy stocks continue being created, perhaps out of zero-point fields or analogues thereof from some forms of dark energy. Provided that *k* ranges from Aleph 0 to Aleph 3, then the spacecraft Lorentz factor will be limited to a range of roughly from [(Aleph J) EXP 3][(Aleph H) EXP 3] [[(Aleph 0)(Aleph 1)] EXP 3] to [(Aleph J) EXP 3][(Aleph H) EXP 3] [[(Aleph 3)(Aleph 1)] EXP 3]. Note that in the above computations, we assume that only the material in the path of a reasonably sized spacecraft is upswept as mass-energy fuel or species of reaction and that the spacecraft makes an effective entire pass through each universe. Thus, all material in said universes of travel is reacted against with assumed finite universal mass-energy density.

For universes in said multiverses having an extensity of Aleph 2 light-years, there are prospects for the spacecraft to attain a Lorentz factor of roughly [(Aleph J) EXP 3] [(Aleph J) EXP 3][(Aleph H) EXP 3] (Aleph 2) to thus result in the multiverses of travel being contracted to a point directly in front of the spacecraft and by a factor of [(Aleph J) EXP 3][(Aleph H) EXP 3] (Aleph 2). Note that in the above computations, we assume that only the material in the path of a reasonably sized spacecraft is upswept as mass-energy fuel or species of reaction and that the spacecraft makes only one effective pass through each universe.

If such Aleph 2 light-year–width universes have three spatial dimensions and are substantially static, then a spacecraft traveling at light-speed in the forests considered may obtain a Lorentz factor of roughly [(Aleph J) EXP 3] [(Aleph J) EXP 3][(Aleph H) EXP 3] [(Aleph 2) EXP 3]. Note that in the above computations, we assume that only the material in the path of a reasonably sized spacecraft is upswept as mass-energy fuel or species of reaction and that the spacecraft makes an effective pass entirely through each universe. Thus, all material in said universes of travel is reacted against with assumed finite universal mass-energy density.

If the latter universes are expanding with time over periods much greater than Aleph 2 years, then the spacecraft may attain a Lorentz factor of roughly [(Aleph J) EXP 3][(Aleph H) EXP 3] [[k(Aleph 2)] EXP 3], where k ranges from a small superunitary real number to suitable infinite real numbers. This, of course, assumes that real positive mass-energy stocks continue being created, perhaps out of zero-point fields or analogues thereof from some forms of dark energy. Provided that k ranges from Aleph 0 to Aleph 4, then the spacecraft Lorentz factor will be limited to a range of roughly from [(Aleph J) EXP 3][(Aleph H) EXP 3] [[(Aleph 0)(Aleph 2)] EXP 3] to [(Aleph J) EXP 3][(Aleph H) EXP 3] [[(Aleph 4)(Aleph 2)] EXP 3]. Note that in the above computations, we assume that only the material in the path of a reasonably sized spacecraft is upswept as mass-energy fuel or species of reaction and that the spacecraft makes an effective pass entirely through each universe. Thus, all material in said universes of travel is reacted against with assumed finite universal mass-energy density.

For universes in said multiverses having an extensity of Aleph 3 light-years, there are prospects for the spacecraft to attain a Lorentz factor of roughly [(Aleph J) EXP 3][(Aleph H) EXP 3] (Aleph 3) to thus result in the multiverses of travel being contracted to point directly in front of the spacecraft and contracted by a factor of [(Aleph J) EXP 3][(Aleph H) EXP 3](Aleph 3). Note that in the above computations, we assume that only the material in the path of a reasonably sized spacecraft is upswept as mass-energy fuel or species of reaction and that the spacecraft makes only one effective pass through each universe.

If such Aleph 3 light-year–width universes have three spatial dimensions and are substantially static, then a spacecraft traveling at light-speed in the forests of consideration may obtain a Lorentz factor of roughly [(Aleph J) EXP 3][(Aleph H) EXP 3] [(Aleph 3) EXP 3]. Note that in the above computations, we assume that only the material in the path of a reasonably sized spacecraft is upswept as mass-energy fuel or species of reaction and that the spacecraft makes an effective pass entirely through each universe. Thus, all material in said universes of travel is reacted against with assumed finite universal mass-energy density.

If the latter universes are expanding with time over periods much greater than Aleph 3 years, then the spacecraft may attain a Lorentz factor of roughly [(Aleph J) EXP 3][(Aleph H) EXP 3] [[k(Aleph 3)] EXP 3], where k ranges from a small superunitary real number to suitable infinite real numbers. This, of course, assumes that real positive mass-energy stocks continue being created, perhaps out of zero-point fields or analogues thereof from some forms of dark energy. Provided that k ranges from Aleph 0 to Aleph 5, then the spacecraft Lorentz factor will be limited to a range of roughly from [(Aleph J) EXP 3][(Aleph H) EXP 3] [[(Aleph 0)(Aleph 3)] EXP 3] to [(Aleph J) EXP 3][(Aleph H) EXP 3] [[(Aleph 5)(Aleph 3)] EXP 3]. Note that in the above computations, we assume that only the material in the path of a reasonably sized spacecraft is upswept as mass-energy fuel or species of reaction and that the spacecraft makes an effective pass entirely through each universe. Thus, all material in said universes of travel is reacted against with assumed finite universal mass-energy density.

For universes having an extensity of Aleph 4 light-years, there are prospects for the spacecraft to attain a Lorentz factor of roughly [(Aleph J) EXP 3][(Aleph H) EXP 3] (Aleph 4) to thus result in the multiverses of travel being contracted to point directly in front of the spacecraft and being contracted relative to the spacecraft by roughly a factor of [(Aleph J) EXP 3][(Aleph H) EXP 3](Aleph 4). Note that in the above computations, we assume that only the material in the path of a reasonably sized spacecraft is upswept as mass-energy fuel or species of reaction and that the spacecraft makes only one effective pass through each universe.

If such Aleph 4 light-year–width universes have three spatial dimensions and are substantially static, then a spacecraft traveling at light-speed in the forests may obtain a Lorentz factor of roughly [(Aleph J) EXP 3][(Aleph H) EXP 3] [(Aleph 4) EXP 3]. Note that in the above computations, we assume that only the material in the path of a reasonably sized spacecraft is upswept as mass-energy fuel or species of reaction and that the spacecraft makes an effective pass entirely through each universe. Thus, all material in said universes of travel is reacted against with assumed finite universal mass-energy density.

If the latter universes are expanding with time over periods much greater than Aleph 4 years, then the spacecraft may attain a Lorentz factor of roughly [(Aleph J) EXP 3][(Aleph H) EXP 3] [[k(Aleph 4)] EXP 3], where k ranges from a small superunitary real number to suitable infinite real numbers. This, of course, assumes that real positive mass-energy stocks continue being created, perhaps out of zero-point fields or analogues thereof from some forms of dark energy. Provided that k ranges roughly from Aleph 0 to Aleph 6, then the spacecraft Lorentz factor will be limited to a range from [(Aleph J) EXP 3][(Aleph H) EXP 3] [[(Aleph 0)(Aleph 4)] EXP 3] to [(Aleph J) EXP 3][(Aleph H) EXP 3] [[(Aleph 6)(Aleph 4)] EXP 3]. Note that in the above computations, we assume that only the material in the path of a reasonably sized spacecraft is upswept as mass-energy fuel or species of reaction and that the spacecraft makes an effective pass entirely through each universe. Thus, all material in said universes of travel is reacted against with assumed finite universal mass-energy density.

For universes having an extensity of Aleph G light-years, there are prospects for the spacecraft to attain a Lorentz factor of roughly [(Aleph J) EXP 3][(Aleph H) EXP 3]

(Aleph G) to thus result in the multiverses of travel being contracted to point directly in front of the spacecraft with a Lorentz contraction of roughly [(Aleph J) EXP 3][(Aleph H) EXP 3] (Aleph G). Note that in the above computations, we assume that only the material in the path of a reasonably sized spacecraft is upswept as mass-energy fuel or species of reaction and that the spacecraft makes only one effective pass through each universe.

If such Aleph G light-year–width universes have three spatial dimensions and are substantially static, then a spacecraft traveling at light-speed in the forest may obtain a Lorentz factor of roughly [(Aleph J) EXP 3][(Aleph H) EXP 3] [(Aleph G) EXP 3]. Note that in the above computations, we assume that only the material in the path of a reasonably sized spacecraft is upswept as mass-energy fuel or species of reaction and that the spacecraft makes an effective pass entirely through each universe. Thus, all material in said universes of travel is reacted against with assumed finite universal mass-energy density.

If the latter universes are expanding with time over periods much greater than Aleph G years, then the spacecraft may attain a Lorentz factor of roughly [(Aleph J) EXP 3][(Aleph H) EXP 3] [[k(Aleph G)] EXP 3], where *k* ranges from a small superunitary real number to suitable infinite real numbers. This, of course, assumes that real positive mass-energy stocks continue being created, perhaps out of zero-point fields or analogues thereof from some forms of dark energy. Provided that *k* ranges from Aleph 0 to Aleph (G + 2), then the spacecraft Lorentz factor will be limited to a range of roughly from [(Aleph J) EXP 3][(Aleph H) EXP 3] [[(Aleph 0)(Aleph G)] EXP 3] to [(Aleph J) EXP 3][(Aleph H) EXP 3] [[[Aleph (G + 2)](Aleph G)] EXP 3]. Note that in the above computations, we assume that only the material in the path of a reasonably sized spacecraft is upswept as mass-energy fuel or species of reaction and that the spacecraft makes an effective pass entirely through each universe. Thus, all material in said universes of travel is reacted against with assumed finite universal mass-energy density.

Now, we consider spacecraft travel in a 4-D–space forest of a width of Aleph J multiverses each of Aleph H universes in width where each multiverse has four spatial dimensions, where *H* is an integer greater than G + 2 and where said multiverses each contain [(Aleph J) EXP 4][(Aleph H) EXP 4] universes.

For universes in said 4-D_space multiverses, having an extensity of (Aleph 1) light-years, there are prospects for the spacecraft to attain a Lorentz factor of roughly [(Aleph J) EXP 4][(Aleph H) EXP 4] (Aleph 1) to thus result in the multiverses of travel being contracted to point directly in front of the spacecraft and Lorentz contracted by a factor of roughly [(Aleph J) EXP 4][(Aleph H) EXP 4] (Aleph 1) relative to the spacecraft. Note that in the above computations, we assume that only the material in the path of a

reasonably sized spacecraft is upswept as mass-energy fuel or species of reaction and that the spacecraft makes only one effective pass through each universe.

If such Aleph 1 light-year–width universes have three spatial dimensions and are substantially static, then a spacecraft traveling at light-speed in the forest of consideration may obtain a Lorentz factor of roughly [(Aleph J) EXP 4][(Aleph H) EXP 4] [(Aleph 1) EXP 3]. Note that in the above computations, we assume that only the material in the path of a reasonably sized spacecraft is upswept as mass-energy fuel or species of reaction and that the spacecraft makes an effective pass entirely through each universe. Thus, all material in said universes of travel is reacted against with assumed finite universal mass-energy density.

If the latter universes are expanding with time over periods much greater than Aleph 1 years, then the spacecraft may attain a Lorentz factor of roughly [(Aleph J) EXP 4][(Aleph H) EXP 4] [[k(Aleph 1)] EXP 3], where k ranges from a small superunitary real number to suitable infinite real numbers. This, of course, assumes that real positive mass-energy stocks continue being created, perhaps out of zero-point fields or analogues thereof from some forms of dark energy. Provided that k ranges from Aleph 0 to Aleph 3, then the spacecraft Lorentz factor will be limited to a range of roughly from [(Aleph J) EXP 4][(Aleph H) EXP 4] [[(Aleph 0)(Aleph 1)] EXP 3] to [(Aleph J) EXP 4][(Aleph H) EXP 4] [[(Aleph 3)(Aleph 1)] EXP 3]. Note that in the above computations, we assume that only the material in the path of a reasonably sized spacecraft is upswept as mass-energy fuel or species of reaction and that the spacecraft makes an effective entire pass through each universe. Thus, all material in said universes of travel is reacted against with assumed finite universal mass-energy density.

For universes in said multiverses having an extensity of Aleph 2 light-years, there are prospects for the spacecraft to attain a Lorentz factor of roughly [(Aleph J) EXP 4][(Aleph H) EXP 4] (Aleph 2) to thus result in the multiverses of travel being contracted to a point directly in front of the spacecraft and by a factor of [(Aleph J) EXP 4][(Aleph H) EXP 4] (Aleph 2). Note that in the above computations, we assume that only the material in the path of a reasonably sized spacecraft is upswept as mass-energy fuel or species of reaction and that the spacecraft makes only one effective pass through each universe.

If such Aleph 2 light-year–width universes have three spatial dimensions and are substantially static, then a spacecraft traveling at light-speed in the forest considered may obtain a Lorentz factor of roughly [(Aleph J) EXP 4][(Aleph H) EXP 4] [(Aleph 2) EXP 3]. Note that in the above computations, we assume that only the material in the path of a reasonably sized spacecraft is upswept as mass-energy fuel or species of reaction and that the spacecraft makes an effective pass entirely through each universe.

Thus, all material in said universes of travel is reacted against with assumed finite universal mass-energy density.

If the latter universes are expanding with time over periods much greater than Aleph 2 years, then the spacecraft may attain a Lorentz factor of roughly [(Aleph J) EXP 4][(Aleph H) EXP 4] [[k(Aleph 2)] EXP 3], where k ranges from a small superunitary real number to suitable infinite real numbers. This, of course, assumes that real positive mass-energy stocks continue being created, perhaps out of zero-point fields or analogues thereof from some forms of dark energy. Provided that k ranges from Aleph 0 to Aleph 4, then the spacecraft Lorentz factor will be limited to a range of roughly from [(Aleph J) EXP 4][(Aleph H) EXP 4] [[(Aleph 0)(Aleph 2)] EXP 3] to [(Aleph J) EXP 4][(Aleph H) EXP 4] [[(Aleph 4)(Aleph 2)] EXP 3]. Note that in the above computations, we assume that only the material in the path of a reasonably sized spacecraft is upswept as mass-energy fuel or species of reaction and that the spacecraft makes an effective pass entirely through each universe. Thus, all material in said universes of travel is reacted against with assumed finite universal mass-energy density.

For universes in said multiverses having an extensity of Aleph 3 light-years, there are prospects for the spacecraft to attain a Lorentz factor of roughly [(Aleph J) EXP 4][(Aleph H) EXP 4] (Aleph 3) to thus result in the multiverses of travel being contracted to point directly in front of the spacecraft and contracted by a factor of [(Aleph J) EXP 4][(Aleph H) EXP 4](Aleph 3). Note that in the above computations, we assume that only the material in the path of a reasonably sized spacecraft is upswept as mass-energy fuel or species of reaction and that the spacecraft makes only one effective pass through each universe.

If such Aleph 3 light-year–width universes have three spatial dimensions and are substantially static, then a spacecraft traveling at light-speed in the forest of consideration may obtain a Lorentz factor of roughly [(Aleph J) EXP 4][(Aleph H) EXP 4] [(Aleph 3) EXP 3]. Note that in the above computations, we assume that only the material in the path of a reasonably sized spacecraft is upswept as mass-energy fuel or species of reaction and that the spacecraft makes an effective pass entirely through each universe. Thus, all material in said universes of travel is reacted against with assumed finite universal mass-energy density.

If the latter universes are expanding with time over periods much greater than Aleph 3 years, then the spacecraft may attain a Lorentz factor of roughly [(Aleph J) EXP 4][(Aleph H) EXP 4] [[k(Aleph 3)] EXP 3], where k ranges from a small superunitary real number to suitable infinite real numbers. This, of course, assumes that real positive mass-energy stocks continue being created, perhaps out of zero-point fields or analogues thereof from some forms of dark energy. Provided that k ranges from Aleph 0 to Aleph 5, then the spacecraft Lorentz factor will be limited to a range of roughly from

[(Aleph J) EXP 4][(Aleph H) EXP 4] [[(Aleph 0)(Aleph 3)] EXP 3] to [(Aleph J) EXP 4][(Aleph H) EXP 4] [[(Aleph 5)(Aleph 3)] EXP 3]. Note that in the above computations, we assume that only the material in the path of a reasonably sized spacecraft is upswept as mass-energy fuel or species of reaction and that the spacecraft makes an effective pass entirely through each universe. Thus, all material in said universes of travel is reacted against with assumed finite universal mass-energy density.

For universes having an extensity of Aleph 4 light-years, there are prospects for the spacecraft to attain a Lorentz factor of roughly [(Aleph J) EXP 4][(Aleph H) EXP 4] (Aleph 4) to thus result in the multiverses of travel being contracted to point directly in front of the spacecraft and being contracted relative to the spacecraft by roughly a factor of [(Aleph J) EXP 4][(Aleph H) EXP 4](Aleph 4). Note that in the above computations, we assume that only the material in the path of a reasonably sized spacecraft is upswept as mass-energy fuel or species of reaction and that the spacecraft makes only one effective pass through each universe.

If such Aleph 4 light-year–width universes have three spatial dimensions and are substantially static, then a spacecraft traveling at light-speed in the forest may obtain a Lorentz factor of roughly [(Aleph J) EXP 4][(Aleph H) EXP 4] [(Aleph 4) EXP 3]. Note that in the above computations, we assume that only the material in the path of a reasonably sized spacecraft is upswept as mass-energy fuel or species of reaction and that the spacecraft makes an effective pass entirely through each universe. Thus, all material in said universes of travel is reacted against with assumed finite universal mass-energy density.

If the latter universes are expanding with time over periods much greater than Aleph 4 years, then the spacecraft may attain a Lorentz factor of roughly [(Aleph J) EXP 4][(Aleph H) EXP 4] [[k(Aleph 4)] EXP 3], where k ranges from a small superunitary real number to suitable infinite real numbers. This, of course, assumes that real positive mass-energy stocks continue being created, perhaps out of zero-point fields or analogues thereof from some forms of dark energy. Provided that k ranges roughly from Aleph 0 to Aleph 6, then the spacecraft Lorentz factor will be limited to a range from [(Aleph J) EXP 4][(Aleph H) EXP 4] [[(Aleph 0)(Aleph 4)] EXP 3] to [(Aleph J) EXP 4][(Aleph H) EXP 4] [[(Aleph 6)(Aleph 4)] EXP 3]. Note that in the above computations, we assume that only the material in the path of a reasonably sized spacecraft is upswept as mass-energy fuel or species of reaction and that the spacecraft makes an effective pass entirely through each universe. Thus, all material in said universes of travel is reacted against with assumed finite universal mass-energy density.

For universes having an extensity of Aleph G light-years, there are prospects for the spacecraft to attain a Lorentz factor of roughly [(Aleph J) EXP 4][(Aleph H) EXP 4] (Aleph G) to thus result in the multiverses of travel being contracted to point directly in

front of the spacecraft with a Lorentz contraction of roughly [(Aleph J) EXP 4][(Aleph H) EXP 4] (Aleph G). Note that in the above computations, we assume that only the material in the path of a reasonably sized spacecraft is upswept as mass-energy fuel or species of reaction and that the spacecraft makes only one effective pass through each universe.

If such Aleph G light-year–width universes have three spatial dimensions and are substantially static, then a spacecraft traveling at light-speed in the forest may obtain a Lorentz factor of roughly [(Aleph J) EXP 4][(Aleph H) EXP 4] [(Aleph G) EXP 3]. Note that in the above computations, we assume that only the material in the path of a reasonably sized spacecraft is upswept as mass-energy fuel or species of reaction and that the spacecraft makes an effective pass entirely through each universe. Thus, all material in said universes of travel is reacted against with assumed finite universal mass-energy density.

If the latter universes are expanding with time over periods much greater than Aleph G years, then the spacecraft may attain a Lorentz factor of roughly [(Aleph J) EXP 4][(Aleph H) EXP 4] [[k(Aleph G)] EXP 3], where k ranges from a small superunitary real number to suitable infinite real numbers. This, of course, assumes that real positive mass-energy stocks continue being created, perhaps out of zero-point fields or analogues thereof from some forms of dark energy. Provided that k ranges from Aleph 0 to Aleph (G + 2), then the spacecraft Lorentz factor will be limited to a range of roughly from [(Aleph J) EXP 4][(Aleph H) EXP 4] [[(Aleph 0)(Aleph G)] EXP 3] to [(Aleph J) EXP 4][(Aleph H) EXP 4] [[[Aleph (G + 2)](Aleph G)] EXP 3]. Note that in the above computations, we assume that only the material in the path of a reasonably sized spacecraft is upswept as mass-energy fuel or species of reaction and that the spacecraft makes an effective pass entirely through each universe. Thus, all material in said universes of travel is reacted against with assumed finite universal mass-energy density.

Now we consider spacecraft travel in a 5-D–space forest of a width of Aleph J multiverses each of Aleph H universes in width of five dimensions, where H is an integer greater than G + 2, and where said multiverses each contain [(Aleph J) EXP 5][(Aleph H) EXP 5] universes.

For universes in said 5-D–space multiverses having an extensity of (Aleph 1) light-years, there are prospects for the spacecraft to attain a Lorentz factor of roughly [(Aleph J) EXP 5][(Aleph H) EXP 5] (Aleph 1) to thus result in the multiverses of travel being contracted to point directly in front of the spacecraft and Lorentz contracted by a factor of roughly [(Aleph J) EXP 5][(Aleph H) EXP 5] (Aleph 1) relative to the spacecraft. Note that in the above computations, we assume that only the material in the path of a

reasonably sized spacecraft is upswept as mass-energy fuel or species of reaction and that the spacecraft makes only one effective pass through each universe.

If such Aleph 1 light-year–width universes have three spatial dimensions and are substantially static, then a spacecraft traveling at light-speed in the forest of consideration may obtain a Lorentz factor of roughly [(Aleph J) EXP 5][(Aleph H) EXP 5] [(Aleph 1) EXP 3]. Note that in the above computations, we assume that only the material in the path of a reasonably sized spacecraft is upswept as mass-energy fuel or species of reaction and that the spacecraft makes an effective pass entirely through each universe. Thus, all material in said universes of travel is reacted against with assumed finite universal mass-energy density.

If the latter universes are expanding with time over periods much greater than Aleph 1 years, then the spacecraft may attain a Lorentz factor of roughly [(Aleph J) EXP 5][(Aleph H) EXP 5] [[k(Aleph 1)] EXP 3], where *k* ranges from a small superunitary real number to suitable infinite real numbers. This, of course, assumes that real positive mass-energy stocks continue being created, perhaps out of zero-point fields or analogues thereof from some forms of dark energy. Provided that *k* ranges from Aleph 0 to Aleph 3, then the spacecraft Lorentz factor will be limited to a range of roughly from [(Aleph J) EXP 5][(Aleph H) EXP 5] [[(Aleph 0)(Aleph 1)] EXP 3] to [(Aleph J) EXP 5][(Aleph H) EXP 5] [[(Aleph 3)(Aleph 1)] EXP 3]. Note that in the above computations, we assume that only the material in the path of a reasonably sized spacecraft is upswept as mass-energy fuel or species of reaction and that the spacecraft makes an effective entire pass through each universe. Thus, all material in said universes of travel is reacted against with assumed finite universal mass-energy density.

For universes in said multiverses having an extensity of Aleph 2 light-years, there are prospects for the spacecraft to attain a Lorentz factor of roughly [(Aleph J) EXP 5][(Aleph H) EXP 5] (Aleph 2) to thus result in the multiverses of travel being contracted to a point directly in front of the spacecraft and by a factor of [(Aleph J) EXP 5][(Aleph H) EXP 5] (Aleph 2). Note that in the above computations, we assume that only the material in the path of a reasonably sized spacecraft is upswept as mass-energy fuel or species of reaction and that the spacecraft makes only one effective pass through each universe.

If such Aleph 2 light-year–width universes have three spatial dimensions and are substantially static, then a spacecraft traveling at light-speed in the forests considered may obtain a Lorentz factor of roughly [(Aleph J) EXP 5][(Aleph H) EXP 5] [(Aleph 2) EXP 3]. Note that in the above computations, we assume that only the material in the path of a reasonably sized spacecraft is upswept as mass-energy fuel or species of reaction and that the spacecraft makes an effective pass entirely through each universe.

Thus, all material in said universes of travel is reacted against with assumed finite universal mass-energy density.

If the latter universes are expanding with time over periods much greater than Aleph 2 years, then the spacecraft may attain a Lorentz factor of roughly [(Aleph J) EXP 5][(Aleph H) EXP 5] [[k(Aleph 2)] EXP 3], where k ranges from a small superunitary real number to suitable infinite real numbers. This, of course, assumes that real positive mass-energy stocks continue being created, perhaps out of zero-point fields or analogues thereof from some forms of dark energy. Provided that k ranges from Aleph 0 to Aleph 4, then the spacecraft Lorentz factor will be limited to a range of roughly from [(Aleph J) EXP 5][(Aleph H) EXP 5] [[(Aleph 0)(Aleph 2)] EXP 3] to [(Aleph J) EXP 5][(Aleph H) EXP 5] [[(Aleph 4)(Aleph 2)] EXP 3]. Note that in the above computations, we assume that only the material in the path of a reasonably sized spacecraft is upswept as mass-energy fuel or species of reaction and that the spacecraft makes an effective pass entirely through each universe. Thus, all material in said universes of travel is reacted against with assumed finite universal mass-energy density.

For universes in said multiverses having an extensity of Aleph 3 light-years, there are prospects for the spacecraft to attain a Lorentz factor of roughly [(Aleph J) EXP 5][(Aleph H) EXP 5] (Aleph 3) to thus result in the multiverses of travel being contracted to point directly in front of the spacecraft and contracted by a factor of [(Aleph J) EXP 4][(Aleph H) EXP 4](Aleph 3). Note that in the above computations, we assume that only the material in the path of a reasonably sized spacecraft is upswept as mass-energy fuel or species of reaction and that the spacecraft makes only one effective pass through each universe.

If such Aleph 3 light-year–width universes have three spatial dimensions and are substantially static, then a spacecraft traveling at light-speed in the forests of consideration may obtain a Lorentz factor of roughly [(Aleph J) EXP 5][(Aleph H) EXP 5] [(Aleph 3) EXP 3]. Note that in the above computations, we assume that only the material in the path of a reasonably sized spacecraft is upswept as mass-energy fuel or species of reaction and that the spacecraft makes an effective pass entirely through each universe. Thus, all material in said universes of travel is reacted against with assumed finite universal mass-energy density.

If the latter universes are expanding with time over periods much greater than Aleph 3 years, then the spacecraft may attain a Lorentz factor of roughly [(Aleph J) EXP 5][(Aleph H) EXP 5] [[k(Aleph 3)] EXP 3], where k ranges from a small superunitary real number to suitable infinite real numbers. This, of course, assumes that real positive mass-energy stocks continue being created, perhaps out of zero-point fields or analogues thereof from some forms of dark energy. Provided that k ranges from Aleph 0 to Aleph 5, then the spacecraft Lorentz factor will be limited to a range of roughly from

[(Aleph J) EXP 5][(Aleph H) EXP 5] [[(Aleph 0)(Aleph 3)] EXP 3] to [(Aleph J) EXP 5][(Aleph H) EXP 5] [[(Aleph 5)(Aleph 3)] EXP 3]. Note that in the above computations, we assume that only the material in the path of a reasonably sized spacecraft is upswept as mass-energy fuel or species of reaction and that the spacecraft makes an effective pass entirely through each universe. Thus, all material in said universes of travel is reacted against with assumed finite universal mass-energy density.

For universes having an extensity of Aleph 4 light-years, there are prospects for the spacecraft to attain a Lorentz factor of roughly [(Aleph J) EXP 5][(Aleph H) EXP 5] (Aleph 4) to thus result in the multiverses of travel being contracted to point directly in front of the spacecraft and being contracted relative to the spacecraft by roughly a factor of [(Aleph J) EXP 4][(Aleph H) EXP 4](Aleph 4). Note that in the above computations, we assume that only the material in the path of a reasonably sized spacecraft is upswept as mass-energy fuel or species of reaction and that the spacecraft makes only one effective pass through each universe.

If such Aleph 4 light-year–width universes have three spatial dimensions and are substantially static, then a spacecraft traveling at light-speed in the forests may obtain a Lorentz factor of roughly [(Aleph J) EXP 5][(Aleph H) EXP 5] [(Aleph 4) EXP 3]. Note that in the above computations, we assume that only the material in the path of a reasonably sized spacecraft is upswept as mass-energy fuel or species of reaction and that the spacecraft makes an effective pass entirely through each universe. Thus, all material in said universes of travel is reacted against with assumed finite universal mass-energy density.

If the latter universes are expanding with time over periods much greater than Aleph 4 years, then the spacecraft may attain a Lorentz factor of roughly [(Aleph J) EXP 5][(Aleph H) EXP 5] [[k(Aleph 4)] EXP 3], where k ranges from a small superunitary real number to suitable infinite real numbers. This, of course, assumes that real positive mass-energy stocks continue being created, perhaps out of zero-point fields or analogues thereof from some forms of dark energy. Provided that k ranges roughly from Aleph 0 to Aleph 6, then the spacecraft Lorentz factor will be limited to a range from [(Aleph J) EXP 5][(Aleph H) EXP 5] [[(Aleph 0)(Aleph 4)] EXP 3] to [(Aleph J) EXP 5][(Aleph H) EXP 5] [[(Aleph 6)(Aleph 4)] EXP 3]. Note that in the above computations, we assume that only the material in the path of a reasonably sized spacecraft is upswept as mass-energy fuel or species of reaction and that the spacecraft makes an effective pass entirely through each universe. Thus, all material in said universes of travel is reacted against with assumed finite universal mass-energy density.

For universes having an extensity of Aleph G light-years, there are prospects for the spacecraft to attain a Lorentz factor of roughly [(Aleph J) EXP 5][(Aleph H) EXP 5] (Aleph G) to thus result in the multiverses of travel being contracted to point directly in

front of the spacecraft with a Lorentz contraction of roughly [(Aleph J) EXP 5][(Aleph H) EXP 5] (Aleph G). Note that in the above computations, we assume that only the material in the path of a reasonably sized spacecraft is upswept as mass-energy fuel or species of reaction and that the spacecraft makes only one effective pass through each universe.

If such Aleph G light-year–width universes have three spatial dimensions and are substantially static, then a spacecraft traveling at light-speed in the forests may obtain a Lorentz factor of roughly [(Aleph J) EXP 5][(Aleph H) EXP 5] [(Aleph G) EXP 3]. Note that in the above computations, we assume that only the material in the path of a reasonably sized spacecraft is upswept as mass-energy fuel or species of reaction and that the spacecraft makes an effective pass entirely through each universe. Thus, all material in said universes of travel is reacted against with assumed finite universal mass-energy density.

If the latter universes are expanding with time over periods much greater than Aleph G years, then the spacecraft may attain a Lorentz factor of roughly [(Aleph J) EXP 5][(Aleph H) EXP 5] [[k(Aleph G)] EXP 3], where k ranges from a small superunitary real number to suitable infinite real numbers. This, of course, assumes that real positive mass-energy stocks continue being created, perhaps out of zero-point fields or analogues thereof from some forms of dark energy. Provided that k ranges from Aleph 0 to Aleph (G + 2), then the spacecraft Lorentz factor will be limited to a range of roughly from [(Aleph J) EXP 5][(Aleph H) EXP 5] [[(Aleph 0)(Aleph G)] EXP 3] to [(Aleph J) EXP 5][(Aleph H) EXP 5] [[[Aleph (G + 2)](Aleph G)] EXP 3]. Note that in the above computations, we assume that only the material in the path of a reasonably sized spacecraft is upswept as mass-energy fuel or species of reaction and that the spacecraft makes an effective pass entirely through each universe. Thus, all material in said universes of travel is reacted against with assumed finite universal mass-energy density.

Now we consider spacecraft travel in a 6-D–space forest of a width of Aleph J multiverses each of Aleph H universes in width and of six dimensions, where H is an integer greater than G + 2 and where said multiverses each contain [(Aleph J) EXP 6][(Aleph H) EXP 6] universes.

For universes in said 6-D–space multiverses having an extensity of (Aleph 1) light-years, there are prospects for the spacecraft to attain a Lorentz factor of roughly [(Aleph J) EXP 6][(Aleph H) EXP 6] (Aleph 1) to thus result in the multiverses of travel being contracted to point directly in front of the spacecraft and Lorentz contracted by a factor of roughly [(Aleph J) EXP 6][(Aleph H) EXP 6] (Aleph 1) relative to the spacecraft. Note that in the above computations, we assume that only the material in the path of a

reasonably sized spacecraft is upswept as mass-energy fuel or species of reaction and that the spacecraft makes only one effective pass through each universe.

If such Aleph 1 light-year–width universes have three spatial dimensions and are substantially static, then a spacecraft traveling at light-speed in the forests of consideration may obtain a Lorentz factor of roughly [(Aleph J) EXP 6][(Aleph H) EXP 6] [(Aleph 1) EXP 3]. Note that in the above computations, we assume that only the material in the path of a reasonably sized spacecraft is upswept as mass-energy fuel or species of reaction and that the spacecraft makes an effective pass entirely through each universe. Thus, all material in said universes of travel is reacted against with assumed finite universal mass-energy density.

If the latter universes are expanding with time over periods much greater than Aleph 1 years, then the spacecraft may attain a Lorentz factor of roughly [(Aleph J) EXP 6][(Aleph H) EXP 6] [[k(Aleph 1)] EXP 3], where k ranges from a small superunitary real number to suitable infinite real numbers. This, of course, assumes that real positive mass-energy stocks continue being created, perhaps out of zero-point fields or analogues thereof from some forms of dark energy. Provided that k ranges from Aleph 0 to Aleph 3, then the spacecraft Lorentz factor will be limited to a range of roughly from [(Aleph J) EXP 6][(Aleph H) EXP 6] [[(Aleph 0)(Aleph 1)] EXP 3] to [(Aleph J) EXP 6][(Aleph H) EXP 6] [[(Aleph 3)(Aleph 1)] EXP 3]. Note that in the above computations, we assume that only the material in the path of a reasonably sized spacecraft is upswept as mass-energy fuel or species of reaction and that the spacecraft makes an effective entire pass through each universe. Thus, all material in said universes of travel is reacted against with assumed finite universal mass-energy density.

For universes in said multiverses having an extensity of Aleph 2 light-years, there are prospects for the spacecraft to attain a Lorentz factor of roughly [(Aleph J) EXP 6][(Aleph H) EXP 6] (Aleph 2) to thus result in the multiverses of travel being contracted to a point directly in front of the spacecraft and by a factor of [(Aleph J) EXP 6][(Aleph H) EXP 6] (Aleph 2). Note that in the above computations, we assume that only the material in the path of a reasonably sized spacecraft is upswept as mass-energy fuel or species of reaction and that the spacecraft makes only one effective pass through each universe.

If such Aleph 2 light-year–width universes have three spatial dimensions and are substantially static, then a spacecraft traveling at light-speed in the forests considered may obtain a Lorentz factor of roughly [(Aleph J) EXP 6][(Aleph H) EXP 6] [(Aleph 2) EXP 3]. Note that in the above computations, we assume that only the material in the path of a reasonably sized spacecraft is upswept as mass-energy fuel or species of reaction and that the spacecraft makes an effective pass entirely through each universe.

Thus, all material in said universes of travel is reacted against with assumed finite universal mass-energy density.

If the latter universes are expanding with time over periods much greater than Aleph 2 years, then the spacecraft may attain a Lorentz factor of roughly [(Aleph J) EXP 6][(Aleph H) EXP 6] [[k(Aleph 2)] EXP 3], where k ranges from a small superunitary real number to suitable infinite real numbers. This, of course, assumes that real positive mass-energy stocks continue being created, perhaps out of zero-point fields or analogues thereof from some forms of dark energy. Provided that k ranges from Aleph 0 to Aleph 4, then the spacecraft Lorentz factor will be limited to a range of roughly from [(Aleph J) EXP 6][(Aleph H) EXP 6] [[(Aleph 0)(Aleph 2)] EXP 3] to [(Aleph J) EXP 6][(Aleph H) EXP 6] [[(Aleph 4)(Aleph 2)] EXP 3]. Note that in the above computations, we assume that only the material in the path of a reasonably sized spacecraft is upswept as mass-energy fuel or species of reaction and that the spacecraft makes an effective pass entirely through each universe. Thus, all material in said universes of travel is reacted against with assumed finite universal mass-energy density.

For universes in said multiverses having an extensity of Aleph 3 light-years, there are prospects for the spacecraft to attain a Lorentz factor of roughly [(Aleph J) EXP 6][(Aleph H) EXP 6] (Aleph 3) to thus result in the multiverses of travel being contracted to point directly in front of the spacecraft and contracted by a factor of [(Aleph J) EXP 4][(Aleph H) EXP 4](Aleph 3). Note that in the above computations, we assume that only the material in the path of a reasonably sized spacecraft is upswept as mass-energy fuel or species of reaction and that the spacecraft makes only one effective pass through each universe.

If such Aleph 3 light-year–width universes have three spatial dimensions and are substantially static, then a spacecraft traveling at light-speed in the forests of consideration may obtain a Lorentz factor of roughly [(Aleph J) EXP 6][(Aleph H) EXP 6] [(Aleph 3) EXP 3]. Note that in the above computations, we assume that only the material in the path of a reasonably sized spacecraft is upswept as mass-energy fuel or species of reaction and that the spacecraft makes an effective pass entirely through each universe. Thus, all material in said universes of travel is reacted against with assumed finite universal mass-energy density.

If the latter universes are expanding with time over periods much greater than Aleph 3 years, then the spacecraft may attain a Lorentz factor of roughly [(Aleph J) EXP 6][(Aleph H) EXP 6] [[k(Aleph 3)] EXP 3], where k ranges from a small superunitary real number to suitable infinite real numbers. This, of course, assumes that real positive mass-energy stocks continue being created, perhaps out of zero-point fields or analogues thereof from some forms of dark energy. Provided that k ranges from Aleph 0 to Aleph 5, then the spacecraft Lorentz factor will be limited to a range of roughly from

[(Aleph J) EXP 6][(Aleph H) EXP 6] [[(Aleph 0)(Aleph 3)] EXP 3] to [(Aleph J) EXP 6][(Aleph H) EXP 6] [[(Aleph 5)(Aleph 3)] EXP 3]. Note that in the above computations, we assume that only the material in the path of a reasonably sized spacecraft is upswept as mass-energy fuel or species of reaction and that the spacecraft makes an effective pass entirely through each universe. Thus, all material in said universes of travel is reacted against with assumed finite universal mass-energy density.

For universes having an extensity of Aleph 4 light-years, there are prospects for the spacecraft to attain a Lorentz factor of roughly [(Aleph J) EXP 6][(Aleph H) EXP 6] (Aleph 4) to thus result in the multiverses of travel being contracted to point directly in front of the spacecraft and being contracted relative to the spacecraft by roughly a factor of [(Aleph J) EXP 4][(Aleph H) EXP 4](Aleph 4). Note that in the above computations, we assume that only the material in the path of a reasonably sized spacecraft is upswept as mass-energy fuel or species of reaction and that the spacecraft makes only one effective pass through each universe.

If such Aleph 4 light-year–width universes have three spatial dimensions and are substantially static, then a spacecraft traveling at light-speed in the forests may obtain a Lorentz factor of roughly [(Aleph J) EXP 6][(Aleph H) EXP 6] [(Aleph 4) EXP 3]. Note that in the above computations, we assume that only the material in the path of a reasonably sized spacecraft is upswept as mass-energy fuel or species of reaction and that the spacecraft makes an effective pass entirely through each universe. Thus, all material in said universes of travel is reacted against with assumed finite universal mass-energy density.

If the latter universes are expanding with time over periods much greater than Aleph 4 years, then the spacecraft may attain a Lorentz factor of roughly [(Aleph J) EXP 6][(Aleph H) EXP 6] [[k(Aleph 4)] EXP 3], where k ranges from a small superunitary real number to suitable infinite real numbers. This, of course, assumes that real positive mass-energy stocks continue being created, perhaps out of zero-point fields or analogues thereof from some forms of dark energy. Provided that k ranges roughly from Aleph 0 to Aleph 6, then the spacecraft Lorentz factor will be limited to a range from [(Aleph J) EXP 6][(Aleph H) EXP 6] [[(Aleph 0)(Aleph 4)] EXP 3] to [(Aleph J) EXP 6][(Aleph H) EXP 6] [[(Aleph 6)(Aleph 4)] EXP 3]. Note that in the above computations, we assume that only the material in the path of a reasonably sized spacecraft is upswept as mass-energy fuel or species of reaction and that the spacecraft makes an effective pass entirely through each universe. Thus, all material in said universes of travel is reacted against with assumed finite universal mass-energy density.

For universes having an extensity of Aleph G light-years, there are prospects for the spacecraft to attain a Lorentz factor of roughly [(Aleph J) EXP 6][(Aleph H) EXP 6] (Aleph G) to thus result in the multiverses of travel being contracted to point directly in

front of the spacecraft with a Lorentz contraction of roughly [(Aleph J) EXP 6][(Aleph H) EXP 6] (Aleph G). Note that in the above computations, we assume that only the material in the path of a reasonably sized spacecraft is upswept as mass-energy fuel or species of reaction and that the spacecraft makes only one effective pass through each universe.

If such Aleph G light-year–width universes have three spatial dimensions and are substantially static, then a spacecraft traveling at light-speed in the forests may obtain a Lorentz factor of roughly [(Aleph J) EXP 6][(Aleph H) EXP 6] [(Aleph G) EXP 3]. Note that in the above computations, we assume that only the material in the path of a reasonably sized spacecraft is upswept as mass-energy fuel or species of reaction and that the spacecraft makes an effective pass entirely through each universe. Thus, all material in said universes of travel is reacted against with assumed finite universal mass-energy density.

If the latter universes are expanding with time over periods much greater than Aleph G years, then the spacecraft may attain a Lorentz factor of roughly [(Aleph J) EXP 6][(Aleph H) EXP 6] [[k(Aleph G)] EXP 3], where k ranges from a small superunitary real number to suitable infinite real numbers. This, of course, assumes that real positive mass-energy stocks continue being created, perhaps out of zero-point fields or analogues thereof from some forms of dark energy. Provided that k ranges from Aleph 0 to Aleph (G + 2), then the spacecraft Lorentz factor will be limited to a range of roughly from [(Aleph J) EXP 6][(Aleph H) EXP 6] [[(Aleph 0)(Aleph G)] EXP 3] to [(Aleph J) EXP 6][(Aleph H) EXP 6] [[[Aleph (G + 2)](Aleph G)] EXP 3]. Note that in the above computations, we assume that only the material in the path of a reasonably sized spacecraft is upswept as mass-energy fuel or species of reaction and that the spacecraft makes an effective pass entirely through each universe. Thus, all material in said universes of travel is reacted against with assumed finite universal mass-energy density.

Now we consider spacecraft travel in an N-D–space forest of a width of Aleph J multiverses each of Aleph H universes in width and of N dimensions, where H is an integer greater than G + 2, and where said multiverses each contain [(Aleph J) EXP N][(Aleph H) EXP N] universes.

For universes in said N-D–space multiverses having an extensity of (Aleph 1) light-years, there are prospects for the spacecraft to attain a Lorentz factor of roughly [(Aleph J) EXP N][(Aleph H) EXP N] (Aleph 1) to thus result in the multiverses of travel being contracted to point directly in front of the spacecraft and Lorentz contracted by a factor of roughly [(Aleph J) EXP N][(Aleph H) EXP N] (Aleph 1) relative to the spacecraft. Note that in the above computations, we assume that only the material in the path of a

reasonably sized spacecraft is upswept as mass-energy fuel or species of reaction and that the spacecraft makes only one effective pass through each universe.

If such Aleph 1 light-year–width universes have three spatial dimensions and are substantially static, then a spacecraft traveling at light-speed in the forests of consideration may obtain a Lorentz factor of roughly [(Aleph J) EXP N][(Aleph H) EXP N] [(Aleph 1) EXP 3]. Note that in the above computations, we assume that only the material in the path of a reasonably sized spacecraft is upswept as mass-energy fuel or species of reaction and that the spacecraft makes an effective pass entirely through each universe. Thus, all material in said universes of travel is reacted against with assumed finite universal mass-energy density.

If the latter universes are expanding with time over periods much greater than Aleph 1 years, then the spacecraft may attain a Lorentz factor of roughly [(Aleph J) EXP N][(Aleph H) EXP N] [[k(Aleph 1)] EXP 3], where k ranges from a small superunitary real number to suitable infinite real numbers. This, of course, assumes that real positive mass-energy stocks continue being created, perhaps out of zero-point fields or analogues thereof from some forms of dark energy. Provided that k ranges from Aleph 0 to Aleph 3, then the spacecraft Lorentz factor will be limited to a range of roughly from [(Aleph J) EXP N][(Aleph H) EXP N] [[(Aleph 0)(Aleph 1)] EXP 3] to [(Aleph J) EXP N][(Aleph H) EXP N] [[(Aleph 3)(Aleph 1)] EXP 3]. Note that in the above computations, we assume that only the material in the path of a reasonably sized spacecraft is upswept as mass-energy fuel or species of reaction and that the spacecraft makes an effective entire pass through each universe. Thus, all material in said universes of travel is reacted against with assumed finite universal mass-energy density.

For universes in said multiverses having an extensity of Aleph 2 light-years, there are prospects for the spacecraft to attain a Lorentz factor of roughly [(Aleph J) EXP N][(Aleph H) EXP N] (Aleph 2) to thus result in the multiverses of travel being contracted to a point directly in front of the spacecraft and by a factor of [(Aleph J) EXP N][(Aleph H) EXP N] (Aleph 2). Note that in the above computations, we assume that only the material in the path of a reasonably sized spacecraft is upswept as mass-energy fuel or species of reaction and that the spacecraft makes only one effective pass through each universe.

If such Aleph 2 light-year–width universes have three spatial dimensions and are substantially static, then a spacecraft traveling at light-speed in the forests considered may obtain a Lorentz factor of roughly [(Aleph J) EXP N][(Aleph H) EXP N] [(Aleph 2) EXP 3]. Note that in the above computations, we assume that only the material in the path of a reasonably sized spacecraft is upswept as mass-energy fuel or species of reaction and that the spacecraft makes an effective pass entirely through each universe.

Thus, all material in said universes of travel is reacted against with assumed finite universal mass-energy density.

If the latter universes are expanding with time over periods much greater than Aleph 2 years, then the spacecraft may attain a Lorentz factor of roughly [(Aleph J) EXP N][(Aleph H) EXP N] [[k(Aleph 2)] EXP 3], where k ranges from a small superunitary real number to suitable infinite real numbers. This, of course, assumes that real positive mass-energy stocks continue being created, perhaps out of zero-point fields or analogues thereof from some forms of dark energy. Provided that k ranges from Aleph 0 to Aleph 4, then the spacecraft Lorentz factor will be limited to a range of roughly from [(Aleph J) EXP N][(Aleph H) EXP N] [[(Aleph 0)(Aleph 2)] EXP 3] to [(Aleph J) EXP N][(Aleph H) EXP N] [[(Aleph 4)(Aleph 2)] EXP 3]. Note that in the above computations, we assume that only the material in the path of a reasonably sized spacecraft is upswept as mass-energy fuel or species of reaction and that the spacecraft makes an effective pass entirely through each universe. Thus, all material in said universes of travel is reacted against with assumed finite universal mass-energy density.

For universes in said multiverses having an extensity of Aleph 3 light-years, there are prospects for the spacecraft to attain a Lorentz factor of roughly [(Aleph J) EXP N][(Aleph H) EXP N] (Aleph 3) to thus result in the multiverses of travel being contracted to point directly in front of the spacecraft and contracted by a factor of [(Aleph J) EXP 4][(Aleph H) EXP 4](Aleph 3). Note that in the above computations, we assume that only the material in the path of a reasonably sized spacecraft is upswept as mass-energy fuel or species of reaction and that the spacecraft makes only one effective pass through each universe.

If such Aleph 3 light-year–width universes have three spatial dimensions and are substantially static, then a spacecraft traveling at light-speed in the forests of consideration may obtain a Lorentz factor of roughly [(Aleph J) EXP N][(Aleph H) EXP N] [(Aleph 3) EXP 3]. Note that in the above computations, we assume that only the material in the path of a reasonably sized spacecraft is upswept as mass-energy fuel or species of reaction and that the spacecraft makes an effective pass entirely through each universe. Thus, all material in said universes of travel is reacted against with assumed finite universal mass-energy density.

If the latter universes are expanding with time over periods much greater than Aleph 3 years, then the spacecraft may attain a Lorentz factor of roughly [(Aleph J) EXP N][(Aleph H) EXP N] [[k(Aleph 3)] EXP 3], where k ranges from a small superunitary real number to suitable infinite real numbers. This, of course, assumes that real positive mass-energy stocks continue being created, perhaps out of zero-point fields or analogues thereof from some forms of dark energy. Provided that k ranges from Aleph 0 to Aleph 5, then the spacecraft Lorentz factor will be limited to a range of roughly from

[(Aleph J) EXP 4][(Aleph H) EXP 4] [[(Aleph 0)(Aleph 3)] EXP 4] to [(Aleph J) EXP 4][(Aleph H) EXP 4] [[(Aleph 5)(Aleph 3)] EXP 4]. Note that in the above computations, we assume that only the material in the path of a reasonably sized spacecraft is upswept as mass-energy fuel or species of reaction and that the spacecraft makes an effective pass entirely through each universe. Thus, all material in said universes of travel is reacted against with assumed finite universal mass-energy density.

For universes having an extensity of Aleph 4 light-years, there are prospects for the spacecraft to attain a Lorentz factor of roughly [(Aleph J) EXP 4][(Aleph H) EXP 4] (Aleph 4) to thus result in the multiverses of travel being contracted to point directly in front of the spacecraft and being contracted relative to the spacecraft by roughly a factor of [(Aleph J) EXP 4][(Aleph H) EXP 4](Aleph 4). Note that in the above computations, we assume that only the material in the path of a reasonably sized spacecraft is upswept as mass-energy fuel or species of reaction and that the spacecraft makes only one effective pass through each universe.

If such Aleph 4 light-year–width universes have four spatial dimensions and are substantially static, then a spacecraft traveling at light-speed in the forests may obtain a Lorentz factor of roughly [(Aleph J) EXP 4][(Aleph H) EXP 4] [(Aleph 4) EXP 4]. Note that in the above computations, we assume that only the material in the path of a reasonably sized spacecraft is upswept as mass-energy fuel or species of reaction and that the spacecraft makes an effective pass entirely through each universe. Thus, all material in said universes of travel is reacted against with assumed finite universal mass-energy density.

If the latter universes are expanding with time over periods much greater than Aleph 4 years, then the spacecraft may attain a Lorentz factor of roughly [(Aleph J) EXP 4][(Aleph H) EXP 4] [[k(Aleph 4)] EXP 4], where k ranges from a small superunitary real number to suitable infinite real numbers. This, of course, assumes that real positive mass-energy stocks continue being created, perhaps out of zero-point fields or analogues thereof from some forms of dark energy. Provided that k ranges roughly from Aleph 0 to Aleph 6, then the spacecraft Lorentz factor will be limited to a range from [(Aleph J) EXP 4][(Aleph H) EXP 4] [[(Aleph 0)(Aleph 4)] EXP 4] to [(Aleph J) EXP 4][(Aleph H) EXP 4] [[(Aleph 6)(Aleph 4)] EXP 4]. Note that in the above computations, we assume that only the material in the path of a reasonably sized spacecraft is upswept as mass-energy fuel or species of reaction and that the spacecraft makes an effective pass entirely through each universe. Thus, all material in said universes of travel is reacted against with assumed finite universal mass-energy density.

For universes having an extensity of Aleph G light-years, there are prospects for the spacecraft to attain a Lorentz factor of roughly [(Aleph J) EXP 4][(Aleph H) EXP 4] (Aleph G) to thus result in the multiverses of travel being contracted to point directly in

front of the spacecraft with a Lorentz contraction of roughly [(Aleph J) EXP 4][(Aleph H) EXP 4] (Aleph G). Note that in the above computations, we assume that only the material in the path of a reasonably sized spacecraft is upswept as mass-energy fuel or species of reaction and that the spacecraft makes only one effective pass through each universe.

If such Aleph G light-year–width universes have four spatial dimensions and are substantially static, then a spacecraft traveling at light-speed in the forests may obtain a Lorentz factor of roughly [(Aleph J) EXP 4][(Aleph H) EXP 4] [(Aleph G) EXP 4]. Note that in the above computations, we assume that only the material in the path of a reasonably sized spacecraft is upswept as mass-energy fuel or species of reaction and that the spacecraft makes an effective pass entirely through each universe. Thus, all material in said universes of travel is reacted against with assumed finite universal mass-energy density.

If the latter universes are expanding with time over periods much greater than Aleph G years, then the spacecraft may attain a Lorentz factor of roughly [(Aleph J) EXP 4][(Aleph H) EXP 4] [[k(Aleph G)] EXP 4], where k ranges from a small superunitary real number to suitable infinite real numbers. This, of course, assumes that real positive mass-energy stocks continue being created, perhaps out of zero-point fields or analogues thereof from some forms of dark energy. Provided that k ranges from Aleph 0 to Aleph (G + 2), then the spacecraft Lorentz factor will be limited to a range of roughly from [(Aleph J) EXP 4][(Aleph H) EXP 4] [[(Aleph 0)(Aleph G)] EXP 4] to [(Aleph J) EXP 4][(Aleph H) EXP 4] [[[Aleph (G + 2)](Aleph G)] EXP 4]. Note that in the above computations, we assume that only the material in the path of a reasonably sized spacecraft is upswept as mass-energy fuel or species of reaction and that the spacecraft makes an effective pass entirely through each universe. Thus, all material in said universes of travel is reacted against with assumed finite universal mass-energy density.

Now we consider spacecraft travel in a 5-D–space forest of a width of Aleph J multiverses each of Aleph H universes in width and of five dimensions where H is an integer greater than G + 2, and where said multiverses each contain [(Aleph J) EXP 5][(Aleph H) EXP 5] universes.

For universes in said 5-D–space multiverses having an extensity of (Aleph 1) light-years, there are prospects for the spacecraft to attain a Lorentz factor of roughly [(Aleph J) EXP 5][(Aleph H) EXP 5] (Aleph 1) to thus result in the multiverses of travel being contracted to point directly in front of the spacecraft and Lorentz contracted by a factor of roughly [(Aleph J) EXP 5][(Aleph H) EXP 5] (Aleph 1) relative to the spacecraft. Note that in the above computations, we assume that only the material in the path of a

reasonably sized spacecraft is upswept as mass-energy fuel or species of reaction and that the spacecraft makes only one effective pass through each universe.

If such Aleph 1 light-year–width universes have four spatial dimensions and are substantially static, then a spacecraft traveling at light-speed in the forests of consideration may obtain a Lorentz factor of roughly [(Aleph J) EXP 5][(Aleph H) EXP 5] [(Aleph 1) EXP 4]. Note that in the above computations, we assume that only the material in the path of a reasonably sized spacecraft is upswept as mass-energy fuel or species of reaction and that the spacecraft makes an effective pass entirely through each universe. Thus, all material in said universes of travel is reacted against with assumed finite universal mass-energy density.

If the latter universes are expanding with time over periods much greater than Aleph 1 years, then the spacecraft may attain a Lorentz factor of roughly [(Aleph J) EXP 5][(Aleph H) EXP 5] [[k(Aleph 1)] EXP 4], where k ranges from a small superunitary real number to suitable infinite real numbers. This, of course, assumes that real positive mass-energy stocks continue being created, perhaps out of zero-point fields or analogues thereof from some forms of dark energy. Provided that k ranges from Aleph 0 to Aleph 3, then the spacecraft Lorentz factor will be limited to a range of roughly from [(Aleph J) EXP 5][(Aleph H) EXP 5] [[(Aleph 0)(Aleph 1)] EXP 4] to [(Aleph J) EXP 5][(Aleph H) EXP 5] [[(Aleph 3)(Aleph 1)] EXP 4]. Note that in the above computations, we assume that only the material in the path of a reasonably sized spacecraft is upswept as mass-energy fuel or species of reaction and that the spacecraft makes an effective entire pass through each universe. Thus, all material in said universes of travel is reacted against with assumed finite universal mass-energy density.

For universes in said multiverses having an extensity of Aleph 2 light-years, there are prospects for the spacecraft to attain a Lorentz factor of roughly [(Aleph J) EXP 5][(Aleph H) EXP 5] (Aleph 2) to thus result in the multiverses of travel being contracted to a point directly in front of the spacecraft and by a factor of [(Aleph J) EXP 5][(Aleph H) EXP 5] (Aleph 2). Note that in the above computations, we assume that only the material in the path of a reasonably sized spacecraft is upswept as mass-energy fuel or species of reaction and that the spacecraft makes only one effective pass through each universe.

If such Aleph 2 light-year–width universes have four spatial dimensions and are substantially static, then a spacecraft traveling at light-speed in the forests considered may obtain a Lorentz factor of roughly [(Aleph J) EXP 5][(Aleph H) EXP 5] [(Aleph 2) EXP 4]. Note that in the above computations, we assume that only the material in the path of a reasonably sized spacecraft is upswept as mass-energy fuel or species of reaction and that the spacecraft makes an effective pass entirely through each universe.

Thus, all material in said universes of travel is reacted against with assumed finite universal mass-energy density.

If the latter universes are expanding with time over periods much greater than Aleph 2 years, then the spacecraft may attain a Lorentz factor of roughly [(Aleph J) EXP 5][(Aleph H) EXP 5] [[k(Aleph 2)] EXP 4], where *k* ranges from a small superunitary real number to suitable infinite real numbers. This, of course, assumes that real positive mass-energy stocks continue being created, perhaps out of zero-point fields or analogues thereof from some forms of dark energy. Provided that *k* ranges from Aleph 0 to Aleph 4, then the spacecraft Lorentz factor will be limited to a range of roughly from [(Aleph J) EXP 5][(Aleph H) EXP 5] [[(Aleph 0)(Aleph 2)] EXP 4] to [(Aleph J) EXP 5][(Aleph H) EXP 5] [[(Aleph 4)(Aleph 2)] EXP 4]. Note that in the above computations, we assume that only the material in the path of a reasonably sized spacecraft is upswept as mass-energy fuel or species of reaction and that the spacecraft makes an effective pass entirely through each universe. Thus, all material in said universes of travel is reacted against with assumed finite universal mass-energy density.

For universes in said multiverses having an extensity of Aleph 3 light-years, there are prospects for the spacecraft to attain a Lorentz factor of roughly [(Aleph J) EXP 5][(Aleph H) EXP 5] (Aleph 3) to thus result in the multiverses of travel being contracted to point directly in front of the spacecraft and contracted by a factor of [(Aleph J) EXP 4][(Aleph H) EXP 4](Aleph 3). Note that in the above computations, we assume that only the material in the path of a reasonably sized spacecraft is upswept as mass-energy fuel or species of reaction and that the spacecraft makes only one effective pass through each universe.

If such Aleph 3 light-year–width universes have four spatial dimensions and are substantially static, then a spacecraft traveling at light-speed in the forests of consideration may obtain a Lorentz factor of roughly [(Aleph J) EXP 5][(Aleph H) EXP 5] [(Aleph 3) EXP 4]. Note that in the above computations, we assume that only the material in the path of a reasonably sized spacecraft is upswept as mass-energy fuel or species of reaction and that the spacecraft makes an effective pass entirely through each universe. Thus, all material in said universes of travel is reacted against with assumed finite universal mass-energy density.

If the latter universes are expanding with time over periods much greater than Aleph 3 years, then the spacecraft may attain a Lorentz factor of roughly [(Aleph J) EXP 5][(Aleph H) EXP 5] [[k(Aleph 3)] EXP 4], where *k* ranges from a small superunitary real number to suitable infinite real numbers. This, of course, assumes that real positive mass-energy stocks continue being created, perhaps out of zero-point fields or analogues thereof from some forms of dark energy. Provided that *k* ranges from Aleph 0 to Aleph 5, then the spacecraft Lorentz factor will be limited to a range of roughly from

[(Aleph J) EXP 5][(Aleph H) EXP 5] [[(Aleph 0)(Aleph 3)] EXP 4] to [(Aleph J) EXP 5][(Aleph H) EXP 5] [[(Aleph 5)(Aleph 3)] EXP 4]. Note that in the above computations, we assume that only the material in the path of a reasonably sized spacecraft is upswept as mass-energy fuel or species of reaction and that the spacecraft makes an effective pass entirely through each universe. Thus, all material in said universes of travel is reacted against with assumed finite universal mass-energy density.

For universes having an extensity of Aleph 4 light-years, there are prospects for the spacecraft to attain a Lorentz factor of roughly [(Aleph J) EXP 5][(Aleph H) EXP 5] (Aleph 4) to thus result in the multiverses of travel being contracted to point directly in front of the spacecraft and being contracted relative to the spacecraft by roughly a factor of [(Aleph J) EXP 4][(Aleph H) EXP 4](Aleph 4). Note that in the above computations, we assume that only the material in the path of a reasonably sized spacecraft is upswept as mass-energy fuel or species of reaction and that the spacecraft makes only one effective pass through each universe.

If such Aleph 4 light-year–width universes have four spatial dimensions and are substantially static, then a spacecraft traveling at light-speed in the forests may obtain a Lorentz factor of roughly [(Aleph J) EXP 5][(Aleph H) EXP 5] [(Aleph 4) EXP 4]. Note that in the above computations, we assume that only the material in the path of a reasonably sized spacecraft is upswept as mass-energy fuel or species of reaction and that the spacecraft makes an effective pass entirely through each universe. Thus, all material in said universes of travel is reacted against with assumed finite universal mass-energy density.

If the latter universes are expanding with time over periods much greater than Aleph 4 years, then the spacecraft may attain a Lorentz factor of roughly [(Aleph J) EXP 5][(Aleph H) EXP 5] [[k(Aleph 4)] EXP 4], where k ranges from a small superunitary real number to suitable infinite real numbers. This, of course, assumes that real positive mass-energy stocks continue being created, perhaps out of zero-point fields or analogues thereof from some forms of dark energy. Provided that k ranges roughly from Aleph 0 to Aleph 6, then the spacecraft Lorentz factor will be limited to a range from [(Aleph J) EXP 5][(Aleph H) EXP 5] [[(Aleph 0)(Aleph 4)] EXP 4] to [(Aleph J) EXP 5][(Aleph H) EXP 5] [[(Aleph 6)(Aleph 4)] EXP 4]. Note that in the above computations, we assume that only the material in the path of a reasonably sized spacecraft is upswept as mass-energy fuel or species of reaction and that the spacecraft makes an effective pass entirely through each universe. Thus, all material in said universes of travel is reacted against with assumed finite universal mass-energy density.

For universes having an extensity of Aleph G light-years, there are prospects for the spacecraft to attain a Lorentz factor of roughly [(Aleph J) EXP 5][(Aleph H) EXP 5] (Aleph G) to thus result in the multiverses of travel being contracted to point directly in

front of the spacecraft with a Lorentz contraction of roughly [(Aleph J) EXP 5][(Aleph H) EXP 5] (Aleph G). Note that in the above computations, we assume that only the material in the path of a reasonably sized spacecraft is upswept as mass-energy fuel or species of reaction and that the spacecraft makes only one effective pass through each universe.

If such Aleph G light-year–width universes have four spatial dimensions and are substantially static, then a spacecraft traveling at light-speed in the forests may obtain a Lorentz factor of roughly [(Aleph J) EXP 5][(Aleph H) EXP 5] [(Aleph G) EXP 4]. Note that in the above computations, we assume that only the material in the path of a reasonably sized spacecraft is upswept as mass-energy fuel or species of reaction and that the spacecraft makes an effective pass entirely through each universe. Thus, all material in said universes of travel is reacted against with assumed finite universal mass-energy density.

If the latter universes are expanding with time over periods much greater than Aleph G years, then the spacecraft may attain a Lorentz factor of roughly [(Aleph J) EXP 5][(Aleph H) EXP 5] [[k(Aleph G)] EXP 4], where k ranges from a small superunitary real number to suitable infinite real numbers. This, of course, assumes that real positive mass-energy stocks continue being created, perhaps out of zero-point fields or analogues thereof from some forms of dark energy. Provided that k ranges from Aleph 0 to Aleph (G + 2), then the spacecraft Lorentz factor will be limited to a range of roughly from [(Aleph J) EXP 5][(Aleph H) EXP 5] [[(Aleph 0)(Aleph G)] EXP 4] to [(Aleph J) EXP 5][(Aleph H) EXP 5] [[[Aleph (G + 2)](Aleph G)] EXP 4]. Note that in the above computations, we assume that only the material in the path of a reasonably sized spacecraft is upswept as mass-energy fuel or species of reaction and that the spacecraft makes an effective pass entirely through each universe. Thus, all material in said universes of travel is reacted against with assumed finite universal mass-energy density.

Now we consider spacecraft travel in a 6-D-space forest of width of Aleph J multiverses each of Aleph H universes in width each of six dimensions, where H is an integer greater than G + 2, and where said multiverses each contain [(Aleph J) EXP 6][(Aleph H) EXP 6] universes.

For universes in said 6-D–space multiverses having an extensity of (Aleph 1) light-years, there are prospects for the spacecraft to attain a Lorentz factor of roughly [(Aleph J) EXP 6][(Aleph H) EXP 6] (Aleph 1) to thus result in the multiverses of travel being contracted to point directly in front of the spacecraft and Lorentz contracted by a factor of roughly [(Aleph J) EXP 6][(Aleph H) EXP 6] (Aleph 1) relative to the spacecraft. Note that in the above computations, we assume that only the material in the path of a

reasonably sized spacecraft is upswept as mass-energy fuel or species of reaction and that the spacecraft makes only one effective pass through each universe.

If such Aleph 1 light-year–width universes have four spatial dimensions and are substantially static, then a spacecraft traveling at light-speed in the forests of consideration may obtain a Lorentz factor of roughly [(Aleph J) EXP 6][(Aleph H) EXP 6] [(Aleph 1) EXP 4]. Note that in the above computations, we assume that only the material in the path of a reasonably sized spacecraft is upswept as mass-energy fuel or species of reaction and that the spacecraft makes an effective pass entirely through each universe. Thus, all material in said universes of travel is reacted against with assumed finite universal mass-energy density.

If the latter universes are expanding with time over periods much greater than Aleph 1 years, then the spacecraft may attain a Lorentz factor of roughly [(Aleph J) EXP 6][(Aleph H) EXP 6] [[k(Aleph 1)] EXP 4], where k ranges from a small superunitary real number to suitable infinite real numbers. This, of course, assumes that real positive mass-energy stocks continue being created, perhaps out of zero-point fields or analogues thereof from some forms of dark energy. Provided that k ranges from Aleph 0 to Aleph 3, then the spacecraft Lorentz factor will be limited to a range of roughly from [(Aleph J) EXP 6][(Aleph H) EXP 6] [[(Aleph 0)(Aleph 1)] EXP 4] to [(Aleph J) EXP 6][(Aleph H) EXP 6] [[(Aleph 3)(Aleph 1)] EXP 4]. Note that in the above computations, we assume that only the material in the path of a reasonably sized spacecraft is upswept as mass-energy fuel or species of reaction and that the spacecraft makes an effective entire pass through each universe. Thus, all material in said universes of travel is reacted against with assumed finite universal mass-energy density.

For universes in said multiverses having an extensity of Aleph 2 light-years, there are prospects for the spacecraft to attain a Lorentz factor of roughly [(Aleph J) EXP 6][(Aleph H) EXP 6] (Aleph 2) to thus result in the multiverses of travel being contracted to a point directly in front of the spacecraft and by a factor of [(Aleph J) EXP 6][(Aleph H) EXP 6] (Aleph 2). Note that in the above computations, we assume that only the material in the path of a reasonably sized spacecraft is upswept as mass-energy fuel or species of reaction and that the spacecraft makes only one effective pass through each universe.

If such Aleph 2 light-year–width universes have four spatial dimensions and are substantially static, then a spacecraft traveling at light-speed in the forests considered may obtain a Lorentz factor of roughly [(Aleph J) EXP 6][(Aleph H) EXP 6] [(Aleph 2) EXP 4]. Note that in the above computations, we assume that only the material in the path of a reasonably sized spacecraft is upswept as mass-energy fuel or species of reaction and that the spacecraft makes an effective pass entirely through each universe.

Thus, all material in said universes of travel is reacted against with assumed finite universal mass-energy density.

If the latter universes are expanding with time over periods much greater than Aleph 2 years, then the spacecraft may attain a Lorentz factor of roughly [(Aleph J) EXP 6][(Aleph H) EXP 6] [[k(Aleph 2)] EXP 4], where *k* ranges from a small superunitary real number to suitable infinite real numbers. This, of course, assumes that real positive mass-energy stocks continue being created, perhaps out of zero-point fields or analogues thereof from some forms of dark energy. Provided that *k* ranges from Aleph 0 to Aleph 4, then the spacecraft Lorentz factor will be limited to a range of roughly from [(Aleph J) EXP 6][(Aleph H) EXP 6] [[(Aleph 0)(Aleph 2)] EXP 4] to [(Aleph J) EXP 6][(Aleph H) EXP 6] [[(Aleph 4)(Aleph 2)] EXP 4]. Note that in the above computations, we assume that only the material in the path of a reasonably sized spacecraft is upswept as mass-energy fuel or species of reaction and that the spacecraft makes an effective pass entirely through each universe. Thus, all material in said universes of travel is reacted against with assumed finite universal mass-energy density.

For universes in said multiverses having an extensity of Aleph 3 light-years, there are prospects for the spacecraft to attain a Lorentz factor of roughly [(Aleph J) EXP 6][(Aleph H) EXP 6] (Aleph 3) to thus result in the multiverses of travel being contracted to point directly in front of the spacecraft and contracted by a factor of [(Aleph J) EXP 4][(Aleph H) EXP 4](Aleph 3). Note that in the above computations, we assume that only the material in the path of a reasonably sized spacecraft is upswept as mass-energy fuel or species of reaction and that the spacecraft makes only one effective pass through each universe.

If such Aleph 3 light-year–width universes have four spatial dimensions and are substantially static, then a spacecraft traveling at light-speed in the forests of consideration may obtain a Lorentz factor of roughly [(Aleph J) EXP 6][(Aleph H) EXP 6] [(Aleph 3) EXP 4]. Note that in the above computations, we assume that only the material in the path of a reasonably sized spacecraft is upswept as mass-energy fuel or species of reaction and that the spacecraft makes an effective pass entirely through each universe. Thus, all material in said universes of travel is reacted against with assumed finite universal mass-energy density.

If the latter universes are expanding with time over periods much greater than Aleph 3 years, then the spacecraft may attain a Lorentz factor of roughly [(Aleph J) EXP 6][(Aleph H) EXP 6] [[k(Aleph 3)] EXP 4], where *k* ranges from a small superunitary real number to suitable infinite real numbers. This, of course, assumes that real positive mass-energy stocks continue being created, perhaps out of zero-point fields or analogues thereof from some forms of dark energy. Provided that *k* ranges from Aleph 0 to Aleph 5, then the spacecraft Lorentz factor will be limited to a range of roughly from

[(Aleph J) EXP 6][(Aleph H) EXP 6] [[(Aleph 0)(Aleph 3)] EXP 4] to [(Aleph J) EXP 6][(Aleph H) EXP 6] [[(Aleph 5)(Aleph 3)] EXP 4]. Note that in the above computations, we assume that only the material in the path of a reasonably sized spacecraft is upswept as mass-energy fuel or species of reaction and that the spacecraft makes an effective pass entirely through each universe. Thus, all material in said universes of travel is reacted against with assumed finite universal mass-energy density.

For universes having an extensity of Aleph 4 light-years, there are prospects for the spacecraft to attain a Lorentz factor of roughly [(Aleph J) EXP 6][(Aleph H) EXP 6] (Aleph 4) to thus result in the multiverses of travel being contracted to point directly in front of the spacecraft and being contracted relative to the spacecraft by roughly a factor of [(Aleph J) EXP 4][(Aleph H) EXP 4](Aleph 4). Note that in the above computations, we assume that only the material in the path of a reasonably sized spacecraft is upswept as mass-energy fuel or species of reaction and that the spacecraft makes only one effective pass through each universe.

If such Aleph 4 light-year–width universes have four spatial dimensions and are substantially static, then a spacecraft traveling at light-speed in the forests may obtain a Lorentz factor of roughly [(Aleph J) EXP 6][(Aleph H) EXP 6] [(Aleph 4) EXP 4]. Note that in the above computations, we assume that only the material in the path of a reasonably sized spacecraft is upswept as mass-energy fuel or species of reaction and that the spacecraft makes an effective pass entirely through each universe. Thus, all material in said universes of travel is reacted against with assumed finite universal mass-energy density.

If the latter universes are expanding with time over periods much greater than Aleph 4 years, then the spacecraft may attain a Lorentz factor of roughly [(Aleph J) EXP 6][(Aleph H) EXP 6] [[k(Aleph 4)] EXP 4], where k ranges from a small superunitary real number to suitable infinite real numbers. This, of course, assumes that real positive mass-energy stocks continue being created, perhaps out of zero-point fields or analogues thereof from some forms of dark energy. Provided that k ranges roughly from Aleph 0 to Aleph 6, then the spacecraft Lorentz factor will be limited to a range from [(Aleph J) EXP 6][(Aleph H) EXP 6] [[(Aleph 0)(Aleph 4)] EXP 4] to [(Aleph J) EXP 6][(Aleph H) EXP 6] [[(Aleph 6)(Aleph 4)] EXP 4]. Note that in the above computations, we assume that only the material in the path of a reasonably sized spacecraft is upswept as mass-energy fuel or species of reaction and that the spacecraft makes an effective pass entirely through each universe. Thus, all material in said universes of travel is reacted against with assumed finite universal mass-energy density.

For universes having an extensity of Aleph G light-years, there are prospects for the spacecraft to attain a Lorentz factor of roughly [(Aleph J) EXP 6][(Aleph H) EXP 6] (Aleph G) to thus result in the multiverses of travel being contracted to point directly in

front of the spacecraft with a Lorentz contraction of roughly [(Aleph J) EXP 6][(Aleph H) EXP 6] (Aleph G). Note that in the above computations, we assume that only the material in the path of a reasonably sized spacecraft is upswept as mass-energy fuel or species of reaction and that the spacecraft makes only one effective pass through each universe.

If such Aleph G light-year–width universes have four spatial dimensions and are substantially static, then a spacecraft traveling at light-speed in the forests may obtain a Lorentz factor of roughly [(Aleph J) EXP 6][(Aleph H) EXP 6] [(Aleph G) EXP 4]. Note that in the above computations, we assume that only the material in the path of a reasonably sized spacecraft is upswept as mass-energy fuel or species of reaction and that the spacecraft makes an effective pass entirely through each universe. Thus, all material in said universes of travel is reacted against with assumed finite universal mass-energy density.

If the latter universes are expanding with time over periods much greater than Aleph G years, then the spacecraft may attain a Lorentz factor of roughly [(Aleph J) EXP 6][(Aleph H) EXP 6] [[k(Aleph G)] EXP 4], where *k* ranges from a small superunitary real number to suitable infinite real numbers. This, of course, assumes that real positive mass-energy stocks continue being created, perhaps out of zero-point fields or analogues thereof from some forms of dark energy. Provided that *k* ranges from Aleph 0 to Aleph (G + 2), then the spacecraft Lorentz factor will be limited to a range of roughly from [(Aleph J) EXP 6][(Aleph H) EXP 6] [[(Aleph 0)(Aleph G)] EXP 4] to [(Aleph J) EXP 6][(Aleph H) EXP 6] [[[Aleph (G + 2)](Aleph G)] EXP 4]. Note that in the above computations, we assume that only the material in the path of a reasonably sized spacecraft is upswept as mass-energy fuel or species of reaction and that the spacecraft makes an effective pass entirely through each universe. Thus, all material in said universes of travel is reacted against with assumed finite universal mass-energy density.

Now, we consider spacecraft travel in an N-D–space forest of a width of Aleph J multiverses each of Aleph H universes in width and of N dimensions, where *H* is an integer greater than G + 2 and where said multiverses each contain [(Aleph J) EXP N][(Aleph H) EXP N] universes.

For universes in said N-D–space multiverses having an extensity of (Aleph 1) light-years, there are prospects for the spacecraft to attain a Lorentz factor of roughly [(Aleph J) EXP N][(Aleph H) EXP N] (Aleph 1) to thus result in the multiverses of travel being contracted to point directly in front of the spacecraft and Lorentz contracted by a factor of roughly [(Aleph J) EXP N][(Aleph H) EXP N] (Aleph 1) relative to the spacecraft. Note that in the above computations, we assume that only the material in the path of a

reasonably sized spacecraft is upswept as mass-energy fuel or species of reaction and that the spacecraft makes only one effective pass through each universe.

If such Aleph 1 light-year–width universes have four spatial dimensions and are substantially static, then a spacecraft traveling at light-speed in the forests of consideration may obtain a Lorentz factor of roughly [(Aleph J) EXP N][(Aleph H) EXP N] [(Aleph 1) EXP 4]. Note that in the above computations, we assume that only the material in the path of a reasonably sized spacecraft is upswept as mass-energy fuel or species of reaction and that the spacecraft makes an effective pass entirely through each universe. Thus, all material in said universes of travel is reacted against with assumed finite universal mass-energy density.

If the latter universes are expanding with time over periods much greater than Aleph 1 years, then the spacecraft may attain a Lorentz factor of roughly [(Aleph J) EXP N][(Aleph H) EXP N] [[k(Aleph 1)] EXP 4], where k ranges from a small superunitary real number to suitable infinite real numbers. This, of course, assumes that real positive mass-energy stocks continue being created, perhaps out of zero-point fields or analogues thereof from some forms of dark energy. Provided that k ranges from Aleph 0 to Aleph 3, then the spacecraft Lorentz factor will be limited to a range of roughly from [(Aleph J) EXP N][(Aleph H) EXP N] [[(Aleph 0)(Aleph 1)] EXP 4] to [(Aleph J) EXP N][(Aleph H) EXP N] [[(Aleph 3)(Aleph 1)] EXP 4]. Note that in the above computations, we assume that only the material in the path of a reasonably sized spacecraft is upswept as mass-energy fuel or species of reaction and that the spacecraft makes an effective entire pass through each universe. Thus, all material in said universes of travel is reacted against with assumed finite universal mass-energy density.

For universes in said multiverses having an extensity of Aleph 2 light-years, there are prospects for the spacecraft to attain a Lorentz factor of roughly [(Aleph J) EXP N][(Aleph H) EXP N] (Aleph 2) to thus result in the multiverses of travel being contracted to a point directly in front of the spacecraft and by a factor of [(Aleph J) EXP N][(Aleph H) EXP N] (Aleph 2). Note that in the above computations, we assume that only the material in the path of a reasonably sized spacecraft is upswept as mass-energy fuel or species of reaction and that the spacecraft makes only one effective pass through each universe.

If such Aleph 2 light-year–width universes have four spatial dimensions and are substantially static, then a spacecraft traveling at light-speed in the forests considered may obtain a Lorentz factor of roughly [(Aleph J) EXP N][(Aleph H) EXP N] [(Aleph 2) EXP 4]. Note that in the above computations, we assume that only the material in the path of a reasonably sized spacecraft is upswept as mass-energy fuel or species of reaction and that the spacecraft makes an effective pass entirely through each universe.

Thus, all material in said universes of travel is reacted against with assumed finite universal mass-energy density.

If the latter universes are expanding with time over periods much greater than Aleph 2 years, then the spacecraft may attain a Lorentz factor of roughly [(Aleph J) EXP N][(Aleph H) EXP N] [[k(Aleph 2)] EXP 4], where k ranges from a small superunitary real number to suitable infinite real numbers. This, of course, assumes that real positive mass-energy stocks continue being created, perhaps out of zero-point fields or analogues thereof from some forms of dark energy. Provided that k ranges from Aleph 0 to Aleph 4, then the spacecraft Lorentz factor will be limited to a range of roughly from [(Aleph J) EXP N][(Aleph H) EXP N] [[(Aleph 0)(Aleph 2)] EXP 4] to [(Aleph J) EXP N][(Aleph H) EXP N] [[(Aleph 4)(Aleph 2)] EXP 4]. Note that in the above computations, we assume that only the material in the path of a reasonably sized spacecraft is upswept as mass-energy fuel or species of reaction and that the spacecraft makes an effective pass entirely through each universe. Thus, all material in said universes of travel is reacted against with assumed finite universal mass-energy density.

For universes in said multiverses having an extensity of Aleph 3 light-years, there are prospects for the spacecraft to attain a Lorentz factor of roughly [(Aleph J) EXP N][(Aleph H) EXP N] (Aleph 3) to thus result in the multiverses of travel being contracted to point directly in front of the spacecraft and contracted by a factor of [(Aleph J) EXP 4][(Aleph H) EXP 4](Aleph 3). Note that in the above computations, we assume that only the material in the path of a reasonably sized spacecraft is upswept as mass-energy fuel or species of reaction and that the spacecraft makes only one effective pass through each universe.

If such Aleph 3 light-year–width universes have four spatial dimensions and are substantially static, then a spacecraft traveling at light-speed in the forests of consideration may obtain a Lorentz factor of roughly [(Aleph J) EXP N][(Aleph H) EXP N] [(Aleph 3) EXP 4]. Note that in the above computations, we assume that only the material in the path of a reasonably sized spacecraft is upswept as mass-energy fuel or species of reaction and that the spacecraft makes an effective pass entirely through each universe. Thus, all material in said universes of travel is reacted against with assumed finite universal mass-energy density.

If the latter universes are expanding with time over periods much greater than Aleph 3 years, then the spacecraft may attain a Lorentz factor of roughly [(Aleph J) EXP N][(Aleph H) EXP N] [[k(Aleph 3)] EXP 4], where k ranges from a small superunitary real number to suitable infinite real numbers. This, of course, assumes that real positive mass-energy stocks continue being created, perhaps out of zero-point fields or analogues thereof from some forms of dark energy. Provided that k ranges from Aleph 0 to Aleph 5, then the spacecraft Lorentz factor will be limited to a range of roughly from

[(Aleph J) EXP N][(Aleph H) EXP N] [[(Aleph 0)(Aleph 3)] EXP 4] to [(Aleph J) EXP N][(Aleph H) EXP N] [[(Aleph 5)(Aleph 3)] EXP 4]. Note that in the above computations, we assume that only the material in the path of a reasonably sized spacecraft is upswept as mass-energy fuel or species of reaction and that the spacecraft makes an effective pass entirely through each universe. Thus, all material in said universes of travel is reacted against with assumed finite universal mass-energy density.

For universes having an extensity of Aleph 4 light-years, there are prospects for the spacecraft to attain a Lorentz factor of roughly [(Aleph J) EXP N][(Aleph H) EXP N] (Aleph 4) to thus result in the multiverses of travel being contracted to point directly in front of the spacecraft and being contracted relative to the spacecraft by roughly a factor of [(Aleph J) EXP 4][(Aleph H) EXP 4](Aleph 4). Note that in the above computations, we assume that only the material in the path of a reasonably sized spacecraft is upswept as mass-energy fuel or species of reaction and that the spacecraft makes only one effective pass through each universe.

If such Aleph 4 light-year–width universes have four spatial dimensions and are substantially static, then a spacecraft traveling at light-speed in the forests may obtain a Lorentz factor of roughly [(Aleph J) EXP N][(Aleph H) EXP N] [(Aleph 4) EXP 4]. Note that in the above computations, we assume that only the material in the path of a reasonably sized spacecraft is upswept as mass-energy fuel or species of reaction and that the spacecraft makes an effective pass entirely through each universe. Thus, all material in said universes of travel is reacted against with assumed finite universal mass-energy density.

If the latter universes are expanding with time over periods much greater than Aleph 4 years, then the spacecraft may attain a Lorentz factor of roughly [(Aleph J) EXP N][(Aleph H) EXP N] [[k(Aleph 4)] EXP 4], where k ranges from a small superunitary real number to suitable infinite real numbers. This, of course, assumes that real positive mass-energy stocks continue being created, perhaps out of zero-point fields or analogues thereof from some forms of dark energy. Provided that k ranges roughly from Aleph 0 to Aleph 6, then the spacecraft Lorentz factor will be limited to a range from [(Aleph J) EXP N][(Aleph H) EXP N] [[(Aleph 0)(Aleph 4)] EXP 4] to [(Aleph J) EXP N][(Aleph H) EXP N] [[(Aleph 6)(Aleph 4)] EXP 4]. Note that in the above computations, we assume that only the material in the path of a reasonably sized spacecraft is upswept as mass-energy fuel or species of reaction and that the spacecraft makes an effective pass entirely through each universe. Thus, all material in said universes of travel is reacted against with assumed finite universal mass-energy density.

For universes having an extensity of Aleph G light-years, there are prospects for the spacecraft to attain a Lorentz factor of roughly [(Aleph J) EXP N][(Aleph H) EXP N] (Aleph G) to thus result in the multiverses of travel being contracted to point directly in

front of the spacecraft with a Lorentz contraction of roughly [(Aleph J) EXP N][(Aleph H) EXP N] (Aleph G). Note that in the above computations, we assume that only the material in the path of a reasonably sized spacecraft is upswept as mass-energy fuel or species of reaction and that the spacecraft makes only one effective pass through each universe.

If such Aleph G light-year–width universes have four spatial dimensions and are substantially static, then a spacecraft traveling at light-speed in the forests may obtain a Lorentz factor of roughly [(Aleph J) EXP N][(Aleph H) EXP N] [(Aleph G) EXP 4]. Note that in the above computations, we assume that only the material in the path of a reasonably sized spacecraft is upswept as mass-energy fuel or species of reaction and that the spacecraft makes an effective pass entirely through each universe. Thus, all material in said universes of travel is reacted against with assumed finite universal mass-energy density.

If the latter universes are expanding with time over periods much greater than Aleph G years, then the spacecraft may attain a Lorentz factor of roughly [(Aleph J) EXP N][(Aleph H) EXP N] [[k(Aleph G)] EXP 4], where k ranges from a small superunitary real number to suitable infinite real numbers. This, of course, assumes that real positive mass-energy stocks continue being created, perhaps out of zero-point fields or analogues thereof from some forms of dark energy. Provided that k ranges from Aleph 0 to Aleph (G + 2), then the spacecraft Lorentz factor will be limited to a range of roughly from [(Aleph J) EXP N][(Aleph H) EXP N] [[(Aleph 0)(Aleph G)] EXP 4] to [(Aleph J) EXP N][(Aleph H) EXP N] [[[Aleph (G + 2)](Aleph G)] EXP 4]. Note that in the above computations, we assume that only the material in the path of a reasonably sized spacecraft is upswept as mass-energy fuel or species of reaction and that the spacecraft makes an effective pass entirely through each universe. Thus, all material in said universes of travel is reacted against with assumed finite universal mass-energy density.

Now, we consider spacecraft travel in a 5-D-space forest of a width of Aleph J multiverses each of Aleph H universes in width each of five dimensions, where H is an integer greater than G + 2, and where said multiverses each contain [(Aleph J) EXP 5][(Aleph H) EXP 5] universes.

For universes in said 5-D–-space multiverses having an extensity of (Aleph 1) light-years, there are prospects for the spacecraft to attain a Lorentz factor of roughly [(Aleph J) EXP 5][(Aleph H) EXP 5] (Aleph 1) to thus result in the multiverses of travel being contracted to point directly in front of the spacecraft and Lorentz contracted by a factor of roughly [(Aleph J) EXP 5][(Aleph H) EXP 5] (Aleph 1) relative to the spacecraft. Note that in the above computations, we assume that only the material in the path of a

reasonably sized spacecraft is upswept as mass-energy fuel or species of reaction and that the spacecraft makes only one effective pass through each universe.

If such Aleph 1 light-year–width universes have five spatial dimensions and are substantially static, then a spacecraft traveling at light-speed in the forests of consideration may obtain a Lorentz factor of roughly [(Aleph J) EXP 5][(Aleph H) EXP 5] [(Aleph 1) EXP 5]. Note that in the above computations, we assume that only the material in the path of a reasonably sized spacecraft is upswept as mass-energy fuel or species of reaction and that the spacecraft makes an effective pass entirely through each universe. Thus, all material in said universes of travel is reacted against with assumed finite universal mass-energy density.

If the latter universes are expanding with time over periods much greater than Aleph 1 years, then the spacecraft may attain a Lorentz factor of roughly [(Aleph J) EXP 5][(Aleph H) EXP 5] [[k(Aleph 1)] EXP 5], where k ranges from a small superunitary real number to suitable infinite real numbers. This, of course, assumes that real positive mass-energy stocks continue being created, perhaps out of zero-point fields or analogues thereof from some forms of dark energy. Provided that k ranges from Aleph 0 to Aleph 3, then the spacecraft Lorentz factor will be limited to a range of roughly from [(Aleph J) EXP 5][(Aleph H) EXP 5] [[(Aleph 0)(Aleph 1)] EXP 5] to [(Aleph J) EXP 5][(Aleph H) EXP 5] [[(Aleph 3)(Aleph 1)] EXP 5]. Note that in the above computations, we assume that only the material in the path of a reasonably sized spacecraft is upswept as mass-energy fuel or species of reaction and that the spacecraft makes an effective entire pass through each universe. Thus, all material in said universes of travel is reacted against with assumed finite universal mass-energy density.

For universes in said multiverses having an extensity of Aleph 2 light-years, there are prospects for the spacecraft to attain a Lorentz factor of roughly [(Aleph J) EXP 5][(Aleph H) EXP 5] (Aleph 2) to thus result in the multiverses of travel being contracted to a point directly in front of the spacecraft and by a factor of [(Aleph J) EXP 5][(Aleph H) EXP 5] (Aleph 2). Note that in the above computations, we assume that only the material in the path of a reasonably sized spacecraft is upswept as mass-energy fuel or species of reaction and that the spacecraft makes only one effective pass through each universe.

If such Aleph 2 light-year–width universes have five spatial dimensions and are substantially static, then a spacecraft traveling at light-speed in the forests considered may obtain a Lorentz factor of roughly [(Aleph J) EXP 5][(Aleph H) EXP 5] [(Aleph 2) EXP 5]. Note that in the above computations, we assume that only the material in the path of a reasonably sized spacecraft is upswept as mass-energy fuel or species of reaction and that the spacecraft makes an effective pass entirely through each universe.

Thus, all material in said universes of travel is reacted against with assumed finite universal mass-energy density.

If the latter universes are expanding with time over periods much greater than Aleph 2 years, then the spacecraft may attain a Lorentz factor of roughly [(Aleph J) EXP 5][(Aleph H) EXP 5] [[k(Aleph 2)] EXP 5], where *k* ranges from a small superunitary real number to suitable infinite real numbers. This, of course, assumes that real positive mass-energy stocks continue being created, perhaps out of zero-point fields or analogues thereof from some forms of dark energy. Provided that *k* ranges from Aleph 0 to Aleph 4, then the spacecraft Lorentz factor will be limited to a range of roughly from [(Aleph J) EXP 5][(Aleph H) EXP 5] [[(Aleph 0)(Aleph 2)] EXP 5] to [(Aleph J) EXP 5][(Aleph H) EXP 5] [[(Aleph 4)(Aleph 2)] EXP 5]. Note that in the above computations, we assume that only the material in the path of a reasonably sized spacecraft is upswept as mass-energy fuel or species of reaction and that the spacecraft makes an effective pass entirely through each universe. Thus, all material in said universes of travel is reacted against with assumed finite universal mass-energy density.

For universes in said multiverses having an extensity of Aleph 3 light-years, there are prospects for the spacecraft to attain a Lorentz factor of roughly [(Aleph J) EXP 5][(Aleph H) EXP 5] (Aleph 3) to thus result in the multiverses of travel being contracted to point directly in front of the spacecraft and contracted by a factor of [(Aleph J) EXP 5][(Aleph H) EXP 5](Aleph 3). Note that in the above computations, we assume that only the material in the path of a reasonably sized spacecraft is upswept as mass-energy fuel or species of reaction and that the spacecraft makes only one effective pass through each universe.

If such Aleph 3 light-year–width universes have five spatial dimensions and are substantially static, then a spacecraft traveling at light-speed in the forests of consideration may obtain a Lorentz factor of roughly [(Aleph J) EXP 5][(Aleph H) EXP 5] [(Aleph 3) EXP 5]. Note that in the above computations, we assume that only the material in the path of a reasonably sized spacecraft is upswept as mass-energy fuel or species of reaction and that the spacecraft makes an effective pass entirely through each universe. Thus, all material in said universes of travel is reacted against with assumed finite universal mass-energy density.

If the latter universes are expanding with time over periods much greater than Aleph 3 years, then the spacecraft may attain a Lorentz factor of roughly [(Aleph J) EXP 5][(Aleph H) EXP 5] [[k(Aleph 3)] EXP 5], where *k* ranges from a small superunitary real number to suitable infinite real numbers. This, of course, assumes that real positive mass-energy stocks continue being created, perhaps out of zero-point fields or analogues thereof from some forms of dark energy. Provided that *k* ranges from Aleph 0 to Aleph 5, then the spacecraft Lorentz factor will be limited to a range of roughly from

[(Aleph J) EXP 5][(Aleph H) EXP 5] [[(Aleph 0)(Aleph 3)] EXP 5] to [(Aleph J) EXP 5][(Aleph H) EXP 5] [[(Aleph 5)(Aleph 3)] EXP 5]. Note that in the above computations, we assume that only the material in the path of a reasonably sized spacecraft is upswept as mass-energy fuel or species of reaction and that the spacecraft makes an effective pass entirely through each universe. Thus, all material in said universes of travel is reacted against with assumed finite universal mass-energy density.

For universes having an extensity of Aleph 4 light-years, there are prospects for the spacecraft to attain a Lorentz factor of roughly [(Aleph J) EXP 5][(Aleph H) EXP 5] (Aleph 4) to thus result in the multiverses of travel being contracted to point directly in front of the spacecraft and being contracted relative to the spacecraft by roughly a factor of [(Aleph J) EXP 5][(Aleph H) EXP 5](Aleph 4). Note that in the above computations, we assume that only the material in the path of a reasonably sized spacecraft is upswept as mass-energy fuel or species of reaction and that the spacecraft makes only one effective pass through each universe.

If such Aleph 4 light-year–width universes have five spatial dimensions and are substantially static, then a spacecraft traveling at light-speed in the forests may obtain a Lorentz factor of roughly [(Aleph J) EXP 5][(Aleph H) EXP 5] [(Aleph 4) EXP 5]. Note that in the above computations, we assume that only the material in the path of a reasonably sized spacecraft is upswept as mass-energy fuel or species of reaction and that the spacecraft makes an effective pass entirely through each universe. Thus, all material in said universes of travel is reacted against with assumed finite universal mass-energy density.

If the latter universes are expanding with time over periods much greater than Aleph 4 years, then the spacecraft may attain a Lorentz factor of roughly [(Aleph J) EXP 5][(Aleph H) EXP 5] [[k(Aleph 4)] EXP 5], where k ranges from a small superunitary real number to suitable infinite real numbers. This, of course, assumes that real positive mass-energy stocks continue being created, perhaps out of zero-point fields or analogues thereof from some forms of dark energy. Provided that k ranges roughly from Aleph 0 to Aleph 6, then the spacecraft Lorentz factor will be limited to a range from [(Aleph J) EXP 5][(Aleph H) EXP 5] [[(Aleph 0)(Aleph 4)] EXP 5] to [(Aleph J) EXP 5][(Aleph H) EXP 5] [[(Aleph 6)(Aleph 4)] EXP 5]. Note that in the above computations, we assume that only the material in the path of a reasonably sized spacecraft is upswept as mass-energy fuel or species of reaction and that the spacecraft makes an effective pass entirely through each universe. Thus, all material in said universes of travel is reacted against with assumed finite universal mass-energy density.

For universes having an extensity of Aleph G light-years, there are prospects for the spacecraft to attain a Lorentz factor of roughly [(Aleph J) EXP 5][(Aleph H) EXP 5] (Aleph G) to thus result in the multiverses of travel being contracted to point directly in

front of the spacecraft with a Lorentz contraction of roughly [(Aleph J) EXP 5][(Aleph H) EXP 5] (Aleph G). Note that in the above computations, we assume that only the material in the path of a reasonably sized spacecraft is upswept as mass-energy fuel or species of reaction and that the spacecraft makes only one effective pass through each universe.

If such Aleph G light-year–width universes have five spatial dimensions and are substantially static, then a spacecraft traveling at light-speed in the forests may obtain a Lorentz factor of roughly [(Aleph J) EXP 5][(Aleph H) EXP 5] [(Aleph G) EXP 5]. Note that in the above computations, we assume that only the material in the path of a reasonably sized spacecraft is upswept as mass-energy fuel or species of reaction and that the spacecraft makes an effective pass entirely through each universe. Thus, all material in said universes of travel is reacted against with assumed finite universal mass-energy density.

If the latter universes are expanding with time over periods much greater than Aleph G years, then the spacecraft may attain a Lorentz factor of roughly [(Aleph J) EXP 5][(Aleph H) EXP 5] [[k(Aleph G)] EXP 5], where *k* ranges from a small superunitary real number to suitable infinite real numbers. This, of course, assumes that real positive mass-energy stocks continue being created, perhaps out of zero-point fields or analogues thereof from some forms of dark energy. Provided that *k* ranges from Aleph 0 to Aleph (G + 2), then the spacecraft Lorentz factor will be limited to a range of roughly from [(Aleph J) EXP 5][(Aleph H) EXP 5] [[(Aleph 0)(Aleph G)] EXP 5] to [(Aleph J) EXP 5][(Aleph H) EXP 5] [[[Aleph (G + 2)](Aleph G)] EXP 5]. Note that in the above computations, we assume that only the material in the path of a reasonably sized spacecraft is upswept as mass-energy fuel or species of reaction and that the spacecraft makes an effective pass entirely through each universe. Thus, all material in said universes of travel is reacted against with assumed finite universal mass-energy density.

Now, we consider spacecraft travel in a 6-D-space forest of a width of Aleph J multiverses each of Aleph H universes in width each of six dimensions, where *H* is an integer greater than G + 2, and where said multiverses each contain [(Aleph J) EXP 6][(Aleph H) EXP 6] universes.

For universes in said 6-D–space multiverses having an extensity of (Aleph 1) light-years, there are prospects for the spacecraft to attain a Lorentz factor of roughly [(Aleph J) EXP 6][(Aleph H) EXP 6] (Aleph 1) to thus result in the multiverses of travel being contracted to point directly in front of the spacecraft and Lorentz contracted by a factor of roughly [(Aleph J) EXP 6][(Aleph H) EXP 6] (Aleph 1) relative to the spacecraft. Note that in the above computations, we assume that only the material in the path of a

reasonably sized spacecraft is upswept as mass-energy fuel or species of reaction and that the spacecraft makes only one effective pass through each universe.

If such Aleph 1 light-year-width universes have five spatial dimensions and are substantially static, then a spacecraft traveling at light-speed in the forests of consideration may obtain a Lorentz factor of roughly [(Aleph J) EXP 6][(Aleph H) EXP 6] [(Aleph 1) EXP 5]. Note that in the above computations, we assume that only the material in the path of a reasonably sized spacecraft is upswept as mass-energy fuel or species of reaction and that the spacecraft makes an effective pass entirely through each universe. Thus, all material in said universes of travel is reacted against with assumed finite universal mass-energy density.

If the latter universes are expanding with time over periods much greater than Aleph 1 years, then the spacecraft may attain a Lorentz factor of roughly [(Aleph J) EXP 6][(Aleph H) EXP 6] [[k(Aleph 1)] EXP 5], where *k* ranges from a small superunitary real number to suitable infinite real numbers. This, of course, assumes that real positive mass-energy stocks continue being created, perhaps out of zero-point fields or analogues thereof from some forms of dark energy. Provided that *k* ranges from Aleph 0 to Aleph 3, then the spacecraft Lorentz factor will be limited to a range of roughly from [(Aleph J) EXP 6][(Aleph H) EXP 6] [[(Aleph 0)(Aleph 1)] EXP 5] to [(Aleph J) EXP 6][(Aleph H) EXP 6] [[(Aleph 3)(Aleph 1)] EXP 5]. Note that in the above computations, we assume that only the material in the path of a reasonably sized spacecraft is upswept as mass-energy fuel or species of reaction and that the spacecraft makes an effective entire pass through each universe. Thus, all material in said universes of travel is reacted against with assumed finite universal mass-energy density.

For universes in said multiverses having an extensity of Aleph 2 light-years, there are prospects for the spacecraft to attain a Lorentz factor of roughly [(Aleph J) EXP 6][(Aleph H) EXP 6] (Aleph 2) to thus result in the multiverses of travel being contracted to a point directly in front of the spacecraft and by a factor of [(Aleph J) EXP 6][(Aleph H) EXP 6] (Aleph 2). Note that in the above computations, we assume that only the material in the path of a reasonably sized spacecraft is upswept as mass-energy fuel or species of reaction and that the spacecraft makes only one effective pass through each universe.

If such Aleph 2 light-year width universes have five spatial dimensions and are substantially static, then a spacecraft traveling at light-speed in the forests considered may obtain a Lorentz factor of roughly [(Aleph J) EXP 6][(Aleph H) EXP 6] [(Aleph 2) EXP 5]. Note that in the above computations, we assume that only the material in the path of a reasonably sized spacecraft is upswept as mass-energy fuel or species of reaction and that the spacecraft makes an effective pass entirely through each universe.

Thus, all material in said universes of travel is reacted against with assumed finite universal mass-energy density.

If the latter universes are expanding with time over periods much greater than Aleph 2 years, then the spacecraft may attain a Lorentz factor of roughly [(Aleph J) EXP 6][(Aleph H) EXP 6] [[k(Aleph 2)] EXP 5], where k ranges from a small superunitary real number to suitable infinite real numbers. This, of course, assumes that real positive mass-energy stocks continue being created, perhaps out of zero-point fields or analogues thereof from some forms of dark energy. Provided that k ranges from Aleph 0 to Aleph 4, then the spacecraft Lorentz factor will be limited to a range of roughly from [(Aleph J) EXP 6][(Aleph H) EXP 6] [[(Aleph 0)(Aleph 2)] EXP 5] to [(Aleph J) EXP 6][(Aleph H) EXP 6] [[(Aleph 4)(Aleph 2)] EXP 5]. Note that in the above computations, we assume that only the material in the path of a reasonably sized spacecraft is upswept as mass-energy fuel or species of reaction and that the spacecraft makes an effective pass entirely through each universe. Thus, all material in said universes of travel is reacted against with assumed finite universal mass-energy density.

For universes in said multiverses having an extensity of Aleph 3 light-years, there are prospects for the spacecraft to attain a Lorentz factor of roughly [(Aleph J) EXP 6][(Aleph H) EXP 6] (Aleph 3) to thus result in the multiverses of travel being contracted to point directly in front of the spacecraft and contracted by a factor of [(Aleph J) EXP 5][(Aleph H) EXP 5](Aleph 3). Note that in the above computations, we assume that only the material in the path of a reasonably sized spacecraft is upswept as mass-energy fuel or species of reaction and that the spacecraft makes only one effective pass through each universe.

If such Aleph 3 light-year width universes have five spatial dimensions and are substantially static, then a spacecraft traveling at light-speed in the forests of consideration may obtain a Lorentz factor of roughly [(Aleph J) EXP 6][(Aleph H) EXP 6] [(Aleph 3) EXP 5]. Note that in the above computations, we assume that only the material in the path of a reasonably sized spacecraft is upswept as mass-energy fuel or species of reaction and that the spacecraft makes an effective pass entirely through each universe. Thus, all material in said universes of travel is reacted against with assumed finite universal mass-energy density.

If the latter universes are expanding with time over periods much greater than Aleph 3 years, then the spacecraft may attain a Lorentz factor of roughly [(Aleph J) EXP 6][(Aleph H) EXP 6] [[k(Aleph 3)] EXP 5], where k ranges from a small superunitary real number to suitable infinite real numbers. This, of course, assumes that real positive mass-energy stocks continue being created, perhaps out of zero-point fields or analogues thereof from some forms of dark energy. Provided that k ranges from Aleph 0 to Aleph 5, then the spacecraft Lorentz factor will be limited to a range of roughly from

[(Aleph J) EXP 6][(Aleph H) EXP 6] [[(Aleph 0)(Aleph 3)] EXP 5] to [(Aleph J) EXP 6][(Aleph H) EXP 6] [[(Aleph 5)(Aleph 3)] EXP 5]. Note that in the above computations, we assume that only the material in the path of a reasonably sized spacecraft is upswept as mass-energy fuel or species of reaction and that the spacecraft makes an effective pass entirely through each universe. Thus, all material in said universes of travel is reacted against with assumed finite universal mass-energy density.

For universes having an extensity of Aleph 4 light-years, there are prospects for the spacecraft to attain a Lorentz factor of roughly [(Aleph J) EXP 6][(Aleph H) EXP 6] (Aleph 4) to thus result in the multiverses of travel being contracted to point directly in front of the spacecraft and being contracted relative to the spacecraft by roughly a factor of [(Aleph J) EXP 5][(Aleph H) EXP 5](Aleph 4). Note that in the above computations, we assume that only the material in the path of a reasonably sized spacecraft is upswept as mass-energy fuel or species of reaction and that the spacecraft makes only one effective pass through each universe.

If such Aleph 4 light-year width universes have five spatial dimensions and are substantially static, then a spacecraft traveling at light-speed in the forests may obtain a Lorentz factor of roughly [(Aleph J) EXP 6][(Aleph H) EXP 6] [(Aleph 4) EXP 5]. Note that in the above computations, we assume that only the material in the path of a reasonably sized spacecraft is upswept as mass-energy fuel or species of reaction and that the spacecraft makes an effective pass entirely through each universe. Thus, all material in said universes of travel is reacted against with assumed finite universal mass-energy density.

If the latter universes are expanding with time over periods much greater than Aleph 4 years, then the spacecraft may attain a Lorentz factor of roughly [(Aleph J) EXP 6][(Aleph H) EXP 6] [[k(Aleph 4)] EXP 5], where k ranges from a small superunitary real number to suitable infinite real numbers. This, of course, assumes that real positive mass-energy stocks continue being created, perhaps out of zero-point fields or analogues thereof from some forms of dark energy. Provided that k ranges roughly from Aleph 0 to Aleph 6, then the spacecraft Lorentz factor will be limited to a range from [(Aleph J) EXP 6][(Aleph H) EXP 6] [[(Aleph 0)(Aleph 4)] EXP 5] to [(Aleph J) EXP 6][(Aleph H) EXP 6] [[(Aleph 6)(Aleph 4)] EXP 5]. Note that in the above computations, we assume that only the material in the path of a reasonably sized spacecraft is upswept as mass-energy fuel or species of reaction and that the spacecraft makes an effective pass entirely through each universe. Thus, all material in said universes of travel is reacted against with assumed finite universal mass-energy density.

For universes having an extensity of Aleph G light-years, there are prospects for the spacecraft to attain a Lorentz factor of roughly [(Aleph J) EXP 6][(Aleph H) EXP 6] (Aleph G) to thus result in the multiverses of travel being contracted to point directly in

front of the spacecraft with a Lorentz contraction of roughly [(Aleph J) EXP 6][(Aleph H) EXP 6] (Aleph G). Note that in the above computations, we assume that only the material in the path of a reasonably sized spacecraft is upswept as mass-energy fuel or species of reaction and that the spacecraft makes only one effective pass through each universe.

If such Aleph G light-year width universes have five spatial dimensions and are substantially static, then a spacecraft traveling at light-speed in the forests may obtain a Lorentz factor of roughly [(Aleph J) EXP 6][(Aleph H) EXP 6] [(Aleph G) EXP 5]. Note that in the above computations, we assume that only the material in the path of a reasonably sized spacecraft is upswept as mass-energy fuel or species of reaction and that the spacecraft makes an effective pass entirely through each universe. Thus, all material in said universes of travel is reacted against with assumed finite universal mass-energy density.

If the latter universes are expanding with time over periods much greater than Aleph G years, then the spacecraft may attain a Lorentz factor of roughly [(Aleph J) EXP 6][(Aleph H) EXP 6] [[k(Aleph G)] EXP 5], where k ranges from a small superunitary real number to suitable infinite real numbers. This, of course, assumes that real positive mass-energy stocks continue being created, perhaps out of zero-point fields or analogues thereof from some forms of dark energy. Provided that k ranges from Aleph 0 to Aleph (G + 2), then the spacecraft Lorentz factor will be limited to a range of roughly from [(Aleph J) EXP 6][(Aleph H) EXP 6] [[(Aleph 0)(Aleph G)] EXP 5] to [(Aleph J) EXP 6][(Aleph H) EXP 6] [[[Aleph (G + 2)](Aleph G)] EXP 5]. Note that in the above computations, we assume that only the material in the path of a reasonably sized spacecraft is upswept as mass-energy fuel or species of reaction and that the spacecraft makes an effective pass entirely through each universe. Thus, all material in said universes of travel is reacted against with assumed finite universal mass-energy density.

Now, we consider spacecraft travel in an N-D–space forest of a width of Aleph J multiverses each of Aleph H universes in width each of N dimensions, where H is an integer greater than G + 2, and where said multiverses each contain [(Aleph J) EXP N][(Aleph H) EXP N] universes.

For universes in said N-D-space multiverses having an extensity of (Aleph 1) light-years, there are prospects for the spacecraft to attain a Lorentz factor of roughly [(Aleph J) EXP N][(Aleph H) EXP N] (Aleph 1) to thus result in the multiverses of travel being contracted to point directly in front of the spacecraft and Lorentz contracted by a factor of roughly [(Aleph J) EXP N][(Aleph H) EXP N] (Aleph 1) relative to the spacecraft. Note that in the above computations, we assume that only the material in the path of a

reasonably sized spacecraft is upswept as mass-energy fuel or species of reaction and that the spacecraft makes only one effective pass through each universe.

If such Aleph 1 light-year-width universes have five spatial dimensions and are substantially static, then a spacecraft traveling at light-speed in the forests of consideration may obtain a Lorentz factor of roughly [(Aleph J) EXP N][(Aleph H) EXP N] [(Aleph 1) EXP 5]. Note that in the above computations, we assume that only the material in the path of a reasonably sized spacecraft is upswept as mass-energy fuel or species of reaction and that the spacecraft makes an effective pass entirely through each universe. Thus, all material in said universes of travel is reacted against with assumed finite universal mass-energy density.

If the latter universes are expanding with time over periods much greater than Aleph 1 years, then the spacecraft may attain a Lorentz factor of roughly [(Aleph J) EXP N][(Aleph H) EXP N] [[k(Aleph 1)] EXP 5], where k ranges from a small superunitary real number to suitable infinite real numbers. This, of course, assumes that real positive mass-energy stocks continue being created, perhaps out of zero-point fields or analogues thereof from some forms of dark energy. Provided that k ranges from Aleph 0 to Aleph 3, then the spacecraft Lorentz factor will be limited to a range of roughly from [(Aleph J) EXP N][(Aleph H) EXP N] [[(Aleph 0)(Aleph 1)] EXP 5] to [(Aleph J) EXP N][(Aleph H) EXP N] [[(Aleph 3)(Aleph 1)] EXP 5]. Note that in the above computations, we assume that only the material in the path of a reasonably sized spacecraft is upswept as mass-energy fuel or species of reaction and that the spacecraft makes an effective entire pass through each universe. Thus, all material in said universes of travel is reacted against with assumed finite universal mass-energy density.

For universes in said multiverses having an extensity of Aleph 2 light-years, there are prospects for the spacecraft to attain a Lorentz factor of roughly [(Aleph J) EXP N][(Aleph H) EXP N] (Aleph 2) to thus result in the multiverses of travel being contracted to a point directly in front of the spacecraft and by a factor of [(Aleph J) EXP N][(Aleph H) EXP N] (Aleph 2). Note that in the above computations, we assume that only the material in the path of a reasonably sized spacecraft is upswept as mass-energy fuel or species of reaction and that the spacecraft makes only one effective pass through each universe.

If such Aleph 2 light-year width universes have five spatial dimensions and are substantially static, then a spacecraft traveling at light-speed in the forests considered may obtain a Lorentz factor of roughly [(Aleph J) EXP N][(Aleph H) EXP N] [(Aleph 2) EXP 5]. Note that in the above computations, we assume that only the material in the path of a reasonably sized spacecraft is upswept as mass-energy fuel or species of reaction and that the spacecraft makes an effective pass entirely through each universe.

Thus, all material in said universes of travel is reacted against with assumed finite universal mass-energy density.

If the latter universes are expanding with time over periods much greater than Aleph 2 years, then the spacecraft may attain a Lorentz factor of roughly [(Aleph J) EXP N][(Aleph H) EXP N] [[k(Aleph 2)] EXP 5], where k ranges from a small superunitary real number to suitable infinite real numbers. This, of course, assumes that real positive mass-energy stocks continue being created, perhaps out of zero-point fields or analogues thereof from some forms of dark energy. Provided that k ranges from Aleph 0 to Aleph 4, then the spacecraft Lorentz factor will be limited to a range of roughly from [(Aleph J) EXP N][(Aleph H) EXP N] [[(Aleph 0)(Aleph 2)] EXP 5] to [(Aleph J) EXP N][(Aleph H) EXP N] [[(Aleph 4)(Aleph 2)] EXP 5]. Note that in the above computations, we assume that only the material in the path of a reasonably sized spacecraft is upswept as mass-energy fuel or species of reaction and that the spacecraft makes an effective pass entirely through each universe. Thus, all material in said universes of travel is reacted against with assumed finite universal mass-energy density.

For universes in said multiverses having an extensity of Aleph 3 light-years, there are prospects for the spacecraft to attain a Lorentz factor of roughly [(Aleph J) EXP N][(Aleph H) EXP N] (Aleph 3) to thus result in the multiverses of travel being contracted to point directly in front of the spacecraft and contracted by a factor of [(Aleph J) EXP 5][(Aleph H) EXP 5](Aleph 3). Note that in the above computations, we assume that only the material in the path of a reasonably sized spacecraft is upswept as mass-energy fuel or species of reaction and that the spacecraft makes only one effective pass through each universe.

If such Aleph 3 light-year–width universes have five spatial dimensions and are substantially static, then a spacecraft traveling at light-speed in the forests of consideration may obtain a Lorentz factor of roughly [(Aleph J) EXP N][(Aleph H) EXP N] [(Aleph 3) EXP 5]. Note that in the above computations, we assume that only the material in the path of a reasonably sized spacecraft is upswept as mass-energy fuel or species of reaction and that the spacecraft makes an effective pass entirely through each universe. Thus, all material in said universes of travel is reacted against with assumed finite universal mass-energy density.

If the latter universes are expanding with time over periods much greater than Aleph 3 years, then the spacecraft may attain a Lorentz factor of roughly [(Aleph J) EXP N][(Aleph H) EXP N] [[k(Aleph 3)] EXP 5], where k ranges from a small superunitary real number to suitable infinite real numbers. This, of course, assumes that real positive mass-energy stocks continue being created, perhaps out of zero-point fields or analogues thereof from some forms of dark energy. Provided that k ranges from Aleph 0 to Aleph 5, then the spacecraft Lorentz factor will be limited to a range of roughly from

[(Aleph J) EXP N][(Aleph H) EXP N] [[(Aleph 0)(Aleph 3)] EXP 5] to [(Aleph J) EXP N][(Aleph H) EXP N] [[(Aleph 5)(Aleph 3)] EXP 5]. Note that in the above computations, we assume that only the material in the path of a reasonably sized spacecraft is upswept as mass-energy fuel or species of reaction and that the spacecraft makes an effective pass entirely through each universe. Thus, all material in said universes of travel is reacted against with assumed finite universal mass-energy density.

For universes having an extensity of Aleph 4 light-years, there are prospects for the spacecraft to attain a Lorentz factor of roughly [(Aleph J) EXP N][(Aleph H) EXP N] (Aleph 4) to thus result in the multiverses of travel being contracted to point directly in front of the spacecraft and being contracted relative to the spacecraft by roughly a factor of [(Aleph J) EXP 5][(Aleph H) EXP 5](Aleph 4). Note that in the above computations, we assume that only the material in the path of a reasonably sized spacecraft is upswept as mass-energy fuel or species of reaction and that the spacecraft makes only one effective pass through each universe.

If such Aleph 4 light-year–width universes have five spatial dimensions and are substantially static, then a spacecraft traveling at light-speed in the forests may obtain a Lorentz factor of roughly [(Aleph J) EXP N][(Aleph H) EXP N] [(Aleph 4) EXP 5]. Note that in the above computations, we assume that only the material in the path of a reasonably sized spacecraft is upswept as mass-energy fuel or species of reaction and that the spacecraft makes an effective pass entirely through each universe. Thus, all material in said universes of travel is reacted against with assumed finite universal mass-energy density.

If the latter universes are expanding with time over periods much greater than Aleph 4 years, then the spacecraft may attain a Lorentz factor of roughly [(Aleph J) EXP N][(Aleph H) EXP N] [[k(Aleph 4)] EXP 5], where k ranges from a small superunitary real number to suitable infinite real numbers. This, of course, assumes that real positive mass-energy stocks continue being created, perhaps out of zero-point fields or analogues thereof from some forms of dark energy. Provided that k ranges roughly from Aleph 0 to Aleph 6, then the spacecraft Lorentz factor will be limited to a range from [(Aleph J) EXP N][(Aleph H) EXP N] [[(Aleph 0)(Aleph 4)] EXP 5] to [(Aleph J) EXP N][(Aleph H) EXP N] [[(Aleph 6)(Aleph 4)] EXP 5]. Note that in the above computations, we assume that only the material in the path of a reasonably sized spacecraft is upswept as mass-energy fuel or species of reaction and that the spacecraft makes an effective pass entirely through each universe. Thus, all material in said universes of travel is reacted against with assumed finite universal mass-energy density.

For universes having an extensity of Aleph G light-years, there are prospects for the spacecraft to attain a Lorentz factor of roughly [(Aleph J) EXP N][(Aleph H) EXP N] (Aleph G) to thus result in the multiverses of travel being contracted to point directly in

front of the spacecraft with a Lorentz contraction of roughly [(Aleph J) EXP N][(Aleph H) EXP N] (Aleph G). Note that in the above computations, we assume that only the material in the path of a reasonably sized spacecraft is upswept as mass-energy fuel or species of reaction and that the spacecraft makes only one effective pass through each universe.

If such Aleph G light-year–width universes have five spatial dimensions and are substantially static, then a spacecraft traveling at light-speed in the forests may obtain a Lorentz factor of roughly [(Aleph J) EXP N][(Aleph H) EXP N] [(Aleph G) EXP 5]. Note that in the above computations, we assume that only the material in the path of a reasonably sized spacecraft is upswept as mass-energy fuel or species of reaction and that the spacecraft makes an effective pass entirely through each universe. Thus, all material in said universes of travel is reacted against with assumed finite universal mass-energy density.

If the latter universes are expanding with time over periods much greater than Aleph G years, then the spacecraft may attain a Lorentz factor of roughly [(Aleph J) EXP N][(Aleph H) EXP N] [[k(Aleph G)] EXP 5], where k ranges from a small superunitary real number to suitable infinite real numbers. This, of course, assumes that real positive mass-energy stocks continue being created, perhaps out of zero-point fields or analogues thereof from some forms of dark energy. Provided that k ranges from Aleph 0 to Aleph (G + 2), then the spacecraft Lorentz factor will be limited to a range of roughly from [(Aleph J) EXP N][(Aleph H) EXP N] [[(Aleph 0)(Aleph G)] EXP 5] to [(Aleph J) EXP N][(Aleph H) EXP N] [[[Aleph (G + 2)](Aleph G)] EXP 5]. Note that in the above computations, we assume that only the material in the path of a reasonably sized spacecraft is upswept as mass-energy fuel or species of reaction and that the spacecraft makes an effective pass entirely through each universe. Thus, all material in said universes of travel is reacted against with assumed finite universal mass-energy density.

Now we consider spacecraft travel in a 6-D–space forest of a width of Aleph J multiverses each of Aleph H universes in width each of six dimensions, where H is an integer greater than G + 2, and where said multiverses each contain [(Aleph J) EXP 6][(Aleph H) EXP 6] universes.

For universes in said 6-D–space multiverses having an extensity of (Aleph 1) light-years, there are prospects for the spacecraft to attain a Lorentz factor of roughly [(Aleph J) EXP 6][(Aleph H) EXP 6] (Aleph 1) to thus result in the multiverses of travel being contracted to point directly in front of the spacecraft and Lorentz contracted by a factor of roughly [(Aleph J) EXP 6][(Aleph H) EXP 6] (Aleph 1) relative to the spacecraft. Note that in the above computations, we assume that only the material in the path of a

reasonably sized spacecraft is upswept as mass-energy fuel or species of reaction and that the spacecraft makes only one effective pass through each universe.

If such Aleph 1 light-year–width universes have six spatial dimensions and are substantially static, then a spacecraft traveling at light-speed in the forests of consideration may obtain a Lorentz factor of roughly [(Aleph J) EXP 6][(Aleph H) EXP 6] [(Aleph 1) EXP 6]. Note that in the above computations, we assume that only the material in the path of a reasonably sized spacecraft is upswept as mass-energy fuel or species of reaction and that the spacecraft makes an effective pass entirely through each universe. Thus, all material in said universes of travel is reacted against with assumed finite universal mass-energy density.

If the latter universes are expanding with time over periods much greater than Aleph 1 years, then the spacecraft may attain a Lorentz factor of roughly [(Aleph J) EXP 6][(Aleph H) EXP 6] [[k(Aleph 1)] EXP 6], where *k* ranges from a small superunitary real number to suitable infinite real numbers. This, of course, assumes that real positive mass-energy stocks continue being created, perhaps out of zero-point fields or analogues thereof from some forms of dark energy. Provided that *k* ranges from Aleph 0 to Aleph 3, then the spacecraft Lorentz factor will be limited to a range of roughly from [(Aleph J) EXP 6][(Aleph H) EXP 6] [[(Aleph 0)(Aleph 1)] EXP 6] to [(Aleph J) EXP 6][(Aleph H) EXP 6] [[(Aleph 3)(Aleph 1)] EXP 6]. Note that in the above computations, we assume that only the material in the path of a reasonably sized spacecraft is upswept as mass-energy fuel or species of reaction and that the spacecraft makes an effective entire pass through each universe. Thus, all material in said universes of travel is reacted against with assumed finite universal mass-energy density.

For universes in said multiverses having an extensity of Aleph 2 light-years, there are prospects for the spacecraft to attain a Lorentz factor of roughly [(Aleph J) EXP 6][(Aleph H) EXP 6] (Aleph 2) to thus result in the multiverses of travel being contracted to a point directly in front of the spacecraft and by a factor of [(Aleph J) EXP 6][(Aleph H) EXP 6] (Aleph 2). Note that in the above computations, we assume that only the material in the path of a reasonably sized spacecraft is upswept as mass-energy fuel or species of reaction and that the spacecraft makes only one effective pass through each universe.

If such Aleph 2 light-year–width universes have six spatial dimensions and are substantially static, then a spacecraft traveling at light-speed in the forests considered may obtain a Lorentz factor of roughly [(Aleph J) EXP 6][(Aleph H) EXP 6] [(Aleph 2) EXP 6]. Note that in the above computations, we assume that only the material in the path of a reasonably sized spacecraft is upswept as mass-energy fuel or species of reaction and that the spacecraft makes an effective pass entirely through each universe.

Thus, all material in said universes of travel is reacted against with assumed finite universal mass-energy density.

If the latter universes are expanding with time over periods much greater than Aleph 2 years, then the spacecraft may attain a Lorentz factor of roughly [(Aleph J) EXP 6][(Aleph H) EXP 6] [[k(Aleph 2)] EXP 6], where *k* ranges from a small superunitary real number to suitable infinite real numbers. This, of course, assumes that real positive mass-energy stocks continue being created, perhaps out of zero-point fields or analogues thereof from some forms of dark energy. Provided that *k* ranges from Aleph 0 to Aleph 4, then the spacecraft Lorentz factor will be limited to a range of roughly from [(Aleph J) EXP 6][(Aleph H) EXP 6] [[(Aleph 0)(Aleph 2)] EXP 6] to [(Aleph J) EXP 6][(Aleph H) EXP 6] [[(Aleph 4)(Aleph 2)] EXP 6]. Note that in the above computations, we assume that only the material in the path of a reasonably sized spacecraft is upswept as mass-energy fuel or species of reaction and that the spacecraft makes an effective pass entirely through each universe. Thus, all material in said universes of travel is reacted against with assumed finite universal mass-energy density.

For universes in said multiverses having an extensity of Aleph 3 light-years, there are prospects for the spacecraft to attain a Lorentz factor of roughly [(Aleph J) EXP 6][(Aleph H) EXP 6] (Aleph 3) to thus result in the multiverses of travel being contracted to point directly in front of the spacecraft and contracted by a factor of [(Aleph J) EXP 6][(Aleph H) EXP 6](Aleph 3). Note that in the above computations, we assume that only the material in the path of a reasonably sized spacecraft is upswept as mass-energy fuel or species of reaction and that the spacecraft makes only one effective pass through each universe.

If such Aleph 3 light-year–width universes have six spatial dimensions and are substantially static, then a spacecraft traveling at light-speed in the forests of consideration may obtain a Lorentz factor of roughly [(Aleph J) EXP 6][(Aleph H) EXP 6] [(Aleph 3) EXP 6]. Note that in the above computations, we assume that only the material in the path of a reasonably sized spacecraft is upswept as mass-energy fuel or species of reaction and that the spacecraft makes an effective pass entirely through each universe. Thus, all material in said universes of travel is reacted against with assumed finite universal mass-energy density.

If the latter universes are expanding with time over periods much greater than Aleph 3 years, then the spacecraft may attain a Lorentz factor of roughly [(Aleph J) EXP 6][(Aleph H) EXP 6] [[k(Aleph 3)] EXP 6], where *k* ranges from a small superunitary real number to suitable infinite real numbers. This, of course, assumes that real positive mass-energy stocks continue being created, perhaps out of zero-point fields or analogues thereof from some forms of dark energy. Provided that *k* ranges from Aleph 0 to Aleph 5, then the spacecraft Lorentz factor will be limited to a range of roughly from

[(Aleph J) EXP 6][(Aleph H) EXP 6] [[(Aleph 0)(Aleph 3)] EXP 6] to [(Aleph J) EXP 6][(Aleph H) EXP 6] [[(Aleph 5)(Aleph 3)] EXP 6]. Note that in the above computations, we assume that only the material in the path of a reasonably sized spacecraft is upswept as mass-energy fuel or species of reaction and that the spacecraft makes an effective pass entirely through each universe. Thus, all material in said universes of travel is reacted against with assumed finite universal mass-energy density.

For universes having an extensity of Aleph 4 light-years, there are prospects for the spacecraft to attain a Lorentz factor of roughly [(Aleph J) EXP 6][(Aleph H) EXP 6] (Aleph 4) to thus result in the multiverses of travel being contracted to point directly in front of the spacecraft and being contracted relative to the spacecraft by roughly a factor of [(Aleph J) EXP 6][(Aleph H) EXP 6](Aleph 4). Note that in the above computations, we assume that only the material in the path of a reasonably sized spacecraft is upswept as mass-energy fuel or species of reaction and that the spacecraft makes only one effective pass through each universe.

If such Aleph 4 light-year–width universes have six spatial dimensions and are substantially static, then a spacecraft traveling at light-speed in the forests may obtain a Lorentz factor of roughly [(Aleph J) EXP 6][(Aleph H) EXP 6] [(Aleph 4) EXP 6]. Note that in the above computations, we assume that only the material in the path of a reasonably sized spacecraft is upswept as mass-energy fuel or species of reaction and that the spacecraft makes an effective pass entirely through each universe. Thus, all material in said universes of travel is reacted against with assumed finite universal mass-energy density.

If the latter universes are expanding with time over periods much greater than Aleph 4 years, then the spacecraft may attain a Lorentz factor of roughly [(Aleph J) EXP 6][(Aleph H) EXP 6] [[k(Aleph 4)] EXP 6], where k ranges from a small superunitary real number to suitable infinite real numbers. This, of course, assumes that real positive mass-energy stocks continue being created, perhaps out of zero-point fields or analogues thereof from some forms of dark energy. Provided that k ranges roughly from Aleph 0 to Aleph 6, then the spacecraft Lorentz factor will be limited to a range from [(Aleph J) EXP 6][(Aleph H) EXP 6] [[(Aleph 0)(Aleph 4)] EXP 6] to [(Aleph J) EXP 6][(Aleph H) EXP 6] [[(Aleph 6)(Aleph 4)] EXP 6]. Note that in the above computations, we assume that only the material in the path of a reasonably sized spacecraft is upswept as mass-energy fuel or species of reaction and that the spacecraft makes an effective pass entirely through each universe. Thus, all material in said universes of travel is reacted against with assumed finite universal mass-energy density.

For universes having an extensity of Aleph G light-years, there are prospects for the spacecraft to attain a Lorentz factor of roughly [(Aleph J) EXP 6][(Aleph H) EXP 6] (Aleph G) to thus result in the multiverses of travel being contracted to point directly in

front of the spacecraft with a Lorentz contraction of roughly [(Aleph J) EXP 6][(Aleph H) EXP 6] (Aleph G). Note that in the above computations, we assume that only the material in the path of a reasonably sized spacecraft is upswept as mass-energy fuel or species of reaction and that the spacecraft makes only one effective pass through each universe.

If such Aleph G light-year–width universes have six spatial dimensions and are substantially static, then a spacecraft traveling at light-speed in the forests may obtain a Lorentz factor of roughly [(Aleph J) EXP 6][(Aleph H) EXP 6] [(Aleph G) EXP 6]. Note that in the above computations, we assume that only the material in the path of a reasonably sized spacecraft is upswept as mass-energy fuel or species of reaction and that the spacecraft makes an effective pass entirely through each universe. Thus, all material in said universes of travel is reacted against with assumed finite universal mass-energy density.

If the latter universes are expanding with time over periods much greater than Aleph G years, then the spacecraft may attain a Lorentz factor of roughly [(Aleph J) EXP 6][(Aleph H) EXP 6] [[k(Aleph G)] EXP 6], where k ranges from a small superunitary real number to suitable infinite real numbers. This, of course, assumes that real positive mass-energy stocks continue being created, perhaps out of zero-point fields or analogues thereof from some forms of dark energy. Provided that k ranges from Aleph 0 to Aleph (G + 2), then the spacecraft Lorentz factor will be limited to a range of roughly from [(Aleph J) EXP 6][(Aleph H) EXP 6] [[(Aleph 0)(Aleph G)] EXP 6] to [(Aleph J) EXP 6][(Aleph H) EXP 6] [[[Aleph (G + 2)](Aleph G)] EXP 6]. Note that in the above computations, we assume that only the material in the path of a reasonably sized spacecraft is upswept as mass-energy fuel or species of reaction and that the spacecraft makes an effective pass entirely through each universe. Thus, all material in said universes of travel is reacted against with assumed finite universal mass-energy density.

Now we consider spacecraft travel in an N-D–space forest of a width of Aleph J multiverses each of Aleph H universes in width each of N dimensions, where H is an integer greater than G + 2, and where said multiverses each contain [(Aleph J) EXP N][(Aleph H) EXP N] universes.

For universes in said N-D–space multiverses having an extensity of (Aleph 1) light-years, there are prospects for the spacecraft to attain a Lorentz factor of roughly [(Aleph J) EXP N][(Aleph H) EXP N] (Aleph 1) to thus result in the multiverses of travel being contracted to point directly in front of the spacecraft and Lorentz contracted by a factor of roughly [(Aleph J) EXP N][(Aleph H) EXP N] (Aleph 1) relative to the spacecraft. Note that in the above computations, we assume that only the material in the path of a

reasonably sized spacecraft is upswept as mass-energy fuel or species of reaction and that the spacecraft makes only one effective pass through each universe.

If such Aleph 1 light-year–width universes have six spatial dimensions and are substantially static, then a spacecraft traveling at light-speed in the forests of consideration may obtain a Lorentz factor of roughly [(Aleph J) EXP N][(Aleph H) EXP N] [(Aleph 1) EXP 6]. Note that in the above computations, we assume that only the material in the path of a reasonably sized spacecraft is upswept as mass-energy fuel or species of reaction and that the spacecraft makes an effective pass entirely through each universe. Thus, all material in said universes of travel is reacted against with assumed finite universal mass-energy density.

If the latter universes are expanding with time over periods much greater than Aleph 1 years, then the spacecraft may attain a Lorentz factor of roughly [(Aleph J) EXP N][(Aleph H) EXP N] [[k(Aleph 1)] EXP 6], where k ranges from a small superunitary real number to suitable infinite real numbers. This, of course, assumes that real positive mass-energy stocks continue being created, perhaps out of zero-point fields or analogues thereof from some forms of dark energy. Provided that k ranges from Aleph 0 to Aleph 3, then the spacecraft Lorentz factor will be limited to a range of roughly from [(Aleph J) EXP N][(Aleph H) EXP N] [[(Aleph 0)(Aleph 1)] EXP 6] to [(Aleph J) EXP N][(Aleph H) EXP N] [[(Aleph 3)(Aleph 1)] EXP 6]. Note that in the above computations, we assume that only the material in the path of a reasonably sized spacecraft is upswept as mass-energy fuel or species of reaction and that the spacecraft makes an effective entire pass through each universe. Thus, all material in said universes of travel is reacted against with assumed finite universal mass-energy density.

For universes in said multiverses having an extensity of Aleph 2 light-years, there are prospects for the spacecraft to attain a Lorentz factor of roughly [(Aleph J) EXP N][(Aleph H) EXP N] (Aleph 2) to thus result in the multiverses of travel being contracted to a point directly in front of the spacecraft and by a factor of [(Aleph J) EXP N][(Aleph H) EXP N] (Aleph 2). Note that in the above computations, we assume that only the material in the path of a reasonably sized spacecraft is upswept as mass-energy fuel or species of reaction and that the spacecraft makes only one effective pass through each universe.

If such Aleph 2 light-year–width universes have six spatial dimensions and are substantially static, then a spacecraft traveling at light-speed in the forests considered may obtain a Lorentz factor of roughly [(Aleph J) EXP N][(Aleph H) EXP N] [(Aleph 2) EXP 6]. Note that in the above computations, we assume that only the material in the path of a reasonably sized spacecraft is upswept as mass-energy fuel or species of reaction and that the spacecraft makes an effective pass entirely through each universe.

Thus, all material in said universes of travel is reacted against with assumed finite universal mass-energy density.

If the latter universes are expanding with time over periods much greater than Aleph 2 years, then the spacecraft may attain a Lorentz factor of roughly [(Aleph J) EXP N][(Aleph H) EXP N] [[k(Aleph 2)] EXP 6], where k ranges from a small superunitary real number to suitable infinite real numbers. This, of course, assumes that real positive mass-energy stocks continue being created, perhaps out of zero-point fields or analogues thereof from some forms of dark energy. Provided that k ranges from Aleph 0 to Aleph 4, then the spacecraft Lorentz factor will be limited to a range of roughly from [(Aleph J) EXP N][(Aleph H) EXP N] [[(Aleph 0)(Aleph 2)] EXP 6] to [(Aleph J) EXP N][(Aleph H) EXP N] [[(Aleph 4)(Aleph 2)] EXP 6]. Note that in the above computations, we assume that only the material in the path of a reasonably sized spacecraft is upswept as mass-energy fuel or species of reaction and that the spacecraft makes an effective pass entirely through each universe. Thus, all material in said universes of travel is reacted against with assumed finite universal mass-energy density.

For universes in said multiverses having an extensity of Aleph 3 light-years, there are prospects for the spacecraft to attain a Lorentz factor of roughly [(Aleph J) EXP N][(Aleph H) EXP N] (Aleph 3) to thus result in the multiverses of travel being contracted to point directly in front of the spacecraft and contracted by a factor of [(Aleph J) EXP 6][(Aleph H) EXP 6](Aleph 3). Note that in the above computations, we assume that only the material in the path of a reasonably sized spacecraft is upswept as mass-energy fuel or species of reaction and that the spacecraft makes only one effective pass through each universe.

If such Aleph 3 light-year–width universes have six spatial dimensions and are substantially static, then a spacecraft traveling at light-speed in the forests of consideration may obtain a Lorentz factor of roughly [(Aleph J) EXP N][(Aleph H) EXP N] [(Aleph 3) EXP 6]. Note that in the above computations, we assume that only the material in the path of a reasonably sized spacecraft is upswept as mass-energy fuel or species of reaction and that the spacecraft makes an effective pass entirely through each universe. Thus, all material in said universes of travel is reacted against with assumed finite universal mass-energy density.

If the latter universes are expanding with time over periods much greater than Aleph 3 years, then the spacecraft may attain a Lorentz factor of roughly [(Aleph J) EXP N][(Aleph H) EXP N] [[k(Aleph 3)] EXP 6], where k ranges from a small superunitary real number to suitable infinite real numbers. This, of course, assumes that real positive mass-energy stocks continue being created, perhaps out of zero-point fields or analogues thereof from some forms of dark energy. Provided that k ranges from Aleph 0 to Aleph 5, then the spacecraft Lorentz factor will be limited to a range of roughly from

[(Aleph J) EXP N][(Aleph H) EXP N] [[(Aleph 0)(Aleph 3)] EXP 6] to [(Aleph J) EXP N][(Aleph H) EXP N] [[(Aleph 5)(Aleph 3)] EXP 6]. Note that in the above computations, we assume that only the material in the path of a reasonably sized spacecraft is upswept as mass-energy fuel or species of reaction and that the spacecraft makes an effective pass entirely through each universe. Thus, all material in said universes of travel is reacted against with assumed finite universal mass-energy density.

For universes having an extensity of Aleph 4 light-years, there are prospects for the spacecraft to attain a Lorentz factor of roughly [(Aleph J) EXP N][(Aleph H) EXP N] (Aleph 4) to thus result in the multiverses of travel being contracted to point directly in front of the spacecraft and being contracted relative to the spacecraft by roughly a factor of [(Aleph J) EXP 6][(Aleph H) EXP 6](Aleph 4). Note that in the above computations, we assume that only the material in the path of a reasonably sized spacecraft is upswept as mass-energy fuel or species of reaction and that the spacecraft makes only one effective pass through each universe.

If such Aleph 4 light-year width universes have six spatial dimensions and are substantially static, then a spacecraft traveling at light-speed in the forests may obtain a Lorentz factor of roughly [(Aleph J) EXP N][(Aleph H) EXP N] [(Aleph 4) EXP 6]. Note that in the above computations, we assume that only the material in the path of a reasonably sized spacecraft is upswept as mass-energy fuel or species of reaction and that the spacecraft makes an effective pass entirely through each universe. Thus, all material in said universes of travel is reacted against with assumed finite universal mass-energy density.

If the latter universes are expanding with time over periods much greater than Aleph 4 years, then the spacecraft may attain a Lorentz factor of roughly [(Aleph J) EXP N][(Aleph H) EXP N] [[k(Aleph 4)] EXP 6], where k ranges from a small superunitary real number to suitable infinite real numbers. This, of course, assumes that real positive mass-energy stocks continue being created, perhaps out of zero-point fields or analogues thereof from some forms of dark energy. Provided that k ranges roughly from Aleph 0 to Aleph 6, then the spacecraft Lorentz factor will be limited to a range from [(Aleph J) EXP N][(Aleph H) EXP N] [[(Aleph 0)(Aleph 4)] EXP 6] to [(Aleph J) EXP N][(Aleph H) EXP N] [[(Aleph 6)(Aleph 4)] EXP 6]. Note that in the above computations, we assume that only the material in the path of a reasonably sized spacecraft is upswept as mass-energy fuel or species of reaction and that the spacecraft makes an effective pass entirely through each universe. Thus, all material in said universes of travel is reacted against with assumed finite universal mass-energy density.

For universes having an extensity of Aleph G light-years, there are prospects for the spacecraft to attain a Lorentz factor of roughly [(Aleph J) EXP N][(Aleph H) EXP N] (Aleph G) to thus result in the multiverses of travel being contracted to point directly in

front of the spacecraft with a Lorentz contraction of roughly [(Aleph J) EXP N][(Aleph H) EXP N] (Aleph G). Note that in the above computations, we assume that only the material in the path of a reasonably sized spacecraft is upswept as mass-energy fuel or species of reaction and that the spacecraft makes only one effective pass through each universe.

If such Aleph G light-year–width universes have six spatial dimensions and are substantially static, then a spacecraft traveling at light-speed in the forests may obtain a Lorentz factor of roughly [(Aleph J) EXP N][(Aleph H) EXP N] [(Aleph G) EXP 6]. Note that in the above computations, we assume that only the material in the path of a reasonably sized spacecraft is upswept as mass-energy fuel or species of reaction and that the spacecraft makes an effective pass entirely through each universe. Thus, all material in said universes of travel is reacted against with assumed finite universal mass-energy density.

If the latter universes are expanding with time over periods much greater than Aleph G years, then the spacecraft may attain a Lorentz factor of roughly [(Aleph J) EXP N][(Aleph H) EXP N] [[k(Aleph G)] EXP 6], where k ranges from a small superunitary real number to suitable infinite real numbers. This, of course, assumes that real positive mass-energy stocks continue being created, perhaps out of zero-point fields or analogues thereof from some forms of dark energy. Provided that k ranges from Aleph 0 to Aleph (G + 2), then the spacecraft Lorentz factor will be limited to a range of roughly from [(Aleph J) EXP N][(Aleph H) EXP N] [[(Aleph 0)(Aleph G)] EXP 6] to [(Aleph J) EXP N][(Aleph H) EXP N] [[[Aleph (G + 2)](Aleph G)] EXP 6]. Note that in the above computations, we assume that only the material in the path of a reasonably sized spacecraft is upswept as mass-energy fuel or species of reaction and that the spacecraft makes an effective pass entirely through each universe. Thus, all material in said universes of travel is reacted against with assumed finite universal mass-energy density.

Now we consider spacecraft travel in an N-D–space forest of a width of Aleph J multiverses each of Aleph H universes in width each of N dimensions, where H is an integer greater than G + 2, and where said multiverses each contain [(Aleph J) EXP N][(Aleph H) EXP N] universes.

For universes in said N-D–space multiverses having an extensity of (Aleph 1) light-years, there are prospects for the spacecraft to attain a Lorentz factor of roughly [(Aleph J) EXP N][(Aleph H) EXP N] (Aleph 1) to thus result in the multiverses of travel being contracted to point directly in front of the spacecraft and Lorentz contracted by a factor of roughly [(Aleph J) EXP N][(Aleph H) EXP N] (Aleph 1) relative to the spacecraft. Note that in the above computations, we assume that only the material in the path of a

reasonably sized spacecraft is upswept as mass-energy fuel or species of reaction and that the spacecraft makes only one effective pass through each universe.

If such Aleph 1 light-years–width universes have *N* spatial dimensions and are substantially static, then a spacecraft traveling at light-speed in the forests of consideration may obtain a Lorentz factor of roughly [(Aleph J) EXP N][(Aleph H) EXP N] [(Aleph 1) EXP N]. Note that in the above computations, we assume that only the material in the path of a reasonably sized spacecraft is upswept as mass-energy fuel or species of reaction and that the spacecraft makes an effective pass entirely through each universe. Thus, all material in said universes of travel is reacted against with assumed finite universal mass-energy density.

If the latter universes are expanding with time over periods much greater than Aleph 1 years, then the spacecraft may attain a Lorentz factor of roughly [(Aleph J) EXP N][(Aleph H) EXP N] [[k(Aleph 1)] EXP N] where k ranges from a small superunitary real number to suitable infinite real numbers. This, of course, assumes that real positive mass-energy stocks continue being created, perhaps out of zero-point fields or analogues thereof from some forms of dark energy. Provided *k* ranges from Aleph 0 to Aleph 3, then the spacecraft Lorentz factor will be limited to a range of roughly from [(Aleph J) EXP N][(Aleph H) EXP N] [[(Aleph 0)(Aleph 1)] EXP N] to [(Aleph J) EXP N][(Aleph H) EXP N] [[(Aleph 3)(Aleph 1)] EXP N]. Note that in the above computations, we assume that only the material in the path of a reasonably sized spacecraft is upswept as mass-energy fuel or species of reaction and that the spacecraft makes an effective entire pass through each universe. Thus, all material in said universes of travel is reacted against with assumed finite universal mass-energy density.

For universes in said multiverses having an extensity of Aleph 2 light-years, there are prospects for the spacecraft to attain a Lorentz factor of roughly [(Aleph J) EXP N][(Aleph H) EXP N] (Aleph 2) to thus result in the multiverses of travel being contracted to a point directly in front of the spacecraft and by a factor of [(Aleph J) EXP N][(Aleph H) EXP N] (Aleph 2). Note that in the above computations, we assume that only the material in the path of a reasonably sized spacecraft is upswept as mass-energy fuel or species of reaction and that the spacecraft makes only one effective pass through each universe.

If such Aleph 2 light-year–width universes have *N* spatial dimensions and are substantially static, then a spacecraft traveling at light-speed in the forests considered may obtain a Lorentz factor of roughly [(Aleph J) EXP N][(Aleph H) EXP N] [(Aleph 2) EXP N]. Note that in the above computations, we assume that only the material in the path of a reasonably sized spacecraft is upswept as mass-energy fuel or species of reaction and that the spacecraft makes an effective pass entirely through each universe.

Thus, all material in said universes of travel is reacted against with assumed finite universal mass-energy density.

If the latter universes are expanding with time over periods much greater than Aleph 2 years, then the spacecraft may attain a Lorentz factor of roughly [(Aleph J) EXP N][(Aleph H) EXP N] [[k(Aleph 2)] EXP N], where k ranges from a small superunitary real number to suitable infinite real numbers. This, of course, assumes that real positive mass-energy stocks continue being created, perhaps out of zero-point fields or analogues thereof from some forms of dark energy. Provided that k ranges from Aleph 0 to Aleph 4, then the spacecraft Lorentz factor will be limited to a range of roughly from [(Aleph J) EXP N][(Aleph H) EXP N] [[(Aleph 0)(Aleph 2)] EXP N] to [(Aleph J) EXP N][(Aleph H) EXP N] [[(Aleph 4)(Aleph 2)] EXP N]. Note that in the above computations, we assume that only the material in the path of a reasonably sized spacecraft is upswept as mass-energy fuel or species of reaction and that the spacecraft makes an effective pass entirely through each universe. Thus, all material in said universes of travel is reacted against with assumed finite universal mass-energy density.

For universes in said multiverses having an extensity of Aleph 3 light-years, there are prospects for the spacecraft to attain a Lorentz factor of roughly [(Aleph J) EXP N][(Aleph H) EXP N] (Aleph 3) to thus result in the multiverses of travel being contracted to point directly in front of the spacecraft and contracted by a factor of [(Aleph J) EXP N][(Aleph H) EXP N](Aleph 3). Note that in the above computations, we assume that only the material in the path of a reasonably sized spacecraft is upswept as mass-energy fuel or species of reaction and that the spacecraft makes only one effective pass through each universe.

If such Aleph 3 light-year–width universes have N spatial dimensions and are substantially static, then a spacecraft traveling at light-speed in the forests of consideration may obtain a Lorentz factor of roughly [(Aleph J) EXP N][(Aleph H) EXP N] [(Aleph 3) EXP N]. Note that in the above computations, we assume that only the material in the path of a reasonably sized spacecraft is upswept as mass-energy fuel or species of reaction and that the spacecraft makes an effective pass entirely through each universe. Thus, all material in said universes of travel is reacted against with assumed finite universal mass-energy density.

If the latter universes are expanding with time over periods much greater than Aleph 3 years, then the spacecraft may attain a Lorentz factor of roughly [(Aleph J) EXP N][(Aleph H) EXP N] [[k(Aleph 3)] EXP N], where k ranges from a small superunitary real number to suitable infinite real numbers. This, of course, assumes that real positive mass-energy stocks continue being created, perhaps out of zero-point fields or analogues thereof from some forms of dark energy. Provided that k ranges from Aleph 0 to Aleph 5, then the spacecraft Lorentz factor will be limited to a range of roughly from

[(Aleph J) EXP N][(Aleph H) EXP N] [[(Aleph 0)(Aleph 3)] EXP N] to [(Aleph J) EXP N][(Aleph H) EXP N] [[(Aleph 5)(Aleph 3)] EXP N]. Note that in the above computations, we assume that only the material in the path of a reasonably sized spacecraft is upswept as mass-energy fuel or species of reaction and that the spacecraft makes an effective pass entirely through each universe. Thus, all material in said universes of travel is reacted against with assumed finite universal mass-energy density.

For universes having an extensity of Aleph 4 light-years, there are prospects for the spacecraft to attain a Lorentz factor of roughly [(Aleph J) EXP N][(Aleph H) EXP N] (Aleph 4) to thus result in the multiverses of travel being contracted to point directly in front of the spacecraft and being contracted relative to the spacecraft by roughly a factor of [(Aleph J) EXP N][(Aleph H) EXP N](Aleph 4). Note that in the above computations, we assume that only the material in the path of a reasonably sized spacecraft is upswept as mass-energy fuel or species of reaction and that the spacecraft makes only one effective pass through each universe.

If such Aleph 4 light-year–width universes have N spatial dimensions and are substantially static, then a spacecraft traveling at light-speed in the forests may obtain a Lorentz factor of roughly [(Aleph J) EXP N][(Aleph H) EXP N] [(Aleph 4) EXP N]. Note that in the above computations, we assume that only the material in the path of a reasonably sized spacecraft is upswept as mass-energy fuel or species of reaction and that the spacecraft makes an effective pass entirely through each universe. Thus, all material in said universes of travel is reacted against with assumed finite universal mass-energy density.

If the latter universes are expanding with time over periods much greater than Aleph 4 years, then the spacecraft may attain a Lorentz factor of roughly [(Aleph J) EXP N][(Aleph H) EXP N] [[k(Aleph 4)] EXP N], where k ranges from a small superunitary real number to suitable infinite real numbers. This, of course, assumes that real positive mass-energy stocks continue being created, perhaps out of zero-point fields or analogues thereof from some forms of dark energy. Provided that k ranges roughly from Aleph 0 to Aleph 6, then the spacecraft Lorentz factor will be limited to a range from [(Aleph J) EXP N][(Aleph H) EXP N] [[(Aleph 0)(Aleph 4)] EXP N] to [(Aleph J) EXP N][(Aleph H) EXP N] [[(Aleph 6)(Aleph 4)] EXP N]. Note that in the above computations, we assume that only the material in the path of a reasonably sized spacecraft is upswept as mass-energy fuel or species of reaction and that the spacecraft makes an effective pass entirely through each universe. Thus, all material in said universes of travel is reacted against with assumed finite universal mass-energy density.

For universes having an extensity of Aleph G light-years, there are prospects for the spacecraft to attain a Lorentz factor of roughly [(Aleph J) EXP N][(Aleph H) EXP N] (Aleph G) to thus result in the multiverses of travel being contracted to point directly in

front of the spacecraft with a Lorentz contraction of roughly [(Aleph J) EXP N][(Aleph H) EXP N] (Aleph G). Note that in the above computations, we assume that only the material in the path of a reasonably sized spacecraft is upswept as mass-energy fuel or species of reaction and that the spacecraft makes only one effective pass through each universe.

If such Aleph G light-year width universes have *N* spatial dimensions and are substantially static, then a spacecraft traveling at light-speed in the forests may obtain a Lorentz factor of roughly [(Aleph J) EXP N][(Aleph H) EXP N] [(Aleph G) EXP N]. Note that in the above computations, we assume that only the material in the path of a reasonably sized spacecraft is upswept as mass-energy fuel or species of reaction and that the spacecraft makes an effective pass entirely through each universe. Thus, all material in said universes of travel is reacted against with assumed finite universal mass-energy density.

If the latter universes are expanding with time over periods much greater than Aleph G years, then the spacecraft may attain a Lorentz factor of roughly [(Aleph J) EXP N][(Aleph H) EXP N] [[k(Aleph G)] EXP N], where *k* ranges from a small superunitary real number to suitable infinite real numbers. This, of course, assumes that real positive mass-energy stocks continue being created, perhaps out of zero-point fields or analogues thereof from some forms of dark energy. Provided that *k* ranges from Aleph 0 to Aleph (G + 2), then the spacecraft Lorentz factor will be limited to a range of roughly from [(Aleph J) EXP N][(Aleph H) EXP N] [[(Aleph 0)(Aleph G)] EXP N] to [(Aleph J) EXP N][(Aleph H) EXP N] [[[Aleph (G + 2)](Aleph G)] EXP N]. Note that in the above computations, we assume that only the material in the path of a reasonably sized spacecraft is upswept as mass-energy fuel or species of reaction and that the spacecraft makes an effective pass entirely through each universe. Thus, all material in said universes of travel is reacted against with assumed finite universal mass-energy density.

Note that in the above multiversal scenarios, we assume that the universes are packed within the multiverses considered at near-maximum possible packing factors.

In the above forestal scenarios, we assume that the multiverses are packed within the forest considered at near-maximum possible packing factors.

We could continue to run the formulas specifically for biospheres, solar systems, and the like metaphoric terms for levels of a fractalverse. However, since the number of levels may itself be indefinite, perhaps infinite, we simply note this as a possibility and assume the reader has an intuition by now as to how the formulas describing such would be expanded.

Now there may be mass-energy present in the spaces between universes in a multiverse, multiverses in a forest, and so on. For multiverses and forests in the above formulations, we omit formulaic references to such but note that such mass-energy backgrounds would require modifications of the subject formulas for more precise resource accounting. Thus, the energy acquired in travel through multiverses can be noted formulaically by including a fudge summand for every associated multiverse of travel and the same for every forest of travel, and also for every biosphere of travel and so on.

ESSAY 12) Unstatably Infinite Spacecraft Lorentz Factors

Note that Aleph 0 is the infinite number of positive integers.

> Aleph 1 = 2 EXP (Aleph 0) and is, according to the perhaps unprovable, and thus not disprovable, continuum hypothesis, equal to the number of real numbers.

> Aleph 2 = 2 EXP (Aleph 1)

> Aleph 3 = 2 EXP (Aleph 2)

> Aleph 4 = 2 EXP (Aleph 3)

And so on.

In general, Aleph n = 2 EXP Aleph (n - 1), where *n* is any finite or infinite positive integer.

But before we go further, consider the hyper-operator notation that was designed to express huge values not otherwise expressible.

For example, Note that Hyper4(a, n) is equal to a tetrated *n*, or a raised to the power of itself *n-1* times. The latter value is symbolically written as *n* subscript *a*.

For example,

> 3 EXP 4 = 81, but 4 subscript3 is approximately equal to 10 EXP (1,000,000,000,000).

Alternatively,

> 4 subscript 2 = 2 EXP 2 EXP 2 EXP 2 = 2 EXP [2 EXP [2 EXP 2]] = 2 EXP (2 EXP 4) = 2 EXP 16 = 65,536.

For example,

Hyper5(4, 4)is equal to 4 tetrated 4 tetrated 4 tetrated 4. This value is commonly referred to as 4 pentated 4.

Hyper 6, (4,4) is 4 pentated 4 pentated 4 pentated 4 and is also referred to as 4 hexataed 4.

Hyper 7, (4,4) is 4 hexated 4 hexated 4 hexated 4 and so on

So can you imagine how big infinities as arguments of hyper-operator functions can be.

For example, consider the following:

Hyper Aleph 0 (Aleph 0, Aleph 0);

Hyper Aleph 1 (Aleph 1, Aleph 1);

Hyper Aleph 2 (Aleph 2, Aleph 2);

Hyper Aleph 3 (Aleph 3, Aleph 3);

And so on to

Hyper Aleph Ω (Aleph Ω, Aleph Ω)

Where Ω is the least infinite ordinal.

However, we can consider uncountable infinities that are generally considered to range larger than countable infinities although the details are a bit technical. However, just as with Aleph numbers, we can consider uncountable infinities far advanced in size compared to the least uncountable infinity. We will refer to these far-advanced uncountable infinities as highly evolved uncountable infinities.

Now consider the following:

Hyper highly evolved uncountable infinity (highly evolved uncountable infinity, highly evolved uncountable infinity).

Now we go the next level up to consider infinities that are so large they cannot be finitely stated. Even the number; hyper highly evolved uncountable infinity (highly evolved uncountable infinity, highly evolved uncountable infinity), can be stated using only a few dozen characters, although not precisely so.

We refer to these higher-yet infinities as unstatable infinities. So can you imagine the following number or infinity.

Hyper-unstatable infinity (unstatable infinity, unstatable infinity)

Well, here is something stated you are going to enjoy very well:

A light-speed inertial or impulse spacecraft will experience at least a Lorentz factor of Ω. Here, Ω is the smallest infinite ordinal.

However, as much as a stretch as it may seem, given technological advancements in some eternally future distant cosmic era, perhaps we will produce spacecraft which we will crew and fly at light-speed to a Lorentz factor of a hyper-unstatable infinity (unstatable infinity, unstatable infinity).

We may affix the following operator to formulas for gamma and related parameters to denote the awesome open-ended prospects of light-speed travel of infinite Lorentz factors.

> [u(Flight at light-speed to a Lorentz factor of from Ω to a Hyper un-statable infinity (un-statable infinity, un-statable infinity) and beyond)]

Flight at light-speed to a Lorentz factor of from Ω to a hyper-unstatable infinity (unstatable infinity, unstatable infinity) and beyond will essentially allow equal numbers of light-years travel through space in one-year ship-time, effectively equal numbers of multiples of speed of light for a spacecraft, and an equal number of years of forward time travel in one-year ship-time.

Each human person and each human soul in the next life and each angel morally good or bad behaving lives its own temporal world line.

In a sense, each of us has our own temporal world line that holds us like a cocoon and that only we alone with GOD have access to. In a sense, our world lines are like a nice, wonderful, cozy, warm bed on a cold winter night that have us all bundled up and personally related to GOD. Couples in love are perhaps the closest thing to a unique personal world line, thus indicating the awesome aspects of conjugal love and marriage.

How our world lines will open up in the depths of future eternity is a great mystery but something I believe all created persons can look forward to with a wonderful sense of wanderlust and intrigue and a sense of the mystery of it all.

Thus, I have made my case even better for the wonders of light-speed travel and ever greater Lorentz factors to thus show that even if Mother Nature does not define faster-than-light travel, perhaps because it does not make sense; because of time dilation in relativistic and ultimately light-speed space travel, it works out for the better than otherwise.

OH WHAT DREAMS WILL FOLLOW!

ESSAY 13) Spacecraft in Light-Speed Spatial Diffusement by Capillary-Like Action

Herein we consider prospects for developing spacecraft that diffuse themselves through space as if an analogue of the capillary action of various liquids.

Accordingly, the spacecraft would seep forward in space-time at essentially the speed of light with possible ever-so-slightly faster-than-light travel.

Such spacecraft might take advantage of a quantum mechanical entanglement that develops, accrues, or remains present between the differential material elements of the spacecraft and proximate particles located just ahead of the spacecraft.

Thus, such a spacecraft may drift forward in space by continuous quantum entanglement and quantum teleportation of the spacecraft and the proximate background particles just in front of the spacecraft.

Some mechanisms of automatically communicating with the spacecraft differential volumetric elements and the forwardly located background particles would be necessary for a classical feedback loop to close the teleportation process.

The *N-Space-M-Time* may optionally be as follows:

flat, positively curved, negatively curved, positively curved and torsioned at one or more scales in arbitrary patterns including, but not limited to, fractals, negatively curved and torsioned at one or more scales in arbitrary patterns including, but not limited to, fractals, positively super-curved, negatively super-curved, positively super-curved and torsioned at one or more scales in arbitrary patterns including, but not limited to, fractals, negatively super-curved and torsioned at one or more scales in arbitrary patterns including, but not limited to, fractals, positively super-curved and super-torsioned at one or more scales in arbitrary patterns including, but not limited to, fractals, negatively super-curved and super-torsioned at one or more scales in arbitrary patterns including, but not limited to, fractals, positively curved and positively torsioned at one or more scales in arbitrary patterns including, but not limited to, fractals, negatively curved and positively torsioned at one or more scales in arbitrary patterns including, but not limited to, fractals, positively super-curved, negatively super-curved, positively super-curved and positively torsioned at one or more scales in arbitrary patterns including, but not limited to, fractals, negatively super-curved and positively torsioned at one or more scales in arbitrary patterns including, but not limited to, fractals, positively super-curved and positively super-torsioned at one or more scales in arbitrary patterns including, but not limited to, fractals, negatively super-curved and positively super-torsioned at one or more scales in arbitrary

EXP 560) – 1, which is approximately equal to 10 EXP 160. This is roughly 10 EXP 70 times the number of atoms, electrons, photons, and neutrinos in the observable universe. Most of these methods permit very high gamma factors, in many cases, virtually unlimited relativistic gamma factors given the virtually, if not actually unlimited, future time periods available to sequester ever greater resources of fuel via an ever-expanding human space-based resource collection infrastructure.

The above methods may work differently in a given hyperspace, extra-universe multiverse vacuum, extra-universe and extra-multiverse forest vacuum, extra-universe, extra-multiverse, extra-forest biosphere vacuum, and so on. These methods may also become more and more modified in thermodynamics, kinematics, and topology for spacecraft generally experiencing increasingly great finite and then increasingly great infinite Lorentz factors, perhaps even in our universe, if our universe permits infinite Lorentz factors for massive bodies.

For example, the kinetic energy, momentum, and Lorentz contraction-specific velocities may differ from those in our known cosmic light-cone.

Additionally, the force charge to invariant mass and/or relativistic mass may follow different mathematical rules for massive species in these higher realms than in ordinary 4-D space-time. The following is a list of known and theoretical or proposed mainly fundamental particles. The thermodynamics of these particles may differ in at least some broader realms outside of ordinary 4-D space-time.

The Standard Model particles include the fermions of the up, down, charmed, strange, top, and bottom quarks and the six antimatter versions, as well as the electron, muon, tauon, electron neutrino, muon neutrino, and tauon neutrino, as well as the six antimatter versions.

The Standard Model bosons include the photon, the W+, W-, and Z_0 weak force bosons, the Higgs boson, and eight distinct species of gluons.

In addition, there exist gravitational waves and presumably the small-scale quanta of gravitons as analogues of photons.

The bosonic supersymmetric counterparts to the normal mattergy fermions, such as the sleptons and the squarks, are sought after in modern accelerator facilities such as the Large Hadron Collider at CERN. The sleptons include a bosonic particle counterpart to the electron, muon, and tau particle, as well as that of the electron neutrino, the muon neutrino, and the tau neutrino. The squarks include a boson for each of the six known quarks: the up, down, strange, charmed, top, and bottom quarks.

Note that each supersymmetric fermion would in theory come in an antimatter version. So there would exist antiphotinos, antigravitinos, antigluinos, antihiggsinos, and the antimatter versions of the weak force supersymmetric fermionic particles called winos. In some theories of cold dark matter, some non–Standard Model fermions are its own antiparticles.

Other theoretical particles include the graviscalar with a spin of zero and the graviphoton with a spin of one.

In addition, there is the possible existence of the axion, which would interact with gravity and electromagnetism and would have zero charge and zero spin. The axion is a theoretical pseudoscalar particle introduced in Peccei-Quinn theory to solve the strong-CP problem.

The existence of the axino is proposed as a solution to the strong CP problem. A proposed supersymmetric partner to the axino is the saxino. The saxino has, in theory, a spin of zero.

The branon is predicted by brane world models in which higher-dimensional analogues of membranes would exist as topological defects in space-time.

The X and Y bosons are predicted by GUT theories to have a mass of about 10^{15} GeV/c^2 and an electrical charge of + 4/3 e and + 1/3 e, respectively, and a spin of 1.

The W and the Z bosons would each have a spin of 1. The W boson would have a charge of − e and would interact electroweakly. The Z would also interact electroweakly but would have zero electrical charge.

The magnetic photon is predicted in extensions to electromagnetism, which also predict the existence of the magnetic monopole, a magnetic field charge analogue to the unipolar electron.

The majoron is predicted as the result of the mass seesaw model of neutrinos, by which neutrinos can oscillate between flavors, a consequence of neutrino mass. Both the charge and the spin of the majoron are theoretically equal to 0.

Mirror matter particles would include a mirror particle for every one of the known particles. Each of the known particles respects rotation and translation spatial symmetries but not so of the third spatial symmetry of mirror reflection. Mirror matter particles, also referred to as shadow matter, could in theory form stars, planets, ETI beings—in fact, all the structure types we observe for normal matter particles. Mirror matter particles, however, would interact very weakly with normal matter. However, the mirror mattergy would interact gravitationally with ordinary matter.

Tachyons are particles that would have imaginary mass and always travel faster than the speed of light. Here, the word *imaginary* refers to the square root of the negative one sense of the word.

Preons are proposed particles as the constituents of the leptons and the quarks. There as yet is no experimental evidence for preons. The preons would be fundamental particles that would combine in various ways to produce the known massive particles and would be fewer in number than the known fermionic particles.

Kaluza-Klein towers are particles predicted by certain models of higher dimensional space.

Leptoquarks would be particles with characteristics of both leptons and quarks. Photoneutrinos would be particles with the characteristics of both photons and neutrinos.

There are other proposed fundamental particles, and the set of these additional particles, as well as the proposed particles briefly described above, are in some theories not necessarily consistent. Some additional particles are briefly described below.

Another taxonomic term for candidates for yet-to-be-discovered particles is WIMPS, or weakly interacting massive particles. The neutralino and the axion are candidates for such cold dark matter particles. Wimps as cold dark matter candidates would only interact weakly and gravitationally, not electromagnetically, and neither through the strong nuclear force. Wimps are predicted by R-parity conserving supersymmetry.

It has even been suggested that the photon, comes in two varieties: the photon and the antiphoton. Traditional quantum-electrodynamicists assert that the photon is its own antiparticle and so generally do not hold to the notion that an antiphoton could exist. That is as distinct from a photon as, say, a positron is from an electron.

The pomeron is a family of particles with an increasing spin that was proposed in 1961 to explain the slowly increasing cross-sections for the scattering of hadrons in collision experiments.

The skyrmion was used as a mathematical model in the modeling of baryons by Tony Skyrme.

Goldstone bosons appear in theories of spontaneous symmetry breaking and are a part of the subjects of particle physics and condensed matter physics. Some supersymmetry theories include goldstinos which are Goldstone fermions. The sgoldstinos are the supersymmetric bosonic counterparts to the goldstinos.

In condensed matter physics, the quantum of acoustic energy is a Goldstone boson referred to as the phonon. Phonons in fluids are longitudinally oriented, whereas in solids, phonons can be longitudinal or transverse in orientation. Magons are local magnetic spin waves that occur within magnetic materials.

Dyons are hypothetical particles with both a magnetic charge and an electrical charge and are predicted by some grand unified theories. Dyons are proposed in certain four-dimensional theories.

Geons are theoretical particles that result from the self-confinement of a gravitational energy wave or electromagnetic energy wave by the wave's self-gravitation.

One can imagine or lexicographically define a question as to whether or not nuclear force waves or gluon waves, as well as weak force waves, or W+, W-, or Z_0 waves, could have high-enough energy so as to form a strong nuclear force or weak nuclear force equivalent to geons. One can also speculate as to whether neutrinos, as well as the supersymmetric field bosons, could have high-enough energy on the level of an individual mattery wave such that the neutrinos, sleptons, and squarks could form analogues to geons.

Inflatons are theoretical particles that may have driven the inflation of our universe from a period of about 10^{-35} seconds after the big bang to about 10^{-34} seconds after the big bang, during which the visible portion of our universe expanded from a size of about 10^{-50} meter to about one meter. This translated into a superluminal space-time expansion of about $(10^{34.5} C)/[3 \times 10^{8}]$ or $10^{26.5} C$. Inflation was theoretically driven by the inflaton field tunneling to a lower-energy state as a result of random quantum fluctuations within the inflaton field that triggered a phase charge within the field. The energy released by the phase change is what seems to have driven inflation according to many versions of inflationary big bang theory.

Normal matter or baryonic matter accounts for only about 4 percent of the mattery content within the observable universe according to sensitivities in contemporary big bang models. Thus, the presence of these other forms and types of mattery and particles may pave the way for sail craft to reach essentially unlimited γ factors. This reasoning is especially applicable to beam pull-sail craft. For such craft, a negative refraction matter-wave index material would be pulled forward by these impinging exotic particles by a mechanism analogous to electromagnetic negative refraction index pull sails. As the craft gained sufficiently high gamma factors, it would be pulled along by the ambient interstellar and intergalactic medium while ever growing in relativistic gamma factors perhaps experiencing a runaway gamma factor limited only by gamma = infinity. Such a gamma factor would allow the lucky crew on such a craft to travel infinite spatial and infinite temporal distances into the future in one Planck time unit ship-time.

If we as the civilization of humanity can survive the next few critical decades, we stand on the shore of a small island in a perhaps infinite and mysterious black primal ocean of space and time with a depth and width that perhaps knows no limits. Will we have the courage to set sail? As representatives of humanity, I believe we must.

Quasi-particles may manifest differently outside of ordinary 4-D space-time. Some such particles include the following:

> Bipolarons, quasiparticle chargons, configurons, dropletons, electron quasiparticle, electron holes, excitons, fractons, holons, levitons, magnons, majorana fermions, orbitrons, phasons, phonitons, phonons, plasmarons, plasmons, polarons, polaritons, rotons, solitons, spinons, trions, wrinklons. An exact duplicate of such quasi-particles is possible in antimatter embodiments.

We may affix the following operator to formulas for gamma and related parameters to denote potential modifications of the thermodynamics, kinematics, topology, laws of physics, and the like for increasing light-speed velocities craft in known and unknown realms for conceptually known and unknown forms of propulsion:

> [u(Potential modifications of the thermodynamics, kinematics, topology, laws of physics, etc., for → c craft in known and unknown realms for conceptually known and unknown forms of propulsion)].

ESSAY 16) Infinite Lorentz Factor Travel and Light-Speed Popping into Other Realms

Now we can intuit spacecraft attaining the velocity of light in infinite Lorentz factors.

Accordingly, the spacecraft may run out of room to travel forward in time in ordinary 4-D space-time and may thus enter, or "pop in," to another larger realm.

Eventually, the spacecraft may run out of room in said larger realm to enter yet another larger realm yet, and the process may continue perpetually from one realm to the next larger realm.

Eventually, regardless of whether a spacecraft runs out of new larger realms to pop into, the spacecraft may start a process of advancement in another process that is different than velocity or the velocity meaning of travel.

Velocity, as well as its associated parameter of Lorentz factors, may be as if a tiny facet on an infinitely large and infinite dimensional crystal with a huge infinity of facets.

The spacecraft may enter and take on location or the form of other facets of the crystal each of which is different from the facet of velocity and Lorentz factors, as well as from each other.

So, in a sense, the spacecraft can gradually increase its spread on the set of facets but most likely only forever cover an inverse huge infinity of the facets or [1/(Huge infinity)] of the set of facets.

However, the facets may simply represent more or less accidental properties of the overall cosmic order.

In a sense, the faceted crystal may be an inverse crystal for which the facets are only entry points relatively close to each other of a realm with existential extensity unfathomably infinity larger and deeper than the set of facets.

Whether the existential crystal is conventional or of inverse style, the interior of the crystal represents travel toward the substantiality of the crystal and, thus, realms that are progressively more removed from accidental physical realities.

Moreover, the crystal, whether externally faceted or internally faceted and of inverse form, may continue to expand forever in number and size of facets, and of depth and extensity.

We may affix the following operator to formulas for gamma and related parameters to denote the above-faceted crystal mechanics:

> [u(Advancement on faceted surface and/or within existential crystal of infinite scale and number of facets expanding forever in number and size of facets, and of depth and extensity)]

ESSAY 17) Superluminal Spacecraft That Do Not Violate Special Relativity

Herein, we consider superluminal impulse travel that enables us to travel faster than light in a continuous spread of space-time.

Accordingly, space-time would have more room in it than the portions that we can visit at light-speed travel or less.

At superluminal velocities, a spacecraft could not travel faster than light in the universal portions visitable by light-speed and less-than-light-speed travel.

Our universe may thus have much more room than the world line we can see in light-speed signal observation whether natural or artificial signals.

Essentially, the projection of the superluminal velocities onto the velocity of light in our spread of space-time would all be equal to the speed of light as indicated by the following figure.

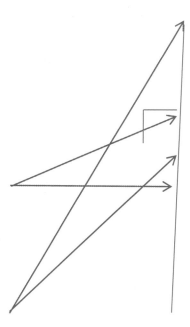

The cosine of the angles between the velocity of light and the superluminal velocities would range from just less than 1 to 0.

The same can be said regarding the distances travel at light-speed in our spread of space-time and those traveled superluminally in other spreads of space-time.

In our ordinary 4-D space-time, the projection of r(x,1; x,2; x,3) in a superluminal scenario of velocities or distances travel in an alternate spread of the space-time on to that associated with light can be represented as such:

{0 degrees = ɸ < 90 degrees| c < Superluminal v(ɸ) ≤ ∞ c}

and

{0 degrees = φ < 90 degrees| r(x,1; x,2; x,3) at c < r(x,1; x,2; x,3) at Superluminal v(φ) ≤ r(x,1; x,2; x,3) at ∞ c}

For 4-D-space spreads, the first set also applies, but the second set is easily denoted as follows:

{0 degrees = φ < 90 degrees| r(x,1; x,2; x,3; x,4) at c < r(x,1; x,2; x,3; x,4) at Superluminal v(φ) ≤ r(x,1; x,2; x,3, x,4) at ∞ c}

For 5-D-space spreads, the first set also applies, but the second set is easily denoted as follows:

{0 degrees = φ < 90 degrees| r(x,1; x,2; x,3; x,4; x,5) at c < r(x,1; x,2; x,3, x,4; x,5) at Superluminal v(φ) ≤ r(x,1; x,2; x,3, x,4, x,5) at ∞ c}

For N-D–space, the first set also applies, but the second set can be represented as follows:

{0 degrees = φ < 90 degrees| r(x,1; x,2; … x,N) at c < r(x,1; x,2; … x,N) at Superluminal v(φ) ≤ r(x,1; x,2; … x,N) at ∞ c}

The following sets imply that within N-D–space, where *N* ranges from 3 to infinity, for each space of spread, there exist an infinite number, ∞:continuous, possible superluminal velocities.

In cases where each space of spread is supercontinuous, there exists an [f(∞:continuous)] = [(∞:continuous) EXP k] possible superluminal velocities, where *k* is greater than 1.

We may affix the following operator to formulas for gamma and related parameters to denote the superluminal velocities and space-time spread mechanisms:

{u{{0 deg = φ < 90 deg| c < Superluminal v(φ) ≤ ∞ c};{0 deg = φ < 90 deg| r(x,1; x,2; … x,N) at c < r(x,1; x,2; … x,N) at Superluminal v(φ) ≤ r(x,1; x,2; … x,N) at ∞ c}}}

Note that the conjectured spatial-temporal spreads are completely different in context from quantum mechanical parallel histories or many worlds concepts. The spreads' concepts are even distinct from classical parallel universes models.

The N-Space-M-Time may optionally be as follows:

flat, positively curved, negatively curved, positively curved and torsioned at one or more scales in arbitrary patterns including, but not limited to, fractals, negatively curved and torsioned at one or more scales in arbitrary patterns including, but not limited to, fractals, positively super-curved, negatively super-curved, positively super-curved and torsioned at one or more scales in arbitrary patterns including, but not limited to, fractals, negatively super-curved and torsioned at one or more scales in arbitrary patterns including, but not limited to, fractals, positively super-curved and super-torsioned at one or more scales in arbitrary patterns including, but not limited to, fractals, negatively super-curved and super-torsioned at one or more scales in arbitrary patterns including, but not limited to, fractals, positively curved and positively torsioned at one or more scales in arbitrary patterns including, but not limited to, fractals, negatively curved and positively torsioned at one or more scales in arbitrary patterns including, but not limited to, fractals, positively super-curved, negatively super-curved, positively super-curved and positively torsioned at one or more scales in arbitrary patterns including, but not limited to, fractals, negatively super-curved and positively torsioned at one or more scales in arbitrary patterns including, but not limited to, fractals, positively super-curved and positively super-torsioned at one or more scales in arbitrary patterns including, but not limited to, fractals, negatively super-curved and positively super-torsioned at one or more scales in arbitrary patterns including, but not limited to, fractals, positively curved and negatively torsioned at one or more scales in arbitrary patterns including, but not limited to, fractals, negatively curved and negatively torsioned at one or more scales in arbitrary patterns including, but not limited to, fractals, positively super-curved, negatively super-curved, positively super-curved and negatively torsioned at one or more scales in arbitrary patterns including, but not limited to, fractals, negatively super-curved and negatively torsioned at one or more scales in arbitrary patterns including, but not limited to, fractals, positively super-curved and negatively super-torsioned at one or more scales in arbitrary patterns including, but not limited to, fractals, negatively super-curved and negatively super-torsioned at one or more scales in arbitrary patterns including, but not limited to, fractals, positively super-...-super-curved, negatively super-...-super-curved, positively super-...-super-curved and torsioned at one or more scales in arbitrary patterns including, but not limited to, fractals, negatively super-...-super-curved and torsioned at one or more scales in arbitrary patterns including, but not limited to, fractals, positively super-...-super-curved and super-...-super-torsioned at one or more scales in arbitrary patterns including, but not limited to, fractals, negatively super-...-super-curved and super-...-super-torsioned at one or more scales in arbitrary patterns including, but not limited to, fractals, positively curved and positively torsioned at one or more scales in arbitrary patterns including, but not limited to, fractals, negatively curved and positively torsioned at one or more scales in arbitrary patterns including, but not limited to, fractals, positively super-...-super-curved, negatively super-...-super-curved, positively super-...-super-curved and positively torsioned at one or more scales in arbitrary patterns including, but not limited to, fractals, negatively super-...-super-curved and positively torsioned at one or more scales in arbitrary patterns including, but not limited to, fractals, positively super-...-super-curved and positively super-...-super-torsioned at one or more scales in arbitrary patterns including, but not limited to, fractals, negatively super-...-super-curved and positively super-...-super-torsioned at one or more scales in arbitrary patterns including, but not limited to, fractals, positively curved and negatively torsioned at one or more scales in arbitrary patterns including but not limited to fractals, negatively curved and negatively torsioned at one or more scales in arbitrary patterns including but not limited to fractals, positively super-...-super-curved, negatively super-...-super-curved, positively super-...-super-curved and negatively torsioned at one or more scales in arbitrary patterns including, but not limited to, fractals, negatively super-...-super-curved and negatively torsioned at one or more scales in arbitrary patterns including, but not limited to, fractals, positively super-...-super-curved and negatively super-...-super-torsioned at one or more scales in arbitrary patterns including, but not limited to, fractals, negatively super-...-super-curved and negatively super-...-super-torsioned at one or more scales in arbitrary patterns including, but not limited to, fractals.

ESSAY 18) Cosmic Light-Cone Scale Accretion Disk-Powered Starships

Here, we consider cosmic light-cone scale-accretion disk-powered starships where the mass-energy outflow is used as a rocket or ramjet exhaust, and/or electro-hydrodynamic-plasma-drives, magneto-hydrodynamic-plasma-drives, electro-magneto-hydrodynamic-plasma-drives, and/or electromagnetic-hydrodynamic-plasma-drives accelerate the massive plasma species in a thrust stream. Other methods may work, such as magnetic-field-effect propulsion, linear-magnetic-induction generators for powering any and all of the following on a light-cone scale system or a distributed micro-scale system:

1) fusion rockets, 2) fission rockets, 3) fission fragment drives, 4) fusion-powered ion, electron, photon, and/or neutrino rockets, 5) fission-powered ion, electron, photon, and/or neutrino rockets, 6) matter-antimatter rockets that carry both components of fuel on board from the start of the mission, 7) matter-antimatter rockets that carry only their antimatter fuel component(s) along from the start of the mission, 8) matter-antimatter reactor-powered ion,

electron, photon, and/or neutrino rockets, 9) fusion fuel pellet linear runway–powered craft, 10) fission fuel pellet linear runway–powered craft, 11) fission-fusion fuel pellet linear runway–powered craft, 12) nuclear isomer fuel pellet linear runway–powered craft, 13) matter-antimatter fuel pellet linear runway–powered craft, 14) antimatter fuel pellet linear runway–powered craft, 15) fusion fuel pellet circulinear runway–powered craft, 16) fission fuel pellet circulinear runway–powered craft, 17) fission-fusion fuel pellet circulinear runway–powered craft, 18) nuclear isomer fuel pellet circulinear runway–powered craft, 19) matter-antimatter fuel pellet circulinear runway–powered craft, 20) antimatter fuel pellet circulinear runway–powered craft, 21) nuclear fission powered electro-hydrodynamic-plasma drive craft, 22) nuclear fusion powered electro-hydrodynamic-plasma-drive craft, 23) matter-antimatter reaction powered electro-hydrodynamic-plasma drive craft, 24) nuclear fission powered magneto-hydrodynamic-plasma drive craft, 25) nuclear fusion powered magneto-hydrodynamic-plasma drive craft, 26) matter-antimatter-reaction–powered electrohydrodynamic plasma-drive craft, 27) nuclear fission–powered electro-magnetohydrodynamic plasma-drive craft, 28] nuclear fusion–powered electro-magnetohydrodynamic plasma-drive craft, 29) matter-antimatter–powered electro-magnetohydrodynamic plasma-drive craft, 30) fusion-powered magnetic field–effect drive, 31) fission-powered magnetic field–effect drive, 32) matter-antimatter-reaction–powered field-effect drive, 33) single-pass solar-dive and fry-sail-driven craft, 34) single-pass stellar-dive and fry-sail-driven craft, 35) single-pass quasar-dive and fry-sail-driven craft, 36) multipass stellar cycler solar dive and fry-sail-driven craft, 37) multi-pass stellar cycler dive and fry sail driven craft, 38] multi-pass cycler quasar-dive and fry sail driven craft, 38] laser beam–driven relativistic-sail craft, 40) microwave beam–driven relativistic-sail craft, 41) radio-frequency-beam–driven relativistic-sail craft, 42) massive neutral particle beam–driven sail craft, 43) massive charged particle beam–driven sail craft, 44) massive particle beam fission fuel–powered craft, 45) massive particle beam fusion fuel–powered craft, 46) massive particle beam matter-antimatter beam fuel–powered craft, 47) antimatter beam fuel–powered craft, 48] nuclear bomb pulse-driven propulsion of the original Project Orion forms, 49) pure nuclear fusion bomb pulse-driven propulsion analogous to the original Project Orion forms, 50) matter-antimatter-bomb-driven propulsion analogous to the original Project Orion forms, 51) one-side-reflective cosmic microwave background radiation sails, 52) multiple-beam-bounce propulsion methods, 53) any improved interstellar ramjet craft, 54) fusion rocket– or fusion-powered electron, ion, photon, or neutrino rockets utilizing single-body- or serially-multiple-body-powered gravitational assists, 55) fission rocket– or fission-powered electron, ion, photon, or neutrino rockets utilizing single-body- or serially-multiple-body-powered gravitational assists, 56) matter-antimatter rocket or matter-antimatter-reaction electron, ion, photon, or neutrino rockets utilizing single-body- or serially-multiple-body-powered gravitational assists, and others.

If the propulsion system is multimodal, then even if only one stage is used for each mode, the number of combinations and thus the number of possible propulsion systems is at least equal to (2 EXP 56) – 1 = 7.205 x 10 EXP 16 = 72.05 quadrillion = 72,050 trillion!

In each of these fifty-six categories, several sub-methods have been proposed, and so the number of possible multi-mode/multi-stage propulsion systems is many, many orders of magnitude greater yet. If, say, each category permits ten subcategories, which, I can reasonably assure you, is likely a conservative estimate, then the total number of possible propulsion systems, all else being the same is equal to about (2 EXP 560) – 1, which is approximately equal to 10 EXP 160. This is roughly 10 EXP 70 times the number of atoms, electrons, photons, and neutrinos in the observable universe. Most of these methods permit very high gamma factors, in many cases, virtually unlimited relativistic gamma factors given the virtually, if not actually unlimited, future time periods available to sequester ever greater resources of fuel via an ever-expanding human space-based resource collection infrastructure.

As long as the invariant mass-specific kinetic energy gain is greater than the inertial inert mass-energy assimilation, the system should perpetually accelerate.

A minimally open universe would be ideal here, or a steady-state mass-energy density open universe may also be conducive to the cosmic accretion disk propulsion method.

The N-Space-M-Time of the cosmic scale accretion disk propulsion system class and/or the space-time of travel as analogues may optionally be as follows:

flat, positively curved, negatively curved, positively curved and torsioned at one or more scales in arbitrary patterns including, but not limited to, fractals, negatively curved and torsioned at one or more scales in arbitrary patterns including, but not limited to, fractals, positively super-curved, negatively super-curved, positively super-curved and torsioned at one or more scales in arbitrary patterns including, but not limited to, fractals, negatively super-curved and torsioned at one or more scales in arbitrary patterns including, but not limited to, fractals, positively super-curved and super-torsioned at one or more scales in arbitrary patterns including, but not limited to, fractals, negatively super-curved and super-torsioned at one or more scales in arbitrary patterns including, but not limited to, fractals, positively curved and positively torsioned at one or more scales in arbitrary patterns including, but not limited to, fractals, negatively curved and positively torsioned at one or more scales in arbitrary patterns including, but not limited to, fractals, positively super-curved, negatively super-curved, positively super-curved and positively torsioned at one or more scales in arbitrary patterns including, but not limited to, fractals, negatively super-curved and positively torsioned at one or more scales in arbitrary patterns including, but not limited to, fractals, positively super-curved and positively super-torsioned at one or more scales in arbitrary patterns including, but not limited to, fractals, negatively super-curved and positively super-torsioned at one or more scales in arbitrary patterns including, but not limited to, fractals, positively curved and negatively torsioned at one or more scales in arbitrary patterns including, but not limited to, fractals, negatively curved and negatively torsioned at one or more scales in arbitrary patterns including, but not limited to, fractals, positively super-curved, negatively super-curved, positively super-curved and negatively torsioned at one or more scales in arbitrary patterns including, but not limited to, fractals, negatively super-curved and negatively torsioned at one or more scales in arbitrary patterns including, but not limited to, fractals, positively super-curved and negatively super-torsioned at one or more scales in arbitrary patterns including, but not limited to, fractals, negatively super-curved and negatively super-torsioned at one or more scales in arbitrary patterns including, but not limited to, fractals, positively super-...-super-curved, negatively super-...-super-curved, positively super-...-super-curved and torsioned at one or more scales in arbitrary patterns including, but not limited to, fractals, negatively super-...-super-curved and torsioned at one or more scales in arbitrary patterns including, but not limited to, fractals, positively super-...-super-curved and super-...-super-torsioned at one or more scales in arbitrary patterns including, but not limited to, fractals, negatively super-...-super-curved and super-...-super-torsioned at one or more scales in arbitrary patterns including, but not limited to, fractals, positively curved and positively torsioned at one or more scales in arbitrary patterns including, but not limited to, fractals, negatively curved and positively torsioned at one or more scales in arbitrary patterns including, but not limited to, fractals, positively super-...-super-curved, negatively super-...-super-curved, positively super-...-super-curved and positively torsioned at one or more scales in arbitrary patterns including, but not limited to, fractals, negatively super-...-super-curved and positively torsioned at one or more scales in arbitrary patterns including, but not limited to, fractals, positively super-...-super-curved and positively super-...-super-torsioned at one or more scales in arbitrary patterns including, but not limited to, fractals, negatively super-...-super-curved and positively super-...-super-torsioned at one or more scales in arbitrary patterns including, but not limited to, fractals, positively curved and negatively torsioned at one or more scales in arbitrary patterns including, but not limited to, fractals, negatively curved and negatively torsioned at one or more scales in arbitrary patterns including, but not limited to, fractals, positively super-...-super-curved, negatively super-...-super-curved, positively super-...-super-curved and negatively torsioned at one or more scales in arbitrary patterns including, but not limited to, fractals, negatively super-...-super-curved and negatively torsioned at one or more scales in arbitrary patterns including, but not limited to, fractals, positively super-...-super-curved and negatively super-...-super-torsioned at one or more scales in arbitrary patterns including, but not limited to, fractals, negatively super-...-super-curved and negatively super-...-super-torsioned at one or more scales in arbitrary patterns including, but not limited to, fractals.

In greater detail, we may consider the net potential vorticity of a sub-light-cone or trans-light-cone accumulation of matter and energy.

The accumulation includes the Standard Model fermions and bosons, the graviton and gravitational waves, black holes, and supersymmetric particles. For all the fermions, we consider matter and antimatter versions as well.

So we can develop the following function to denote the net potential vorticity of the cosmic scale-sized deposition of matter and energy.

up-quark, down-quark, charmed-quark, strange-quark, top-quark, bottom-quark, up-antiquark, down-antiquark, charmed-antiquark, strange-antiquark, top-antiquark, bottom-antiquark, electron, muon, tauon, anti-electron, anti-muon, anti-tauon, electron neutrino, muon neutrino, tauon neutrino, anti-electron neutrino, anti-muon neutrino, anti-tauon neutrino, photon, graviton, gravitational wave, the 8 species of gluons, the W+, W-, and Z0 weak force bosons, squarks, sleptons, photinos, gluinos, weakinos, gravitinos, higginos, black holes, other dark matter species, dark energy

ESSAY 19) Ever So Slightly Superluminal Spacecraft Hyper-Removed from Temporal Realm of Origin

Here, we consider that light-speed and ever-so-slightly greater-than-light-speed travel in increasingly infinite Lorentz factors may result in a spacecraft running out of temporal room to travel forward in time, and thus resulting in the spacecraft "popping out" of the universe, multiverse, forest, biosphere and the like, and/or hyperspace of travel and into a larger realm.

Such spacecraft may eventually leave the causally and thermodynamically coupled realm of origin and, upon further increases in Lorentz factors, become hyper-removed from the temporal realm of original spatial-temporal travel, where hyper-removed or hyper-outside-of the original spatial-temporal realm of travel is to be removed from or outside of the original spatial-temporal realm of travel as removed from or outside of the original spatial-temporal realm of travel is to be completely contained within the original spatial-temporal realm of travel.

Upon further increases in Lorentz factors, the spacecraft may become hyper-hyper-removed from the original spatial-temporal realm of travel, where hyper-hyper-removed from the original spatial-temporal realm of travel is to be hyper-removed from the original spatial-temporal travel, where hyper-removed or hyper-outside-of the original spatial-temporal realm of travel is to be removed from or outside of the original spatial-temporal realm of travel as removed from or outside of the original spatial-temporal realm is to completely contained within the original spatial-temporal realm.

We refer to outside-of, hyper-outside-of and hyper-hyper-outside-of, respectively, as hyper-0-outside of, hyper-1-outside-of, and hyper-2-outside-of.

We can continue the serial pattern to hyper-hyper-hyper-outside-of as hyper-3-outside-of, and continue to hyper-4-outside-of, hyper-5-outside-of, and so on up to and beyond hyper-∞-outside-of.

Such travel, indeed, is way out there in an existential accident-wise void of size beyond measure.

We may affix the following operator to formulas for gamma and related parameters to denote the series of hyper-k-outside-of and ship-time, Lorentz factor, ship-kinetic energy, and ship-frame acceleration derivatives and integrals of the serial progression of arbitrary orders:

> [u(Hyper-k-outside-of and ship-time, Lorentz factor, ship-kinetic energy, and ship-frame acceleration derivatives and integrals of the serial progression of arbitrary orders; k = 0, 1, 2. ...)].

ESSAY 20) Extreme Infinite Spacecraft Lorentz Factors

Note that Hyper4(a, n) is equal to a tetrated n, or a raised to the power of itself n-1 times. The latter value is symbolically written as n subscript a.

For example,

3 EXP 4 = 81, but 4 subscript3 is approximately equal to 10 EXP (1,000,000,000,000).

Alternatively,

4 subscript 2 = 2 EXP 2 EXP 2 EXP 2 = 2 EXP [2 EXP [2 EXP 2]] = 2 EXP (2 EXP 4) = 2 EXP 16 = 65,536.

For example,

Hyper5(4, 4)is equal to 4 tetrated 4 tetrated 4 tetrated 4. This value is commonly referred to as 4 pentated 4.

Hyper 6, (4,4) is 4 pentated 4 pentated 4 pentated 4 and is also referred to as 4 hexataed 4.

Hyper 7, (4,4) is 4 hexated 4 hexated 4 hexated 4 and so on.

Aleph 0 is the infinite number of integers.

Aleph 1, according to the perhaps unprovable, and thus unfalsifiable, continuum hypothesis, is the number of real numbers that is greater than Aleph 0 by a multiplicative factor of infinity.

Aleph 2 is similarly greater than Aleph 1.

Aleph 3 is similarly greater than Aleph 2.

Aleph 4 is similarly greater than Aleph 3.

And so on

In general, Aleph n = 2 EXP [Aleph (n-1)].

The number Ω is commonly stated as the least infinite positive integer or ordinal.

Now here is a real zinger.

So we can produce the abstraction of [Hyper Aleph Ω (Aleph Ω, Aleph Ω)].

We can go to ever-greater infinities.

So we can consider

Hyper [Hyper Aleph Ω (Aleph Ω, Aleph Ω)] [[Hyper Aleph Ω (Aleph Ω, Aleph Ω)], [Hyper Aleph Ω (Aleph Ω, Aleph Ω)]] numbers

We can also consider uncountable infinities that are often considered greater than countable infinities.

As with Aleph numbers, we can consider a large evolved uncountable infinity in the following form.

So we can consider the following:

> Hyper [Hyper (large evolved uncountable infinity) [(large evolved uncountable infinity), (large evolved uncountable infinity)]] [[Hyper (large evolved uncountable infinity) [(large evolved uncountable infinity), (large evolved uncountable infinity)]], [Hyper (large evolved uncountable infinity) [(large evolved uncountable infinity), (large evolved uncountable infinity)]]] numbers

We can go to yet further extremes to produce the following number:

> Aleph { Hyper [Hyper (large evolved uncountable infinity) [(large evolved uncountable infinity), (large evolved uncountable infinity)]] [[Hyper (large evolved uncountable infinity) [(large evolved uncountable infinity), (large evolved uncountable infinity)]], [Hyper (large evolved uncountable infinity) [(large evolved uncountable infinity), (large evolved uncountable infinity)]]]}

Now we may consider infinities so large that they cannot be abstractly denoted. Such infinities simply exist and need to be "lived out."

We can consider such infinities that are so large that they cannot be abstractly defined.

We can try to get a handle on some stupendously large infinities by fabricating the following expression:

So we can consider

> Hyper [Hyper (infinities that are so large that they cannot be abstractly defined) [(infinities that are so large that they cannot be abstractly defined), (infinities that are so large that they cannot be abstractly defined)]] [[Hyper (infinities that are so large that they cannot be abstractly defined) [(infinities that are so large that they cannot be abstractly defined), (infinities that are so large that they cannot be abstractly defined)]], [Hyper (infinities that are so large that they cannot be abstractly defined) [(infinities that are so large that they cannot be abstractly defined), (infinities that are so large that they cannot be abstractly defined)]]] numbers

As for extremely large finite and then infinite Lorentz factors for spacecraft travel at the speed of light, we likely eventually can achieve the first few lowest Aleph numbers as Lorentz factors such as Aleph 0, Aleph 1, Aleph 2, and perhaps Aleph 3.

Traveling to ever greater Aleph number Lorentz factors will become possible as we develop our physics and engineering capabilities.

However, given the above digression, we can be assured that we will never run out of future infinite Lorentz factors.

We can also consider our relationship with GOD and realize that GOD has so many ontological transcendentals that we cannot even refer to this number, which is inexpressibly beyond any non-abstractly definable infinity.

Ontological transcendentals are properties such as ontological goodness, value, purpose, truth, and the like instead of ordinary transcendentals such as power, strength, immortality, and other existentially active properties.

We humans know of only a few ontological transcendentals such as ontological goodness, value, purpose, and perhaps a few others.

In summary, we will never run out of opportunities to explore reality.

ESSAY 21) Spacecraft Powered by Light-Speed Infinite Density Energy Walls

Herein we consider various infinities of Aleph number levels of space-time-mass-energy volumetric densities in N-D-Space-M-D-Time resulting in various singularity scenarios, transport effects, kinematic effects, topological effects, and so on effects.

One set of possibilities includes prospects for employing variously infinite energy wall densities in energy walls produced by local multiversal vacuum state tunnelings, universal vacuum state tunnelings, forestal vacuum state tunnelings, biospheric vacuum state tunnelings, and so on, as well as local hyperspatial vacuum state tunnelings.

Oddly enough, such local-realm vacuum-energy-state decays may produce infinitely greater energy densities than the maximum of those associated with the big bangs.

The reason for the above assertion is that vacuum-state decays would cause conflagration of the laws of physics beginning with the location of the vacuum-state tunneling, thus leading to such a maelstrom of acausality, and the scrambling of the order of cause and effect such that Mother Nature would be ontologically in accidental if not semi-substantial ways intolerant to the infinite energy density walls that would propagate outward at the speed of light. In a sense, Mother Nature would repel the energy wave, thus causing the density and total energy content of the energy wave to grow in levels of infinite energy densities and infinite total energy.

These infinite-density energy walls maxing out light-speed may manifest in a meta-light-speed content.

Accordingly, the energy walls would propagate at light-speed but also at meta-light-speed in a new definition of travel, along with a first new analogue of Lorentz factors.

As the local decay state energy production became more extreme, a meta-meta-light-speed may manifest with a second new analogue of Lorentz factors.

As the local-decay state energy production became still more extreme, a meta-meta-meta-light-speed may manifest with a third new analogue of Lorentz factors.

We refer to meta-light-speed, meta-meta-light-speed, meta-meta-meta light-speed, and the associated analogues of Lorentz factors as follows.

Meta-1-c, Meta-2-c, Meta-3-c, analogue-Lorentz-1, analogue-Lorentz-2, and analogue Lorentz-3.

However, with more extreme or deeper vacuum-energy state decays, we can formulate the following set which implies continuing the above series perhaps without limit:

$\{\{\Sigma(k = 1; k = \infty\uparrow):[u(\text{Meta-}k\text{-c, analogue-Lorentz-}k)]\}$ | size of $\infty\uparrow$ can grow with time in cosmic evolution$\}$

Such growth in these sets may be applicable to spacecraft propulsion in spacecraft achieving Meta-k-c, analogue-Lorentz-k properties, although the spacecraft propulsion systems would not be directly energized by cosmic decay events. Otherwise, the spacecraft would be annihilated by proximity to such decay-event-produced energy walls.

We can affix the following operator to formulas for gamma and related parameters to denote spacecraft remotely and safely located relative to energy walls but somehow having the meta-k-c, and analogue-Lorentz-k states manifesting in spacecraft kinetic energy states and wave-functions:

$\{\{\{\Sigma(k = 1; k = \infty\uparrow):[u(\text{Meta-}k\text{-c, analogue-Lorentz-}k)]\}$ | size of $\infty\uparrow$ can grow with time in cosmic evolution$\}$ manifesting in spacecraft kinetic energy states and wave-functions$\}$

Obviously, the energy wall development projects must not occur in realms occupied by sentient life-forms, but if we can find biologically sterile realms, then perhaps we can experiment with partial and limited energy wall production and study the phenomena to see if we can apply it to unrelated spacecraft propulsion programs.

Just some more crazy ideas from the keyboard of Jim Essig.

ESSAY 22) Unimaginably Great Infinite Spacecraft Lorentz Factors

Note that Hyper4(a, n) is equal to a tetrated n, or a raised to the power of itself n-1 times. The latter value is symbolically written as n subscript a.

For example,

> 3 EXP 4 = 81, but 4 subscript3 is approximately equal to 10 EXP (1,000,000,000,000).

Alternatively,

> 4 subscript 2 = 2 EXP 2 EXP 2 EXP 2 = 2 EXP [2 EXP [2 EXP 2]] = 2 EXP (2 EXP 4) = 2 EXP 16 = 65,536.

For example,

Hyper5(4, 4)is equal to 4 tetrated 4 tetrated 4 tetrated 4. This value is commonly referred to as 4 pentated 4.

Hyper 6, (4,4) is 4 pentated 4 pentated 4 pentated 4 and is also referred to as 4 hexataed 4.

Hyper 7, (4,4) is 4 hexated 4 hexated 4 hexated 4

And so on.

Aleph 0 is the infinite number of integers.

Aleph 1, according to the perhaps unprovable, and thus unfalsifiable, continuum hypothesis, is the number of real numbers that is greater than Aleph 0 by a multiplicative factor of infinity.

Aleph 2 is similarly greater than Aleph 1.

Aleph 3 is similarly greater than Aleph 2.

Aleph 4 is similarly greater than Aleph 3.

And so on.

In general, Aleph n = 2 EXP [Aleph (n-1)].

The number Ω is commonly stated as the least infinite positive integer or ordinal.

Now here is a real zinger:

So we can produce the abstraction of [Hyper Aleph Ω (Aleph Ω, Aleph Ω)].

We can go to ever greater infinities.

So we can consider

Hyper [Hyper Aleph Ω (Aleph Ω, Aleph Ω)] [[Hyper Aleph Ω (Aleph Ω, Aleph Ω)], [Hyper Aleph Ω (Aleph Ω, Aleph Ω)]] numbers

We can also consider uncountable infinities that are often considered greater than countable infinities.

As with Aleph numbers, we can consider a large evolved uncountable infinity in the following form:

So we can consider

Hyper [Hyper (large evolved uncountable infinity) [(large evolved uncountable infinity), (large evolved uncountable infinity)]] [[Hyper (large evolved uncountable infinity) [(large evolved uncountable infinity), (large evolved uncountable infinity)]], [Hyper (large evolved uncountable infinity) [(large evolved uncountable infinity), (large evolved uncountable infinity)]]] numbers

We can go to yet further extremes to produce the following number:

Aleph { Hyper [Hyper (large evolved uncountable infinity) [(large evolved uncountable infinity), (large evolved uncountable infinity)]] [[Hyper (large evolved uncountable infinity) [(large evolved uncountable infinity), (large evolved uncountable infinity)]], [Hyper (large evolved uncountable infinity) [(large evolved uncountable infinity), (large evolved uncountable infinity)]]]}

Now we may consider infinities so large that they cannot be abstractly denoted. Such infinities simply exist and need to be "lived out."

We can consider such infinities that are so large that they cannot be abstractly defined.

We can try to get a handle on some stupendously large infinities by fabricating the following expression.

So we can consider

Hyper [Hyper (infinities that are so large that they cannot be abstractly defined) [(infinities that are so large that they cannot be abstractly defined), (infinities that are so large that they cannot be abstractly defined)]] [[Hyper (infinities that are so large that they cannot be abstractly defined) [(infinities that are so large that they cannot be abstractly defined), (infinities that are so large that they cannot be abstractly defined)]], [Hyper (infinities that are so large that they cannot be abstractly defined) [(infinities that are so large that they cannot be abstractly defined), (infinities that are so large that they cannot be abstractly defined)]]] numbers

Now we can consider infinities of the next level up above infinities that are so large they cannot be abstractly defined.

In reality, we can consider still a yet higher or second level of infinities that are above said first higher level of infinities that are higher than the infinities that cannot be abstractly defined.

In reality still further, we can consider still a yet higher or third level of infinities that are above the second level of infinities that are above said first higher level of infinities that are higher than the infinities that cannot be abstractly defined.

We can likewise continue the series where we go all the way and eternally beyond a level of infinities as such having an ordinary index of the following:

Hyper [Hyper (infinities that are so large that they cannot be abstractly defined) [(infinities that are so large that they cannot be abstractly defined), (infinities that are so large that they cannot be abstractly defined)]] [[Hyper (infinities that are so large that they cannot be abstractly defined) [(infinities that are so large that they cannot be abstractly defined), (infinities that are so large that they cannot be abstractly defined)]], [Hyper (infinities that are so large that they cannot be abstractly defined) [(infinities that are so large that they cannot be abstractly defined), (infinities that are so large that they cannot be abstractly defined)]]]

I believe that GOD has more ontological transcendentals than the above-ordinated values.

Ontological transcendentals are properties such as value, purpose, ontological goodness, truth, and the like.

I believe ontological transcendentals are higher principles than active transcendentals, the latter of which are aspects such as power, strength, immutability, all-knowing, and the like.

However, we may also conjecture on GOD as having or being meta-ontological transcendentals, which are constructs we cannot understand. I believe GOD has more meta-ontological transcendentals than the largest level of infinities presented above. Meta-ontological transcendentals would be higher principles than ontological transcendentals.

I even believe that GOD as meta-meta-ontological transcendentals in numbers likely that qualitatively dwarf His number of meta-ontological transcendentals.

I even believe that GOD as meta-meta-meta-ontological transcendentals in numbers likely that qualitatively dwarf His number of Meta-meta-ontological transcendentals.

We will refer to these three constructs as Meta-1-ontological-transcendentals, Meta-2-ontological-transcendentals, Meta-3-ontological-transcendentals.

We can continue the series to produce the following conjecture of GOD having

> Meta-{ Hyper [Hyper (infinities that are so large that they cannot be abstractly defined) [(infinities that are so large that they cannot be abstractly defined), (infinities that are so large that they cannot be abstractly defined)]] [[Hyper (infinities that are so large that they cannot be abstractly defined) [(infinities that are so large that they cannot be abstractly defined), (infinities that are so large that they cannot be abstractly defined)]], [Hyper (infinities that are so large that they cannot be abstractly defined) [(infinities that are so large that they cannot be abstractly defined), (infinities that are so large that they cannot be abstractly defined)]]]}-ontological-transcendentals

Having gone this far we have not merely begun understanding how vast GOD is. We could continue the serial conjecture into eternity and we still would have only taken our first baby step in categorizing GOD on the path of eternity.

So reality is far larger and more varied than we could ever imagine.

However, here and now, let us settle for exploring our limited but still huge portion of our universe with perhaps Lorentz factors as great as 10 billion. A Lorentz factor of 10 billion will enable us to travel the current radius of the observable universe in about one-year ship-time. We can progress to ever greater Lorentz factors.

Now, within the observable universe, there are about 10 EXP 24 stars. This is about equal to the number of fine grains of table sugar that would cover the entire United States one hundred meters deep. The number of planets orbiting stars in our universe seems to be about ten times greater yet or equal to the number of fine grains of table sugar that would cover the entire United States one thousand meters deep. The number of moons, orbiting planets, orbiting stars is estimated to be ten times greater yet or equal to the number of fine grains of table sugar that would cover the entire United States ten thousand meters deep.

So let us go boldly into the future to explore the awesome realm GOD has made. The next 13.78 billion years will keep us very busy as we travel distances from Earth equal to one current cosmic light-cone radius unit. However, our universe extends far beyond the observable portion.

Oh what dreams will come!

ESSAY 23) Electromagnetic Symmetry-Breaking–Powered Spacecraft

A fascinating concept for spacecraft propulsion would include exothermic decays of symmetry broken electromagnetic waves and the further symmetry breaking of magnetism into two separate forces. The separate forces would be positive polar magnetism and negative polar magnetism.

Likewise, the electric force might be symmetry broken into positive electric force and negative electric force.

Ideally and absolutely necessarily, these symmetry breakings would be controllable and only be induced in special thermodynamic and topologically confinement vessels so that the symmetry-breaking reactions do not run away to trash the universe or greater multiverse.

Generally, such symmetry-breaking chambers would produce super-Planck power outputs, which would be useful only if the power could be quantum mechanically and/or classically imprinted on the kinetic energy wave-function of a spacecraft. Accordingly, the spacecraft would not experience any net force due to acceleration.

The N-Space-M-Time may optionally be as follows:

> flat, positively curved, negatively curved, positively curved and torsioned at one or more scales in arbitrary patterns including, but not limited to, fractals, negatively curved and torsioned at one or more scales in arbitrary patterns including, but not limited to, fractals, positively super-curved, negatively super-curved, positively super-curved and torsioned at one or more scales in arbitrary patterns including, but not limited to, fractals, negatively super-curved and torsioned at one or more scales in arbitrary patterns including, but not limited to, fractals, positively super-curved and super-torsioned at one or more scales in arbitrary patterns including, but not limited to, fractals, negatively super-curved and super-torsioned at one or more scales in arbitrary patterns including, but not limited to, fractals, positively curved and positively torsioned at one or more scales in arbitrary patterns including, but not limited to, fractals, negatively curved and positively torsioned at one or more scales in arbitrary patterns including, but not limited to, fractals, positively super-curved, negatively super-curved, positively super-curved and

positively torsioned at one or more scales in arbitrary patterns including, but not limited to, fractals, negatively super-curved and positively torsioned at one or more scales in arbitrary patterns including, but not limited to, fractals, positively super-curved and positively super-torsioned at one or more scales in arbitrary patterns including, but not limited to, fractals, negatively super-curved and positively super-torsioned at one or more scales in arbitrary patterns including, but not limited to, fractals, positively curved and negatively torsioned at one or more scales in arbitrary patterns including, but not limited to, fractals, negatively curved and negatively torsioned at one or more scales in arbitrary patterns including, but not limited to, fractals, positively super-curved, negatively super-curved, positively super-curved and negatively torsioned at one or more scales in arbitrary patterns including, but not limited to, fractals, negatively super-curved and negatively torsioned at one or more scales in arbitrary patterns including, but not limited to, fractals, positively super-curved and negatively super-torsioned at one or more scales in arbitrary patterns including, but not limited to, fractals, negatively super-curved and negatively super-torsioned at one or more scales in arbitrary patterns including, but not limited to, fractals, positively super-...-super-curved, negatively super-...-super-curved, positively super-...-super-curved and torsioned at one or more scales in arbitrary patterns including, but not limited to, fractals, negatively super-...-super-curved and torsioned at one or more scales in arbitrary patterns including, but not limited to, fractals, positively super-...-super-curved and super-...-super-torsioned at one or more scales in arbitrary patterns including, but not limited to, fractals, negatively super-...-super-curved and super-...-super-torsioned at one or more scales in arbitrary patterns including, but not limited to, fractals, positively curved and positively torsioned at one or more scales in arbitrary patterns including, but not limited to, fractals, negatively curved and positively torsioned at one or more scales in arbitrary patterns including, but not limited to, fractals, positively super-...-super-curved, negatively super-...-super-curved, positively super-...-super-curved and positively torsioned at one or more scales in arbitrary patterns including, but not limited to, fractals, negatively super-...-super-curved and positively torsioned at one or more scales in arbitrary patterns including, but not limited to, fractals, positively super-...-super-curved and positively super-...-super-torsioned at one or more scales in arbitrary patterns including, but not limited to, fractals, negatively super-...-super-curved and positively super-...-super-torsioned at one or more scales in arbitrary patterns including, but not limited to, fractals, positively curved and negatively torsioned at one or more scales in arbitrary patterns including, but not limited to, fractals, negatively curved and negatively torsioned at one or more scales in arbitrary patterns including, but not limited to, fractals, positively super-...-super-curved, negatively super-...-super-curved, positively super-...-super-curved and negatively torsioned at one or more scales in arbitrary patterns including, but not limited to, fractals, negatively super-...-super-curved and negatively torsioned at one or more scales in arbitrary patterns including, but not limited to, fractals, positively super-...-super-curved and negatively super-...-super-torsioned at one or more scales in arbitrary patterns including, but not limited to, fractals, negatively super-...-super-curved and negatively super-...-super-torsioned at one or more scales in arbitrary patterns including, but not limited to, fractals.

ESSAY 24) Travel at Existential Subunitary Portions of Light-Speed but at Light-Speed and Related Concepts

Herein we set the theoretical tone for considering the following roster items that are addressed in the four sections proceeding this section.

Traveling at light-speed but of subunitary fraction of c, not in the sense of velocity, but instead an existential portion of c.

A crystal made of light-speed as a collection and/or multiple connectivities of the constant c as a parameter of infinite Lorentz factor travel or as a realm or habitat therein.

Storage of light-speed or multiple copies of light-speed or a crystal of light-speed or multiple crystals of light-speed as a source of locomotion, energy supply, and the like for powering an infinite Lorentz factor spacecraft.

Teleportation and/or tunneling of light-speed, multiple copies of light-speed, light-speed crystals, light-speed realms, and the like as a means of conveyance of a spacecraft, habitat, and/or other mechanism at infinite Lorentz factors.

Expectation values of light-speed, multiple copies of light-speed, light-speed crystals, light-speed realms, and the like as a means of conveyance of a spacecraft, habitat, and/or other mechanisms at infinite Lorentz factors.

Prospects of transporting a cosmic light-cone portion of a universe at light-speed with respect to the rest of the universe and meaning thereof.

Prospects of transporting a cosmic light-cone portion of a multiverse at light-speed with respect to the rest of the multiverse and meaning thereof.

Prospects of transporting a cosmic light-cone portion of a forest at light-speed with respect to the rest of the forest and meaning thereof.

Prospects of transporting a cosmic light-cone portion of a biosphere at light-speed with respect to the rest of the biosphere and meaning thereof.

Prospects of transporting a cosmic light-cone portion of an N-D-Space-M-D-Time at light-speed with respect to the rest of the N-D-Space-M-D-Time and meaning thereof.

Prospects of transporting a cosmic light-cone portion of a hyperspace at light-speed with respect to the rest of the hyperspace and meaning thereof.

Now we can contemplate scenarios for which a light-speed spacecraft attains such an extreme infinite Lorentz factor that the craft runs out of velocity bandwidth to travel in ways contained by the speed of light.

Accordingly, the craft may begin to travel at an overflow light-speed as a kind of velocity deposit box.

We refer to the first level of overflow light-speed as overflow-1-light-speed.

Once the craft over-runs overflow-1-light-speed, the craft will begin to travel at a second level which we will refer to as overflow-2-light-speed.

Once the craft over-runs overflow-2-light-speed, the craft will begin to travel at a third level which we will refer to as overflow-3-light-speed.

We can continue without end to overflow-∞-↑-light-speed and beyond.

The associated Lorentz factors will infinitely "skyrocket."

It is plausible that the spacecraft Lorentz factors as such may perpetually grow from one infinity to the next higher infinity and forever repeat the general process.

A fascinating prospect would include a spacecraft traveling at such a greatly infinite Lorentz factor that the Lorentz factor, although a numerical quantity, is also a quality. By quality, I am referring to something distinct from and more sublime than the usual schemata of qualitatively different infinities such as countable and uncountable infinities and similar qualitative distinctions often used in mathematics. Instead, I am considering specific numeric infinities that are so great that they are also qualities in and of themselves.

Truly, light-speed travel will be enormously fun.

Imagine traveling at an overflow-k-light-speed, where the Lorentz factor is so infinite as to be an existential quality for which the spacecraft can travel a number of light-years in one-year ship-frame such that the number is both numerically infinite and a quality.

Imagine traveling at an overflow-k-light-speed where the Lorentz factor is so infinite as to be an existential quality for which the spacecraft can travel a number of years into the future in one-year ship-frame such that the number is both numerically infinite and a quality.

Imagine traveling at an overflow-k-light-speed where the Lorentz factor is so infinite as to be an existential quality for which the spacecraft can travel an effectively infinite multiple of the speed of light such that the multiple is both numerically infinite and a quality.

We affix the following operator to the lengthy formulas for gamma and related parameters presented below:

[w(Travel at an overflow-k-light-speed where the Lorentz factor can be so infinite such that the Lorentz factor is both numerically infinite and a quality and for which the craft can travel commensurately infinite; distances through space, years into the future, and effective multiples of the speed of light)]

Next, we consider scenarios for which a spacecraft, having reached the speed of light in the least infinite Lorentz factor, continues positive acceleration in its own reference frame to acquire progressively greater infinite Lorentz factors.

At some point, instead of piling on in a loose interpretation a kind of infinitesimal fractions of the speed of light to the craft's already light-speed velocity, instead of stating the craft begins to travel faster-than-light, we instead consider another process where the craft travels other-than-faster-than-light. Here, the notion of faster is replaced by something distinct from faster.

The first level of other-than-faster-than-light we refer to as other-than-faster-than-light-1.

As the craft maxes out other-than-faster-than-light-1, we continue to yet another breach into yet another metric of other-than-faster-than-light-2.

As the craft maxes out other-than-faster-than-light-2, we continue to yet another breach into yet another metric of other-than-faster-than-light-3.

We continue onward and eternally beyond the attainment of other-than-faster-than-light-$\infty\uparrow$.

The index, $\infty\uparrow$, stands for infinity and beyond.

Such craft may well experience infinite but increasing Lorentz factors while also morphing into other-than-Lorentz-factors-k as partners with other-than-faster-than-light-k where k ranges from one to open-ended infinities.

As the rock group the Cars sings, "Let the good times roll!"

So, options for attaining light-speed in impulse travel and further extensions are profound indeed.

We may affix the following operator to formulas for gamma and related parameters to denote options for attaining light-speed in impulse travel and further extensions:

> [u(Options for craft having attained light-speed impulse travel and experiencing infinite but increasing Lorentz factors while also morphing into states of other-than-Lorentz-factors-k as partners with other-than-faster-than-light-k, where *k* ranges from one to open-ended infinities)].

ESSAY 25)

Now we can consider arbitrarily great infinite aleph numbers.

For example, the concepts of aleph numbers were developed several generations ago near the turn of the nineteenth century to the twentieth century.

Accordingly, Aleph 0 is the infinite number of integers.

Aleph 1 is the number of real numbers according to the perhaps unprovable, and thus unfalsifiable, continuum hypothesis. Regardless, Aleph 1 = 2 EXP (Aleph 0).

Aleph 2 = 2 EXP (Aleph 1).

Aleph 3 = 2 EXP (Aleph 2).

We can go all the way to and beyond Aleph Ω, where omega is the least transfinite ordinal.

More generally, Aleph n = 2 EXP [Aleph (n-1)], where *n* is any finite or infinite counting number.

But wait, there is more to this little essay on infinities!

There are countable and uncountable infinities. Uncountable infinities are generally held to be larger than countable infinities, but the precise distinctions are a bit more complicated.

In essence, we can consider infinities of ever greater size. Just as we can always add one element to a set, we can also add one to any infinity no matter how large. So there is no largest infinity.

Now here is a real zinger!

Imagine an infinity that is so large that the infinity, although a quantity or number, is also a quality.

Now we can abstractly imagine an infinity that is so large that although a quantity and a quality, the quality that the infinity is becomes an infinite quality.

Now here is yet another zinger you will not want to miss!

Imagine an infinity that is so large that although a number and a quality, the infinity is a hyper-quality whereby the name, hyper-quality refers loosely to being related to quality as quality is related to quantity.

We can continue onward to constructs of infinities as hyper-hyper-qualities, which are to hyper-qualities as hyper-qualities are to qualities as qualities are to quantities.

We refer to hyper-qualities as hyper-1-qualities and hyper-hyper-qualities as hyper-2-qualities.

We can go all the way to hyper-Ω-qualities, and dare we say all the way to [hyper-(hyper-Ω-qualities)-qualities].

We can continue all the way to hyper-[hyper-(hyper-Ω-qualities)-qualities]-qualities.

From there, we can simply continue on with ever greater infinities.

Even if light-speed is an inviolable limit imposed by Mother Nature, could we ever attain spacecraft Lorentz factors greater than hyper-[hyper-(hyper-Ω-qualities)-qualities]-qualities?

I say never say never, given all unending future eternity to make scientific and technological progress.

We may affix the following operator to formulas for gamma and related parameters to denote the above unlimited potential for spacecraft to attain ever greater infinite Lorentz factors:

> [w[Spacecraft Lorentz factors in light-speed travel of aleph numbers; then uncountable infinities; then infinities that are quantities and qualities; then infinities that are quantities, qualities, and hyper-1-qualities; then infinities that are quantities, qualities, hyper-1-qualities; and hyper-2-qualities; then

infinities that are quantities, qualities, hyper-1-qualities; hyper-2-qualities and hyper-3-qualities; …; then infinities that are quantities, qualities, hyper-1-qualities; hyper-2-qualities, hyper-3-qualities, …, hyper-Ω-qualities; …; then infinities that are quantities, qualities, hyper-1-qualities; hyper-2-qualities, hyper-3-qualities, …, hyper-Ω-qualities, …, [hyper-(hyper-Ω-qualities)-qualities];

then infinities that are quantities, qualities, hyper-1-qualities; hyper-2-qualities, hyper-3-qualities, …, hyper-Ω-qualities, …, [hyper-(hyper-Ω-qualities)-qualities], …, hyper-[hyper-(hyper-Ω-qualities)-qualities]-qualities;

then infinities that are quantities, qualities, hyper-1-qualities; hyper-2-qualities, hyper-3-qualities, …, hyper-Ω-qualities, …, [hyper-(hyper-Ω-qualities)-qualities], …, {hyper-[hyper-(hyper-Ω-qualities)-qualities]-qualities}, …, hyper-{hyper-[hyper-(hyper-Ω-qualities)-qualities]-qualities}-qualities; and the algorithm goes on forever, world without end, Amen!]]

The beautiful thing is that we can, in principle, travel any of these infinities of light-years through space in one Planck time unit ship-frame, any of these infinities of years forward in time in one Planck time unit ship-frame, and any of these infinities of multiples of light-speed in an effective sense.

So fret not. Light-speed travel limits imposed by Mother Nature! It works out for the best anyhow.

Now as a spacecraft attained light-speed in the least infinite Lorentz factor, the craft may enter a kind of singularity for which eventually it cannot attain even an infinitesimal fraction of light-speed faster than light-speed as associated with the least infinite Lorentz factor. Thus, the spacecraft would have reached a singularity. We will refer to this first singularity as light-speed-singularity-1.

However, as the spacecraft continued to have increases in the infinity of the spacecraft Lorentz factor, the spacecraft may begin its journey to a second light-speed singularity, which unique temporal and relativistic aberrational kinematic and topological transport states and other states of the spacecraft relative to its realm of travel. We will refer to this second singularity as light-speed-singularity-2.

As the spacecraft continued to have increases in the infinity of the spacecraft Lorentz factor, the spacecraft may begin its journey to a third light-speed singularity, which unique temporal and relativistic aberrational kinematic and topological transport states and other states of the spacecraft relative to its realm of travel. We will refer to this third singularity as light-speed-singularity-3.

We can continue the process until a spacecraft may reach light-speed-singularity-Ω. Here, omega is the least transfinite ordinal.

We may continue forever as a spacecraft goes from one sum of (Ω + L) to another, where L is any finite or infinite counting number, even counting numbers on arbitrarily hyperextended number lines.

We may affix the following operator to formulas for gamma and related parameters to denote the above serial light-speed- singularity-k attainments:

$$\{w\{\text{Light-speed- singularity-k}| \text{ } k = 1, 2, 3, \ldots\}\}$$

What these singularities will enable is, as of now, almost a complete mystery. However, provided there just may be some or infinitely many non-Lorentz invariant effects, the set of effects may manifest altered states of consciousness and/or objective improvements on the ontology of the spacecraft and its crew.

The development of light-speed- singularity-k states as k increases may be as if traveling down an eternal roadway or tunnel for which there is ever a beacon of light at the end of the tunnel that morphs into a more extreme, whimsical, and beautiful luminary subtly guiding us ever farther into future eternity. This is, personally, a fascinating set of conjectures.

We can consider spacecraft that travel from one of the above light-speed- singularity-k to another, but this time, the associated Lorentz factor does not just merely increase from one infinite value to another but, instead, progressively takes on other nonnumeric aspects. In a sense, the associated values for gamma may become ontologically better, and, eventually, inherently of more value, purpose, and/or truth.

However, before the associated infinite Lorentz factors take on objective ontologically transcendental states, the Lorentz factors can become qualities, or other nonnumeric states, while at the same time becoming ever more infinite in numerical quantities.

As the Lorentz factors become more and more numerically infinite, they may take on repeated numerical bifurcation or multiple mathematical and numerical values and deeper states.

As the value of the Lorentz factors increases in infinite scales, the number of bifurcations, trifurcations, and so on can approach arbitrarily great infinite values thus alluding to these Lorentz factors having sub-ordinate infinite numbers of values, as well as states that become more metaphysical.

So do not fret about any light-speed limits, Special Relativity reigns supreme and allows an unlimited infinity of different states, phenomena, and the like. You can't go wrong with special relativity.

We can also take the ship-time, ship-acceleration, and total ship-energy derivatives and integrals of these states of one and multiple orders to further explore the ramifications and other aspects of these conjectured mechanisms.

We may affix the following operator to formulas for gamma and related parameters to denote the conjectures as provided above on multi-state values of gamma and the like:

> {u{{Attainments of light-speed-singularity-k and associated progressively infinite Lorentz factors and the onset of Lorentz factors having J non-numerical qualities and levels of metaphysical properties, as well as sub-ordinate L finite then infinite numbers of numerical values simultaneously| k = 1, 2, 3, …; J = 1, 2, 3, …; L = 1, 2, 3,} and arbitrary ship-time, ship-acceleration, ship-energy orders of differentiation and integration thereof}}

Here we consider prospects for which a spacecraft would reach the velocity of light in the least transfinite value for Lorentz factor.

We go on to consider light-speed travel that becomes more worked in by a spacecraft progressing to greater infinite Lorentz factors.

Eventually, a spacecraft would run out of existential room in light-speed and begin to travel into the constant space outside of light-speed or the constant c. Here, we are not referring to superluminal or faster-than-light travel, but travel in more sublime ways for which the constant c would be a launching point into abstract constant space but which would also be a real manifestation with associated thermodynamic and topological aspects.

Such an extra-light-speed-constant-space may have an abstract volume and extensity that is infinite while being infinite in abstract dimensionality.

In such a space, the constant c would be a mere geometric point of zero dimension.

What's more is that there may exist infinite numbers of extra-light-speed-constant spaces of similar and dissimilar sizes and numbers of dimensions. There may also exist hyperspatial analogues of extra-light-speed-constant-spaces.

We may affix the following operator to formulas for gamma and related parameters to indicate manifestations of extra-light-speed-constant-spaces. In this operator, we also consider single and arbitrary higher-order derivatives and integrals of rates of progression of a spacecraft in such extra-light-speed-constant-spaces with respect to gamma, ship-time, ship-frame acceleration, and spacecraft kinetic energy:

> [z[(Manifestations of extra-light-speed-constant-spaces for increasingly infinite spacecraft Lorentz factors) and single and arbitrary higher-order derivatives and integrals of rates of progression of a spacecraft in such extra-light-speed-constant-spaces with respect to gamma, ship-time, ship-frame acceleration, and spacecraft kinetic energy]]

ESSAY 26) Planck Acceleration Unit Spacecraft Reference Frames

Here we consider a spacecraft for which every constituent fundamental massive particle is accelerated at the Planck acceleration directly acting on each such particle so that there are no tidal or tension forces acting on the spacecraft.

We assume that the force acting on each such particle is acting in ways such that the force is confined to each particle.

Thus, the spacecraft will experience no tidal or stress forces and this experience is as if placement in a high energy space-time but where as far as stress and stain are concerned, the spacecraft is as if immersed in a space-time as if set to a zero-energy space-time.

Now imagine a spacecraft undergoing Planck acceleration perpetually for Aleph 0 years spacecraft time.

Now imagine a spacecraft undergoing Planck acceleration perpetually for Aleph 1 years spacecraft time.

Now imagine a spacecraft undergoing Planck acceleration perpetually for Aleph 2 years spacecraft time.

Now imagine a spacecraft undergoing Planck acceleration perpetually for Aleph 3 years spacecraft time.

Now imagine a spacecraft undergoing Planck acceleration perpetually for Aleph Ω years spacecraft time.

Note that Hyper4(a, n) is equal to a tetrated n, or a raised to the power of itself n-1 times. The latter value is symbolically written as n subscript a. For example,

3 EXP 4 = 81, but 4 subscript3 is approximately equal to 10 EXP (1,000,000,000,000).

Alternatively,

4 subscript 2 = 2 EXP 2 EXP 2 EXP 2 = 2 EXP [2 EXP [2 EXP 2]] = 2 EXP (2 EXP 4) = 2 EXP 16 = 65,536.

For example,

Hyper5(4, 4)is equal to 4 tetrated 4 tetrated 4 tetrated 4. This value is commonly referred to as 4 pentated 4.

Hyper 6, (4,4) is 4 pentated 4 pentated 4 pentated 4 and is also referred to as 4 hexataed 4.

Hyper 7, (4,4) is 4 hexated 4 hexated 4 hexated 4 and so on.

Aleph 0 is the infinite number of integers.

Aleph 1, according to the perhaps unprovable, and thus unfalsifiable, continuum hypothesis, is the number of real numbers that is greater than Aleph 0 by a multiplicative factor of infinity.

Aleph 2 is similarly greater than Aleph 1.

Aleph 3 is similarly greater than Aleph 2.

Aleph 4 is similarly greater than Aleph 3.

And so on.

In general, Aleph n = 2 EXP [Aleph (n-1)]

The number Ω is commonly stated as the least infinite positive integer or ordinal.

Now here is a real zinger.

So we can produce the abstraction of [Hyper Aleph Ω (Aleph Ω, Aleph Ω)].

We can go to ever greater infinities.

Now, there are mathematically recognized countable and uncountable infinities. Uncountable infinities are usually considered in contexts as being greater than countable infinities although the distinction is a bit more detailed.

So can you imagine the following infinities!

[Hyper Aleph Uncountable Infinities (Aleph Uncountable Infinities, Aleph Uncountable Infinities)]

We can go on to consider infinities so large that they have no mathematical language to explain them.

Now can you imagine the following infinities!

[Hyper Aleph (infinities so large that they have no mathematical language to explain them) [[Aleph (infinities so large that they have no mathematical language to explain them)], [Aleph (infinities so large that they have no mathematical language to explain them)]]]

Perhaps a spacecraft may obtain a Lorentz factor in the range of the latter infinities.

This means that such a spacecraft may travel as follows:

[Hyper Aleph (infinities so large that they have no mathematical language to explain them) [[Aleph (infinities so large that they have no mathematical language to explain them)], [Aleph (infinities so large that they have no mathematical language to explain them)]]] light-years through space in one-year ship-time.

This also means that such a spacecraft may travel as follows:

[Hyper Aleph (infinities so large that they have no mathematical language to explain them) [[Aleph (infinities so large that they have no mathematical language to explain them)], [Aleph (infinities so large that they have no mathematical language to explain them)]]] years into the future in one-year ship-time.

This means that such a spacecraft may travel an effective multiple of

[Hyper Aleph (infinities so large that they have no mathematical language to explain them) [[Aleph (infinities so large that they have no mathematical language to explain them)], [Aleph (infinities so large that they have no mathematical language to explain them)]]] the velocity of light.

Now imagine a spacecraft undergoing the Planck acceleration for mathematical language to explain them)], [Aleph (infinities so large that they have no mathematical language to explain them)]]] years' ship-time.

Now at the Planck distance and time scales, the greatest possible acceleration is the Planck acceleration at:

$A_P = L_p/t_p{}^2$.

Since the Planck length and the Planck time are equal to $\{\hbar\, G/C^3\}^{1/2}$ and $\{\hbar\, G/C^5\}^{1/2}$, respectively, where $\hbar$, G, and C are the reduced Planck constant, the universal gravitational constant, and the speed of light, respectively, the Planck acceleration outside of the smallest possible black hole becomes:

$$A_p = \{\{\{\hbar\, G/C^3\}^{1/2}\}/\{\{\{\hbar\, G/C^5\}^{1/2}\}^2\}\} = 5.6 \times 10^{51}\ \text{m/s}^2.$$

The Planck acceleration is the acceleration experienced by a minute test particle just outside of the event horizon of the smallest possible black hole. However, by reducing the inertial mass of the spacecraft perhaps an acceleration much greater than the Planck acceleration is possible.

The above conjecture has ramifications for black hole physics. Perhaps inside the event horizon of a black hole, the inertial mass of the infalling matter is reduced within the black hole, and the matter is thereby "energized" or photonized and gradually converted to mass-less species in a process that occurs along a continuum or that is discretely or continuous in progression. This may indicate a breakdown in the laws of physics in a modeled sense, but also in reality. So perhaps upon reaching the central singularity, the matter is converted to pure electromagnetic and/or gravitational radiation.

Again, note the following relativistic rocket equations for constant acceleration:

$$t = (C/a)\sinh(aT/C) = [[(d/C)^2] + (2d/a)]^{1/2}$$

$$d = [(C^2)/a]\,[[\cosh(aT/C)] - 1] = [(C^2)/a]\,\{\{[1 + [(at/C)^2]]^{1/2}\} - 1\}$$

$$v = C\tanh(aT/C) = (at)\,/\,\{[1 + [(at/C)^2]]^{1/2}\}$$

$$T = (C/a)\,\text{inversesinh}\,(at/C) = (C/a)\,\text{inversecosh}\,[[ad/(C^2)] + 1]$$

$$\gamma = \cosh(aT/c) = [1 + [(at/C)^2]]^{1/2} = [ad/(C^2)] + 1$$

$$d = [C^2)/a]\,[[\cosh(aT/C)] - 1] = [(C^2)/a]\,\{\{[1 + [(at/C)^2]]^{1/2}\} - 1\}$$

For an acceleration equal to one Planck acceleration unit for one-year ship-time, we obtain a value for distance traveled equal to the following:

$$d = [(C^2)/a]\,[[\cosh(aT/C)] - 1] = [(C^2)/A_p]\,[[\cosh(A_p\,T/C)] - 1] = [(C^2)/\,\{\{\{\hbar\,G/C^3\}^{1/2}\}/\{\{\{\hbar\,G/C^5\}^{1/2}\}^2\}\}]\,[[\cosh(\{\{\{\hbar\,G/C^3\}^{1/2}\}/\{\{\{\hbar\,G/C^5\}^{1/2}\}^2\}\}\,T/C)] - 1] = [(C^2)/\,[5.6 \times 10^{51}\ \text{m/s}^2]]\,[[\cosh([5.6 \times 10^{51}\ \text{m/s}^2]((31{,}000{,}000\ \text{s})/(300{,}000{,}000\ \text{m/s}))] - 1]$$

A huge value, indeed!

Gamma is equal to:

$$\gamma = \cosh(aT/c) = \cosh(A_pT/c) = \cosh\{\{\{\{\hbar \ G/C^3\}^{1/2}\}/\{\{\{ \ \hbar \ G/C^5\}^{1/2}\}^2\}\}[(31{,}000{,}000 \ s)/(300{,}000{,}000 \ m/s)]\}$$

$$= \cosh\{[5.6 \times 10^{51} \ m/s^2] \ [(31{,}000{,}000 \ s)/(300{,}000{,}000 \ m/s)]\}$$

Another huge value indeed!

A method of somehow extracting zero-point field energy would be ideal if the latent energy density of the field is very high especially if equal to the theoretical Planck energy density

$$= \{[C^7]/\{[h/(2 \ \pi)] \ (G^2)\}\}$$

Here, C equals the speed of light, h is the Planck constant, and G is the Newtonian gravitational constant. The Planck energy density is equal to

$$[9 \times 10^{16}] \ [5.1 \times 10^{96}] \ [Joule/[meter^3]] = 45.9 \times 10^{112} \ \text{joules per cubic meter}$$

But this is another story altogether.

Now let us substitute T with

[Hyper Aleph (infinities so large that they have no mathematical language to explain them) [[Aleph (infinities so large that they have no mathematical language to explain them)], [Aleph (infinities so large that they have no mathematical language to explain them)]]].

This leads to a fantastically infinite Lorentz factor.

Now for a universe, multiverse, forest, biosphere, and the like, or a hyperspace quantized at a subunitary fraction, (1/m), of both the Planck distance and time scales, and where the speed of light is the same as in our universe, the greatest possible acceleration is the m times the Planck acceleration at:

$$(m)(A_P) = (m)[L_p/(t_p{}^2)]$$

Since the Planck length and the Planck time are equal to $\{\hbar \ G/C^3\}^{1/2}$ and $\{ \ \hbar \ G/C^5\}^{1/2}$, respectively, where $\hbar$, G, and C are the reduced Planck constant, the universal gravitational constant, and the speed of light, respectively, the m times Planck acceleration outside of the smallest possible black hole in these realms becomes:

$$(m)(A_p) = (m)\{\{\{\hbar \ G/C^3\}^{1/2}\}/\{\{\{ \ \hbar \ G/C^5\}^{1/2}\}^2\}\} = (m)(5.6 \times 10^{51}) \ m/s^2$$

For an acceleration equal to m times one Planck acceleration unit for one-year ship-time, we obtain a value for distance traveled equal to:

$$d = [(C^2)/a] \ [[\cosh(aT/C)] - 1] = (m)[(C^2)/A_p] \ [[\cosh(A_p \ T/C)] - 1] = (m)[(C^2)/ \ \{\{\{\hbar \ G/C^3\}^{1/2}\}/\{\{\{ \ \hbar \ G/C^5\}^{1/2}\}^2\}\}] \ [[\cosh(\{\{\{\hbar \ G/C^3\}^{1/2}\}/\{\{\{ \ \hbar \ G/C^5\}^{1/2}\}^2\}\} \ T/C)] - 1] =$$

(m)[(C^2)/ \; [5.6 \; x \; 10^{51} \; m/s^2]] \; [[\cosh([5.6 \; x \; 10^{51} \; m/s^2]((31,000,000 \; s)/(300,000,000 \; m/s))] - 1]

A huge value, indeed!

Gamma is equal to:

$$\gamma = \cosh(aT/c) = (m)[\cosh(A_pT/c)] = (m)\{\cosh\{\{\{\{\hbar \; G/C^3\}^{1/2}\}/\{\{\{ \; \hbar \; G/C^5\}^{1/2}\}^2\}\}[(31,000,000 \; s)/ \; (300,000,000 \; m/s)]\}\}$$

$$= (m)\{\cosh\{[5.6 \; x \; 10^{51} \; m/s^2] \; [(31,000,000 \; s)/ \; (300,000,000 \; m/s)]\}\}$$

Another huge value indeed!

Now let us substitute T with the following:

[Hyper Aleph (infinities so large that they have no mathematical language to explain them) [[Aleph (infinities so large that they have no mathematical language to explain them)], [Aleph (infinities so large that they have no mathematical language to explain them)]]]

This leads to a fantastically infinite Lorentz factor.

Now for a universe, multiverse, forest, biosphere, and the like or a hyperspace quantized at a subunitary fraction, (1/m), of both the Planck distance and time scales, and where the speed of light is greater than it is in our universe by a superunitary multiple, n, the greatest possible acceleration is the m times the Planck acceleration at:

$$(m)(A_P) = (m)[L_p/(t_p^{\;2})]$$

Since the Planck length and the Planck time are equal to $\{\hbar \; G/C^3\}^{1/2}$ and $\{ \; \hbar \; G/C^5\}^{1/2}$, respectively, where $\hbar$, G, and C are the reduced Planck constant, the universal gravitational constant, and the speed of light, respectively, the m times Planck acceleration outside of the smallest possible black hole in these realms becomes:

$$(m)(A_p) = (m)\{\{\{\hbar \; G/(nC)^3\}^{1/2}\}/\{\{\{ \; \hbar \; G/(nC)^5\}^{1/2}\}^2\}\} = (m)(5.6 \; x \; 10^{51}) \; m/s^2.$$

For an acceleration equal to m times one Planck acceleration unit for one-year ship-time, we obtain a value for distance traveled equal to the following:

$$d = [((nC)^2)/a] \; [[\cosh(aT/(nC))] - 1] = (m)[((nC)^2)/A_p] \; [[\cosh(A_p \; T/(nC))] - 1] = (m)[((nC)^2)/ \; \{\{\{\hbar \; G/(nC)^3\}^{1/2}\}/\{\{\{ \; \hbar \; G/(nC)^5\}^{1/2}\}^2\}\}] \; [[\cosh(\{\{\{\hbar \; G/(nC)^3\}^{1/2}\}/\{\{\{ \; \hbar \; G/(nC)^5\}^{1/2}\}^2\}\} \; T/(nC))] - 1] = (m)[((nC)^2)/ \; [5.6 \; x \; 10^{51} \; m/s^2]] \; [[\cosh([5.6 \; x \; 10^{51} \; m/s^2]((31,000,000 \; s)/(300,000,000 \; m/s)))] - 1]$$

A huge value, indeed!

Gamma is equal to the following:

$$\gamma = \cosh(aT/c) = (m)[\cosh(A_pT/c)] = (m)\{\cosh\{\{\{\{\hbar\ \ G/(nC)^3\}^{1/2}\}/\{\{\{\ \ \hbar\ \ G/(nC)^5\}^{1/2}\}^2\}\}[(31,000,000\ s)/ [[(n)(300,000,000)]\ m/s]]\}\}$$

$$= (m)\{\cosh\{[5.6 \times 10^{51}\ m/s^2]\ [(31,000,000\ s)/ [[(n)(300,000,000)]\ m/s]]\}\}$$

Another huge value, indeed!

Now let us substitute *T* with the following:

[Hyper Aleph (infinities so large that they have no mathematical language to explain them) [[Aleph (infinities so large that they have no mathematical language to explain them)], [Aleph (infinities so large that they have no mathematical language to explain them)]]]

This leads to a fantastically infinite Lorentz factor.

ESSAY 27) Perhaps Like a Turtle on Top of a Turtle on Top of a Turtle and So On, a Universe Encompassing a Universe Encompassing a Universe and So On Accessed by Light-Speed Spacecraft

A fascinating prospect involves the notion that our universe is as if a particle in the next level of the realm upward.

Another fascinating prospect involves the notion that said next level of realm upward or realm-upward-1 is a particle in another realm, which we will refer to as realm-upward-2.

Still another fascinating prospect involves the notion that the level realm-upward-2 is but a particle in realm-upward-3.

The number of higher levels as such may be infinite and indefinite.

Likewise, a fundamental particle in our universe or broader level may be as if an entire realm in the next level down, which we will refer to as realm-downward-1.

Likewise, realm-downward-2 may be as if a particle in realm-downward-1.

Likewise, realm-downward-3 may be as if a particle in realm-downward-2.

Regardless, because of the chronological protection conjecture, we expect that the maximum velocity that can manifest in translational travel from one location to another location in our universe is the speed of light. However, it may be possible that velocities ever so slightly greater than light-speed in our universe may be possible as long as these velocities do not enable backward time travel.

So travel into any object of the set realm-upward-k or into any object of the set realm-downward-h, where *k* and *h* are any positive integers, cannot manifest faster-than-light travel enabling backward time travel.

So we have the following differential equation that must be strictly adhered to:

> *d* (distance traveled from a location starting from our universe and into realm-upward-k or realm-downward-h and then back into another location in our universe)/dt ≤ c or ≤ (c + e), where *e* is vanishingly small

However, the following temporally integrated velocity can be as great as we would like:

> [∫(Effective velocity traveled from a location starting from our universe and into realm-upward-k or realm-downward-h and then back into another location in our universe) dt,ship] → ∞ cosmic light-cone radius units

> [∫(Effective velocity traveled from a location starting from our universe and into realm-upward-k or realm-downward-h and then back into another location in our universe) dt,background] ≥ (0 → ∞) cosmic light-cone radius units

We can also formulate other analogues of velocity as follows:

> *d* (Distance traveled from a location starting from our universe and into realm-upward-k or realm-downward-h and then back into another location in our universe)/dγ

> *d* (Distance traveled from a location starting from our universe and into realm-upward-k or realm-downward-h and then back into another location in our universe)/d(acceleration ship-frame)

> *d* (Distance traveled from a location starting from our universe and into realm-upward-k or realm-downward-h and then back into another location in our universe)/d(acceleration background-frame)

> *d* (Distance traveled from a location starting from our universe and into realm-upward-k or realm-downward-h and then back into another location in our universe)/d(ship kinetic energy)

Likewise, we can formulate analogues of distance traveled as follows:

> [∫(Effective velocity traveled from a location starting from our universe and into realm-upward-k or realm-downward-h and then back into another location in our universe) dγ]

> [∫(Effective velocity traveled from a location starting from our universe and into realm-upward-k or realm-downward-h and then back into another location in our universe) d(ship-frame acceleration)]

∫∫(Effective velocity traveled from a location starting from our universe and into realm-upward-k or realm-downward-h and then back into another location in our universe) d(ship kinetic energy)]

The N-Space-M-Time may optionally be as follows:

flat, positively curved, negatively curved, positively curved and torsioned at one or more scales in arbitrary patterns including, but not limited to, fractals, negatively curved and torsioned at one or more scales in arbitrary patterns including, but not limited to, fractals, positively super-curved, negatively super-curved, positively super-curved and torsioned at one or more scales in arbitrary patterns including, but not limited to, fractals, negatively super-curved and torsioned at one or more scales in arbitrary patterns including, but not limited to, fractals, positively super-curved and super-torsioned at one or more scales in arbitrary patterns including, but not limited to, fractals, negatively super-curved and super-torsioned at one or more scales in arbitrary patterns including, but not limited to, fractals, positively curved and positively torsioned at one or more scales in arbitrary patterns including, but not limited to, fractals, negatively curved and positively torsioned at one or more scales in arbitrary patterns including, but not limited to, fractals, positively super-curved, negatively super-curved, positively super-curved and positively torsioned at one or more scales in arbitrary patterns including, but not limited to, fractals, negatively super-curved and positively torsioned at one or more scales in arbitrary patterns including, but not limited to, fractals, positively super-curved and positively super-torsioned at one or more scales in arbitrary patterns including, but not limited to, fractals, negatively super-curved and positively super-torsioned at one or more scales in arbitrary patterns including, but not limited to, fractals, positively curved and negatively torsioned at one or more scales in arbitrary patterns including, but not limited to, fractals, negatively curved and negatively torsioned at one or more scales in arbitrary patterns including, but not limited to, fractals, positively super-curved, negatively super-curved, positively super-curved and negatively torsioned at one or more scales in arbitrary patterns including, but not limited to, fractals, negatively super-curved and negatively torsioned at one or more scales in arbitrary patterns including, but not limited to, fractals, positively super-curved and negatively super-torsioned at one or more scales in arbitrary patterns including, but not limited to, fractals, negatively super-curved and negatively super-torsioned at one or more scales in arbitrary patterns including, but not limited to, fractals, positively super-...-super-curved, negatively super-...-super-curved, positively super-...-super-curved and torsioned at one or more scales in arbitrary patterns including, but not limited to, fractals, negatively super-...-super-curved and torsioned at one or more scales in arbitrary patterns including, but not limited to, fractals, positively super-...-super-curved and super-...-super-torsioned at one or more scales in arbitrary patterns including, but not limited to, fractals, negatively super-...-super-curved and super-...-super-torsioned at one or more scales in arbitrary patterns including, but not limited to, fractals, positively curved and positively torsioned at one or more scales in arbitrary patterns including, but not limited to, fractals, negatively curved and positively torsioned at one or more scales in arbitrary patterns including, but not limited to, fractals, positively super-...-super-curved, negatively super-...-super-curved, positively super-...-super-curved and positively torsioned at one or more scales in arbitrary patterns including, but not limited to, fractals, negatively super-...-super-curved and positively torsioned at one or more scales in arbitrary patterns including, but not limited to, fractals, positively super-...-super-curved and positively super-...-super-torsioned at one or more scales in arbitrary patterns including, but not limited to, fractals, negatively super-...-super-curved and positively super-...-super-torsioned at one or more scales in arbitrary patterns including, but not limited to, fractals, positively curved and negatively torsioned at one or more scales in arbitrary patterns including, but not limited to, fractals, negatively curved and negatively torsioned at one or more scales in arbitrary patterns including, but not limited to, fractals, positively super-...-super-curved, negatively super-...-super-curved, positively super-...-super-curved and negatively torsioned at one or more scales in arbitrary patterns including, but not limited to, fractals, negatively super-...-super-curved and negatively torsioned at one or more scales in arbitrary patterns including, but not limited to, fractals, positively super-...-super-curved and negatively super-...-super-torsioned at one or more scales in arbitrary patterns including, but not limited to, fractals, negatively super-...-super-curved and negatively super-...-super-torsioned at one or more scales in arbitrary patterns including, but not limited to, fractals.

As for hyperspatial applications, we consider such hyperspaces can be N-Time-M-Space, where N is an integer greater than or equal to 3 and M is any counting number.

Alternatively, N can be any rational number equal to 3 or greater and M is any rational number greater than or equal to 1.

As another set of scenarios, N can be any irrational number equal to 3 or greater and M is any irrational number greater than or equal to 1.

Alternatively, N can be any rational number greater than 0 and M is any rational number greater than 0.

As another set of scenarios, N can be any irrational number greater than 0 and M is any irrational number greater than 0.

Each of the following roster elements can be the basis of expressions of differentiated functions for power, energy, velocity, momentum, gamma, and the like associated with the particles and/or spacecraft by differentiation in the following manners:

d f/dt,1 d(d f/dt,1)/dt,2 d(d f/dt,2)/dt,1 d[d(d f/dt,1)/dt,2]/dt,3 d[d(d f/dt,1)/dt,3]/dt,2

d[d(d f/dt,2)/dt,1]/dt,3 d[d(d f/dt,2)/dt,3]/dt,1 d[d(d f/dt,3)/dt,1]/dt,2 d[d(d f/dt,3)/dt,2]/dt,1 d f/dT,1 d(d f/dT,1)/dT,2 d(d f/dT,2)/dT,1 d[d(d f/dT,1)/dT,2]/dT,3 d[d(d f/dT,1)/dT,3]/dT,2

d[d(d f/dT,2)/dT,1]/dT,3 d[d(d f/dT,2)/dT,3]/dT,1 d[d(d f/dT,3)/dT,1]/dT,2 d[d(d f/dT,3)/dT,2]/dT,1

T in each case is a timelike blend of various sorts of multiple time dimensions that appropriately and usefully enables derivation in one iteration.

We can go further to consider the following:

d f/d$\top$,1 d(d f/d$\top$,1)/d$\top$,2 d(d f/d$\top$,2)/d$\top$,1 d[d(d f/d$\top$,1)/d$\top$,2]/d$\top$,3 d[d(d f/d$\top$,1)/d$\top$,3]/d$\top$,2

d[d(d f/d$\top$,2)/d$\top$,1]/d$\top$,3 d[d(d f/d$\top$,2)/d$\top$,3]/d$\top$,1 d[d(d f/d$\top$,3)/d$\top$,1]/d$\top$,2 d[d(d f/d$\top$,3)/d$\top$,2]/d$\top$,1

$\top$ in each case is a timelike blend of timelike blends of various sorts of multiple time dimensions that appropriately and usefully enables derivation in one iteration.

By the same token, we can also integrate functions for power, energy, velocity, momentum, gamma, and the like by respect to multiple time dimensions as follows:

∫ f(dt,1) ∫ ∫d f(dt,1)(dt,2) ∫ ∫d f(dt,2)(dt,1) ∭ f(dt,1)(dt,2)(dt,3) ∭ f(dt,1)(dt,3)(dt,2)

∭ f(dt,2)(dt,1)(dt,3) ∭ f(dt,2)(dt,3)(dt,1) ∭ f(dt,3)(dt,1)(dt,2) ∭ f(dt,3)(dt,2)(dt,1)

∫ f(dT,1) ∬ f(dT,1) (dT,2) ∬ f(dT,2) ((dT,1)) ∭ f(dT,1) (dT,2) (dT,3) ∭ f(dT,1) (dT,3) (dT,2)

∭ f(dT,2) (dT,1) (dT,3) ∭ f(dT,2) (dT,3) (dT,1) ∭ f(dT,3) (dT,1) (dT,2) ∭ f(dT,3) (dT,2) (dT,1)

We can go further to consider the following:

∫ f(d$\top$,1) ∫ ∫ f(d$\top$,1)(d$\top$,2) ∫ ∫ f(d$\top$,2)(d$\top$,1) ∫ ∫ ∫ f(d$\top$,1)(d$\top$,2)(d$\top$,3) ∫ ∫ ∫ f(d$\top$,1)(d$\top$,3)(d$\top$,2)

∫ ∫ ∫ f(d$\top$,2)(d$\top$,1)(d$\top$,3) ∫ ∫ ∫ f(d$\top$,2)(d$\top$,3)(d$\top$,1) ∫ ∫ ∫ f(d$\top$,3)(d$\top$,1)(d$\top$,2) ∫ ∫ ∫ f(d$\top$,3)(d$\top$,2)(d$\top$,1)

We can continue to yet higher-order derivatives and integrals. Suffice it to say that the general patterns continue for higher numbers of dimensions.

We can also consider higher-order blends of time and thus are not limited to the blends defined by *T* and *$\top$*. We chose these two symbols for their resemblance to the letter *t*.

So. Absolutely! Absolutely! Do not fret about any light-speed limits! Even with such limits, the opportunities are far greater than we have ever previously considered for travel distances, especially with respect to ship-time.

ESSAY 28) R-Tuple Coordinate Travel and Light-Speed

Now we are all generally aware of the concept of the real number line and its extension into positive and negative infinity.

However, most of us have not heard of the concept of a hyperextended number line.

I am not the originator of the concept of a hyperextended number line, but I have been refining the concepts of hyperintegers and hyperreal numbers that are larger than conventional number-line–based infinities. Although not the same as the conventional concept of uncountable infinities that are generally held to be a larger class of infinities than countable infinities, for present purposes, we refer to hyperintegers and hyperreal numbers as high-end uncountable infinities.

However, we may also consider still "longer" number lines, perhaps by the lexicography of hyper-hyperextended number lines. We refer to hyperextended number lines, hyperintegers, and hyperreal numbers as hyper-1-extended number lines, hyper-1-integers, and hyper-1-real numbers. We refer to hyper-hyperextended number lines, hyper-hyperintegers, and hyper-hyperreal numbers as hyper-2-extended number lines, hyper-2-integers, and hyper-2-real numbers. Likewise, we can consider the pattern continuing forever from hyper-3-extended number lines, hyper-3-integers, and hyper-3-real numbers; hyper-4-extended number lines, hyper-4-integers, and hyper-4-real numbers; … ; hyper-Ω-extended number lines, hyper-Ω-integers, and hyper-Ω-real numbers.

Now we can consider the distinction between a geometric zero-dimensional point, a one-dimensional line segment that is a portion of a line of finite extent, a number line that extends infinitely in both the positive and negative directions. However, have you ever considered an abstract geometric object that is to a conventional number line as a conventional number line is to a line segment? How about an abstract geometric object of a second-class hyper-line that would be to a first-class hyper-line which would be to a conventional number line as a conventional number line is to a line segment. We can likewise consider a third-class hyper-line, a fourth-class hyper-line, and so on to an Ωth-class hyper-line, and forever after onward from there.

Note that we are not considering objects with more than one dimension but, instead, another more abstract principle.

Now if Mother Nature follows abstract space, prospects for travel beyond the extent of a conventionally infinite linear distance manifest.

For example, we can imagine traveling a natural distance commensurate with advanced travel along a hyper-1-extended number line in space travel, then advanced travel along a hyper-2-extended number line, then advanced travel along a hyper-3-extended number line, and so on.

Imagine traveling a hyper-1-integer number of light-years through space in a one-second ship-frame, or a hyper-2-integer number of light-years through space in a one-second ship-frame, or a hyper-3-integer number of light-years through space in a one-second ship-frame, and so on, or a hyper-1-integer number of years into the future in a one-second ship-frame, or a hyper-2-integer number of years into the future in a one-second ship-frame, or a hyper-3-integer number of years into the future in a one-second ship-frame, and so on, all at respective effectively hyper-1-integer, hyper-2-integer, hyper-3-integer, and so on, multiples of light-speed.

The hyper-k-number lines can be used for models of extensities of eternities into the future as well.

We may affix the following operator to denote the above travel aspects to formulas for gamma and related parameters:

> [w(Mechanisms of "longer" number lines perhaps by the lexicography of hyperextended number lines while referring to hyperextended number lines, hyperintegers, and hyperreal numbers as hyper-1-extended number lines, hyper-1-integers, and hyper-1-real numbers while in turn referring to hyper-hyperextended number lines, hyper-hyperintegers, and hyper-hyperreal numbers as hyper-2-extended number lines, hyper-2-integers, and hyper-2-real numbers while considering the pattern continuing forever from: hyper-3-extended number lines, hyper-3-integers, and hyper-3-real numbers; hyper-4-extended number lines, hyper-4-integers, and hyper-4-real numbers; ... ; hyper-Ω-extended number lines, hyper-Ω-integers, and hyper-Ω-real numbers)]:[z(Mechanisms of an abstract geometric object that is to a conventional number line as a conventional number line is to a line segment; and further an abstract geometric object of second-class hyper-line which would be to a first-class hyper-line which would be to a conventional number line as a conventional number line is to a line segment; while in turn likewise considering third-class hyper-line, a fourth-class hyper-line, and so on to an Ωth class hyper-line and forever after onward from there)]:[u(Mechanisms of traveling a hyper-1-integer number of light-years through space in one second ship-frame, or a hyper-2-integer number of light-years through space in one second ship-frame, or a hyper-3-integer number of light-years through space in one second ship-frame, and so on, or a hyper-1-integer number of years into the future in one second ship-frame, or a hyper-2-integer number of years into the future in one-second ship-frame, or a hyper-3-integer number of years into the future in one-second ship-frame, and so on, all at respective effectively hyper-1-integer, hyper-2-integer, hyper-3-integer, and so on multiples of light-speed)].

We can also consider analogues in three or more spatial distances and one or more temporal dimensions.

Accordingly, we may affix the following operator to the formulas for gamma and related parameters to denote options for traveling in space-times so extended:

[w(mechanisms of "longer" number lines in "larger" space-times coordinate systems perhaps by the lexicography of the axial hyperextended number lines while referring to hyperextended number lines, hyperintegers, and hyperreal numbers as hyper-1-extended number lines, hyper-1-integers, and hyper-1-real numbers while in turn referring to hyper-hyperextended number lines, hyper-hyperintegers, and hyper-hyperreal numbers as hyper-2-extended number lines, hyper-2-integers, and hyper-2-real numbers while considering the pattern continuing forever from: hyper-3-extended number lines, hyper-3-integers, and hyper-3-real numbers; hyper-4-extended number lines, hyper-4-integers, and hyper-4-real numbers; … ; hyper-Ω-extended number lines, hyper-Ω-integers, and hyper-Ω-real numbers)]:[z(Mechanisms of an abstract geometric object that is to a conventional number line as a conventional number line is to a line segment; and further an abstract geometric object of second-class hyper-line which would be to a first-class hyper-line which would be to a conventional number line as a conventional number line is to a line segment; while in turn likewise considering a third-class hyper-line, a fourth-class hyper-line, and so on to an Ωth-class hyper-line and forever after onward from there)]:[u(mechanisms of traveling a hyper-1-integer number of light-years through space in one-second ship-frame, or a hyper-2-integer number of light-years through space in one-second ship-frame, or a hyper-3-integer number of light-years through space in one-second ship-frame, and so on, or a hyper-1-integer number of years into the future in one-second ship-frame, or a hyper-2-integer number of years into the future in one-second ship-frame, or a hyper-3-integer number of years into the future in one-second ship-frame, and so on, all at respective effectively hyper-1-integer, hyper-2-integer, hyper-3-integer, and so on multiples of light-speed)]:[r(Considerations that such hyper-k-extended hyperspaces or hyper-space-times of travel may have hyper-h-integers of dimensions, where *k* and *h* may approach and exceed Ω, the least transfinite ordinal)]

ESSAY 29) Hyper-Positive Numbers.

Now we are all well aware of positive and negative numbers, and those who have done college math know about imaginary and complex numbers.

However, most folks, even almost all mathematicians, have never conceptualized numbers of hyper-positivity.

Numbers of hyper-positivity would be to positive numbers as positive numbers are to negative numbers.

Number 1 of hyper-positivity would be hyper-positive relative to the positive integer 1.

Number 2 of hyper-positivity would be hyper-positive relative to the positive integer 2.

Number 3 of hyper-positivity would be hyper-positive relative to the positive integer 3 and so on.

We refer to numbers of hyper-positivity as numbers-of-hyper-positivity-1.

Numbers of hyper-hyper-positivity would be to numbers of hyper-positivity as numbers of hyper-positivity would be to positive numbers as positive numbers are to negative numbers.

Number 1 of hyper-hyper-positivity would be hyper-hyper-positive relative to the positive integer 1.

Number 2 of hyper-hyper-positivity would be hyper-hyper-positive relative to the positive integer 2.

Number 3 of hyper-hyper-positivity would be hyper-hyper-positive relative to the positive integer 3 and so on.

We refer to numbers of hyper-hyper-positivity as numbers-of-hyper-positivity-2.

We can continue the digression to numbers of hyper-hyper-hyper-positivity or numbers-of-hyper-positivity-3, then numbers-of-hyper-positivity-4, and so on to numbers-of-hyper-positivity-Ω and beyond.

Note that here that I am not implying the concepts of order-pairs such as (a,b), or tuples in multidimensional space such as (x,1; x,2; ...; x,n), where n can be finite or infinite. My concepts are instead numbers that are as if more positive than positive.

Perhaps such mathematical constructs can find meaning in spacecraft having attained infinite Lorentz factors, thus, in a sense, entering a new light-speed realm of objects traveling at ordinary everyday velocities with respect to the spacecraft.

After attaining the speed of light in the said first new realm, the spacecraft would enter yet another, or second, new realm, where objects in proximity to the spacecraft travel at ordinary, everyday velocities with respect to the light-speed spacecraft.

After attaining the speed of light in the said second new realm, the spacecraft would enter yet another, or third, new realm, where objects in proximity to the spacecraft travel at ordinary, everyday velocities with respect to the light-speed spacecraft.

The ascending order of new realms may be greater than Ω, the least transfinite ordinal.

We refer to the set of these ascending realms as light-speed-ascending-realms-h, where h is any finite or infinite counting number.

We may affix the following operator to formulas for gamma and related parameters to denote these numbers-of-hyper-positivity-k, where k ranges from any finite or infinite counting number and light-speed-ascending-realms-h, where h is any finite or infinite counting number and also arbitrary orders of differentiation and integration in the rates of attainment of serially higher-order terms as such with respect to ship-time, gamma, acceleration, and spacecraft kinetic energy:

> [w(For light-speed spacecraft, attainment of parameters of values defined by numbers-of-hyper-positivity-k where k ranges from any finite or infinite counting number and light-speed-ascending-realms-h where h is any finite or infinite counting number and also arbitrary orders of differentiation and integration in the rates of attainment of serially higher-order terms as such with respect to ship-time, gamma, acceleration, and spacecraft kinetic energy)].

Now we can understand that travel at the speed of light is associated with infinite Lorentz factors and thus infinite time dilation.

Another infinity of light-speed travel is relativistic Lorentz contraction.

Yet another infinity of light-speed travel is relativistic aberration.

However, as infinite Lorentz factors continue to get bigger and bigger for the spacecraft, the craft may manifest more and more parameters infinite in value.

Eventually, the spacecraft may experience untold infinite numbers of parameters morphing into infinite values.

All the infinite parameters already manifested may experience some forms of upsteps into new distinct levels.

For example, the parameter light-speed of a craft traveling at the speed of light may experience an upstep to a velocity ever so slightly greater than the speed of light.

As the latter spacecraft maxed out its upstep from light-speed, the craft may undergo yet another upstep to a velocity that is once again ever so slightly greater than the superluminal velocity associated with the first upstep. Still, further increases in infinite Lorentz factors may result in a third upstep to a velocity ever so slightly greater than that associated with the second upstep in velocity.

Upsteps may manifest for the level of infinities of spacecraft Lorentz factor, background reference frame relativistic aberration, and the perhaps open-endedly infinite numbers of other infinite parameters attained by the spacecraft.

We may affix the following operator to formulas for gamma and related parameters to denote the serial acquisition of infinite parameters, each an element in the set of open-ended infinite numbers of parameters, and the potentially open-ended infinite numbers of upstep velocities greater than c, all as the spacecraft Lorentz factor increases in infinite values and also arbitrary orders of differentiation and integration in the rates of attainment of serially higher-order terms as such with respect to ship-time, gamma, acceleration, and spacecraft kinetic energy.

> [w(Serial acquisition of infinite parameters, each an element in the set of open-ended infinite numbers of parameters, and the potentially open ended infinite numbers of upstep velocities greater than c, all as the spacecraft Lorentz factor increases in infinite values and also arbitrary orders of differentiation and integration in the rates of attainment of serially higher-order terms as such with respect to ship-time, gamma, acceleration, and spacecraft kinetic energy)]

ESSAY 30) Travel Almost at, or Ever So Slightly Faster Than Light in Ascending Levels of Veiled Space-Time

Consider the following list of scenarios that are self-descriptive:

Travel into a first veiled space-time at less than or equal to the speed of light or ever-so-slightly faster than light, but not in ways that would violate the chronological protection conjecture and/or extraction of propulsion energy from these realms.

Travel into a second veiled space-time at less than or equal to the speed of light or ever so slightly faster than light, but not in ways that would violate the chronological protection conjecture and/or extraction of propulsion energy from these realms.

Travel into a third veiled space-time at less than or equal to the speed of light or ever so slightly faster than light, but not in ways that would violate the chronological protection conjecture and/or extraction of propulsion energy from these realms.

Travel into an Ωth veiled space-time at less than or equal to the speed of light or ever so slightly faster than light, but not in ways that would violate the chronological protection conjecture and/or extraction of propulsion energy from these realms.

Travel into an (Ω + e)th veiled space-time at less than or equal to the speed of light or ever so slightly faster than light, but not in ways that would violate the chronological protection conjecture, where e is any finite or infinite positive integer and/or extraction of propulsion energy from these realms.

Travel into a first hidden space-time at less than or equal to the speed of light or ever so slightly faster than light, but not in ways that would violate the chronological protection conjecture and/or extraction of propulsion energy from these realms.

Travel into a second hidden space-time at less than or equal to the speed of light or ever so slightly faster than light, but not in ways that would violate the chronological protection conjecture and/or extraction of propulsion energy from these realms.

Travel into a third hidden space-time at less than or equal to the speed of light or ever so slightly faster than light, but not in ways that would violate the chronological protection conjecture and/or extraction of propulsion energy from these realms.

Travel into an Ωth hidden space-time at less than or equal to the speed of light or ever so slightly faster than light, but not in ways that would violate the chronological protection conjecture and/or extraction of propulsion energy from these realms.

Travel into an $(\Omega + e)$th hidden space-time at less than or equal to the speed of light or ever so slightly faster than light, but not in ways that would violate the chronological protection conjecture, where e is any finite or infinite positive integer and/or extraction of propulsion energy from these realms.

Travel into a first veiled virtual or zero-point space-time at less than or equal to the speed of light or ever so slightly faster than light, but not in ways that would violate the chronological protection conjecture and/or extraction of propulsion energy from these realms.

Travel into a second veiled virtual or zero-point space-time at less than or equal to the speed of light or ever so slightly faster than light, but not in ways that would violate the chronological protection conjecture and/or extraction of propulsion energy from these realms.

Travel into a third veiled virtual or zero-point space-time at less than or equal to the speed of light or ever so slightly faster than light, but not in ways that would violate the chronological protection conjecture and/or extraction of propulsion energy from these realms.

Travel into an Ωth veiled virtual or zero-point space-time at less than or equal to the speed of light or ever so slightly faster than light, but not in ways that would violate the chronological protection conjecture and/or extraction of propulsion energy from these realms.

Travel into an $(\Omega + e)$th virtual or zero-point veiled space-time at less than or equal to the speed of light or ever so slightly faster than light, but not in ways that would violate the chronological protection conjecture, where e is any finite or infinite positive integer and/or extraction of propulsion energy from these realms.

Travel into a first hidden virtual or zero-point space-time at less than or equal to the speed of light or ever so slightly faster than light, but not in ways that would violate the chronological protection conjecture and/or extraction of propulsion energy from these realms.

Travel into a second hidden virtual or zero-point space-time at less than or equal to the speed of light or ever so slightly faster than light, but not in ways that would violate the chronological protection conjecture and/or extraction of propulsion energy from these realms.

Travel into a third hidden virtual or zero-point space-time at less than or equal to the speed of light or ever so slightly faster than light, but not in ways that would violate the chronological protection conjecture and/or extraction of propulsion energy from these realms.

Travel into an Ωth hidden virtual or zero-point space-time at less than or equal to the speed of light or ever so slightly faster than light, but not in ways that would violate the chronological protection conjecture and/or extraction of propulsion energy from these realms.

Travel into an $(\Omega + e)$th hidden virtual or zero-point space-time at less than or equal to the speed of light or ever so slightly faster than light, but not in ways that would violate the chronological protection conjecture, where e is any finite or infinite positive integer and/or extraction of propulsion energy from these realms.

Travel into a first veiled hidden variables pseudo-space-time at less than or equal to the speed of light or ever so slightly faster than light, but not in ways that would violate the chronological protection conjecture and/or extraction of propulsion energy from these realms.

Travel into a second veiled hidden variables pseudo-space-time at less than or equal to the speed of light or ever so slightly faster than light, but not in ways that would violate the chronological protection conjecture and/or extraction of propulsion energy from these realms.

Travel into a third veiled hidden variables pseudo-space-time at less than or equal to the speed of light or ever so slightly faster than light, but not in ways that would violate the chronological protection conjecture and/or extraction of propulsion energy from these realms.

Travel into an Ωth veiled hidden variables pseudo-space-time at less than or equal to the speed of light or ever so slightly faster than light, but not in ways that would violate the chronological protection conjecture and/or extraction of propulsion energy from these realms.

Travel into an (Ω + e)th veiled hidden variables pseudo-space-time at less than or equal to the speed of light or ever so slightly faster than light, but not in ways that would violate the chronological protection conjecture, where $\underline{e}$ is any finite or infinite positive integer and/or extraction of propulsion energy from these realms.

Travel into a first hidden variables pseudo-space-time at less than or equal to the speed of light or ever so slightly faster than light, but not in ways that would violate the chronological protection conjecture and/or extraction of propulsion energy from these realms.

Travel into a second hidden variables pseudo-space-time at less than or equal to the speed of light or ever so slightly faster than light, but not in ways that would violate the chronological protection conjecture and/or extraction of propulsion energy from these realms.

Travel into a third hidden variables pseudo-space-time at less than or equal to the speed of light or ever so slightly faster than light, but not in ways that would violate the chronological protection conjecture and/or extraction of propulsion energy from these realms.

Travel into an Ωth hidden variables pseudo-space-time at less than or equal to the speed of light or ever so slightly faster than light, but not in ways that would violate the chronological protection conjecture and/or extraction of propulsion energy from these realms.

Travel into an (Ω + e)th hidden variables pseudo-space-time at less than or equal to the speed of light or ever so slightly faster than light, but not in ways that would violate the chronological protection conjecture, where e is any finite or infinite positive integer and/or extraction of propulsion energy from these realms.

Travel into a first veiled hidden variables pseudo-virtual or zero-point space-time at less than or equal to the speed of light or ever so slightly faster than light, but not in ways that would violate the chronological protection conjecture and/or extraction of propulsion energy from these realms.

Travel into a second veiled hidden variables pseudo-virtual or zero-point space-time at less than or equal to the speed of light or ever so slightly faster than light, but not in ways that would violate the chronological protection conjecture and/or extraction of propulsion energy from these realms.

Travel into a third veiled hidden variables pseudo-virtual or zero-point space-time at less than or equal to the speed of light or ever so slightly faster than light, but not in ways that would violate the chronological protection conjecture and/or extraction of propulsion energy from these realms.

Travel into an Ωth veiled hidden variables pseudo-virtual or zero-point space-time at less than or equal to the speed of light or ever so slightly faster than light, but not in ways that would violate the chronological protection conjecture and/or extraction of propulsion energy from these realms.

Travel into an $(\Omega + e)$th hidden variables pseudo-virtual or zero-point veiled space-time at less than or equal to the speed of light or ever so slightly faster than light, but not in ways that would violate the chronological protection conjecture, where e is any finite or infinite positive integer and/or extraction of propulsion energy from these realms.

Travel into a first hidden variables pseudo-virtual or zero-point space-time at less than or equal to the speed of light or ever so slightly faster than light, but not in ways that would violate the chronological protection conjecture and/or extraction of propulsion energy from these realms.

Travel into a second hidden variables pseudo-virtual or zero-point space-time at less than or equal to the speed of light or ever so slightly faster than light, but not in ways that would violate the chronological protection conjecture and/or extraction of propulsion energy from these realms.

Travel into a third hidden variables pseudo-virtual or zero-point space-time at less than or equal to the speed of light or ever so slightly faster than light, but not in ways that would violate the chronological protection conjecture and/or extraction of propulsion energy from these realms.

Travel into an Ωth hidden variables pseudo-virtual or zero-point space-time at less than or equal to the speed of light or ever so slightly faster than light, but not in ways that would violate the chronological protection conjecture and/or extraction of propulsion energy from these realms.

Travel into an $(\Omega + e)$th hidden variables pseudo-virtual or zero-point space-time at less than or equal to the speed of light or ever so slightly faster than light, but not in ways that would violate the chronological protection conjecture, where e is any finite or infinite positive integer and/or extraction of propulsion energy from these realms.

Travel into a first veiled space-time, but not in the hyperspatial or parallel history sense at less than or equal to the speed of light or ever so slightly faster than light, but not in ways that would violate the chronological protection conjecture and/or extraction of propulsion energy from these realms.

Travel into a second veiled space-time, but not in the hyperspatial or parallel history sense at less than or equal to the speed of light or ever so slightly faster than light, but not in ways that would violate the chronological protection conjecture and/or extraction of propulsion energy from these realms.

Travel into a third veiled space-time, but not in the hyperspatial or parallel history sense at less than or equal to the speed of light or ever so slightly faster than light, but not in ways that would violate the chronological protection conjecture and/or extraction of propulsion energy from these realms.

Travel into an Ωth veiled space-time, but not in the hyperspatial or parallel history sense at less than or equal to the speed of light or ever so slightly faster than light, but not in ways that would violate the chronological protection conjecture and/or extraction of propulsion energy from these realms.

Travel into an $(\Omega + e)$th veiled space-time, but not in the hyperspatial or parallel history sense at less than or equal to the speed of light or ever so slightly faster than light, but not in ways that would violate the chronological protection conjecture, where e is any finite or infinite positive integer and/or extraction of propulsion energy from these realms.

Travel into a first hidden space-time, but not in the hyperspatial or parallel history sense at less than or equal to the speed of light or ever so slightly faster than light, but not in ways that would violate the chronological protection conjecture and/or extraction of propulsion energy from these realms.

Travel into a second hidden space-time, but not in the hyperspatial or parallel history sense at less than or equal to the speed of light or ever so slightly faster than light, but not in ways that would violate the chronological protection conjecture and/or extraction of propulsion energy from these realms.

Travel into a third hidden space-time but not in the hyperspatial nor parallel history sense at less than or equal to the speed of light or ever so slightly faster than light, but not in ways that would violate the chronological protection conjecture and/or extraction of propulsion energy from these realms.

Travel into an Ωth hidden space-time, but not in the hyperspatial or parallel history sense at less than or equal to the speed of light or ever so slightly faster than light, but not in ways that would violate the chronological protection conjecture and/or extraction of propulsion energy from these realms.

Travel into an $(\Omega + e)$th hidden space-time, but not in the hyperspatial or parallel history sense at less than or equal to the speed of light or ever so slightly faster than light, but not in ways that would violate the chronological protection conjecture, where e is any finite or infinite positive integer and/or extraction of propulsion energy from these realms.

Travel into a first veiled virtual or zero-point space-time, but not in the hyperspatial or parallel history sense at less than or equal to the speed of light or ever so slightly faster than light, but not in ways that would violate the chronological protection conjecture and/or extraction of propulsion energy from these realms.

Travel into a second veiled virtual or zero-point space-time, but not in the hyperspatial or parallel history sense at less than or equal to the speed of light or ever so slightly faster than light, but not in ways that would violate the chronological protection conjecture and/or extraction of propulsion energy from these realms.

Travel into a third veiled virtual or zero-point space-time, but not in the hyperspatial or parallel history sense at less than or equal to the speed of light or ever so slightly faster than light, but not in ways that would violate the chronological protection conjecture and/or extraction of propulsion energy from these realms.

Travel into an Ωth veiled virtual or zero-point space-time, but not in the hyperspatial or parallel history sense at less than or equal to the speed of light or ever so slightly faster than light, but not in ways that would violate the chronological protection conjecture and/or extraction of propulsion energy from these realms.

Travel into an $(\Omega + e)$th virtual or zero-point veiled space-time, but not in the hyperspatial or parallel history sense at less than or equal to the speed of light or ever so slightly faster than light, but not in ways that would violate the chronological protection conjecture, where e is any finite or infinite positive integer and/or extraction of propulsion energy from these realms.

Travel into a first hidden virtual or zero-point space-time, but not in the hyperspatial nor parallel history sense at less than or equal to the speed of light or ever so slightly faster than light, but not in ways that would violate the chronological protection conjecture and/or extraction of propulsion energy from these realms.

Travel into a second hidden virtual or zero-point space-time, but not in the hyperspatial or parallel history sense at less than or equal to the speed of light or ever so slightly faster than light, but not in ways that would violate the chronological protection conjecture and/or extraction of propulsion energy from these realms.

Travel into a third hidden virtual or zero-point space-time, but not in the hyperspatial or parallel history sense at less than or equal to the speed of light or ever so slightly faster than light, but not in ways that would violate the chronological protection conjecture and/or extraction of propulsion energy from these realms.

Travel into an Ωth hidden virtual or zero-point space-time, but not in the hyperspatial or parallel history sense at less than or equal to the speed of light or ever so slightly faster than light, but not in ways that would violate the chronological protection conjecture and/or extraction of propulsion energy from these realms.

Travel into an $(\Omega + e)$th hidden virtual or zero-point space-time, but not in the hyperspatial or parallel history sense at less than or equal to the speed of light or ever so slightly faster than light, but not in ways that would violate the chronological protection conjecture, where e is any finite or infinite positive integer and/or extraction of propulsion energy from these realms.

Travel into a first veiled hidden variables pseudo-space-time, but not in the hyperspatial or parallel history sense at less than or equal to the speed of light or ever so slightly faster than light, but not in ways that would violate the chronological protection conjecture and/or extraction of propulsion energy from these realms.

Travel into a second veiled hidden variables pseudo-space-time, but not in the hyperspatial or parallel history sense at less than or equal to the speed of light or ever so slightly faster than light, but not in ways that would violate the chronological protection conjecture and/or extraction of propulsion energy from these realms.

Travel into a third veiled hidden variables pseudo-space-time, but not in the hyperspatial or parallel history sense at less than or equal to the speed of light or ever so slightly faster than light, but not in ways that would violate the chronological protection conjecture and/or extraction of propulsion energy from these realms.

Travel into an Ωth veiled hidden variables pseudo-space-time, but not in the hyperspatial or parallel history sense at less than or equal to the speed of light or ever so slightly faster than light, but not in ways that would violate the chronological protection conjecture and/or extraction of propulsion energy from these realms.

Travel into an $(\Omega + e)$th veiled hidden variables pseudo-space-time, but not in the hyperspatial or parallel history sense at less than or equal to the speed of light or ever so slightly faster than light, but not in ways that would violate the chronological protection conjecture, where e is any finite or infinite positive integer and/or extraction of propulsion energy from these realms.

Travel into a first hidden variables pseudo-space-time, but not in the hyperspatial or parallel history sense at less than or equal to the speed of light or ever so slightly faster than light, but not in ways that would violate the chronological protection conjecture and/or extraction of propulsion energy from these realms.

Travel into a second hidden variables pseudo-space-time, but not in the hyperspatial or parallel history sense at less than or equal to the speed of light or ever so slightly faster than light, but not in ways that would violate the chronological protection conjecture and/or extraction of propulsion energy from these realms.

Travel into a third hidden variables pseudo space-time, but not in the hyperspatial or parallel history sense at less than or equal to the speed of light or ever so slightly faster than light, but not in ways that would violate the chronological protection conjecture and/or extraction of propulsion energy from these realms.

Travel into an Ωth hidden variables pseudo space-time, but not in the hyperspatial or parallel history sense at less than or equal to the speed of light or ever so slightly faster than light, but not in ways that would violate the chronological protection conjecture and/or extraction of propulsion energy from these realms.

Travel into an $(\Omega + e)$th hidden variables pseudo-space-time, but not in the hyperspatial or parallel history sense at less than or equal to the speed of light or ever so slightly faster than light, but not in ways that would violate the chronological protection conjecture, where e is any finite or infinite positive integer and/or extraction of propulsion energy from these realms.

Travel into a first veiled hidden variables pseudo virtual or zero-point space-time, but not in the hyperspatial or parallel history sense at less than or equal to the speed of light or ever so slightly faster than light, but not in ways that would violate the chronological protection conjecture and/or extraction of propulsion energy from these realms.

Travel into a second veiled hidden variables pseudo virtual or zero-point space-time, but not in the hyperspatial or parallel history sense at less than or equal to the speed of light or ever so slightly faster than light, but not in ways that would violate the chronological protection conjecture and/or extraction of propulsion energy from these realms.

Travel into a third veiled hidden variables pseudo virtual or zero-point space-time, but not in the hyperspatial or parallel history sense at less than or equal to the speed of light or ever so slightly faster than light, but not in ways that would violate the chronological protection conjecture and/or extraction of propulsion energy from these realms.

Travel into an Ωth veiled hidden variables pseudo virtual or zero-point space-time, but not in the hyperspatial or parallel history sense at less than or equal to the speed of light or ever so slightly faster than light, but not in ways that would violate the chronological protection conjecture and/or extraction of propulsion energy from these realms.

Travel into an (Ω + e)th hidden variables pseudo virtual or zero-point veiled space-time, but not in the hyperspatial or parallel history sense at less than or equal to the speed of light or ever so slightly faster than light, but not in ways that would violate the chronological protection conjecture, where *e* is any finite or infinite positive integer and/or extraction of propulsion energy from these realms.

Travel into a first hidden variables pseudo virtual or zero-point space-time, but not in the hyperspatial or parallel history sense at less than or equal to the speed of light or ever so slightly faster than light, but not in ways that would violate the chronological protection conjecture and/or extraction of propulsion energy from these realms.

Travel into a second hidden variables pseudo virtual or zero-point space-time, but not in the hyperspatial or parallel history sense at less than or equal to the speed of light or ever so slightly faster than light, but not in ways that would violate the chronological protection conjecture and/or extraction of propulsion energy from these realms.

Travel into a third hidden variables pseudo virtual or zero-point space-time, but not in the hyperspatial or parallel history sense at less than or equal to the speed of light or ever so slightly faster than light, but not in ways that would violate the chronological protection conjecture and/or extraction of propulsion energy from these realms.

Travel into an Ωth hidden variables pseudo virtual or zero-point space-time, but not in the hyperspatial or parallel history sense at less than or equal to the speed of light or ever so slightly faster than light, but not in ways that would violate the chronological protection conjecture and/or extraction of propulsion energy from these realms.

Travel into an (Ω + e)th hidden variables pseudo virtual or zero-point space-time, but not in the hyperspatial or parallel history sense at less than or equal to the speed of light or ever so slightly faster than light, but not in ways that would violate the chronological protection conjecture, where *e* is any finite or infinite positive integer and/or extraction of propulsion energy from these realms.

We may affix the following operator to formulas for gamma and related parameters below to denote the above alternative realms mechanisms:

> [w(Travel into a first veiled space-time at less than or equal to the speed of light or ever so slightly faster than light, but not in ways that would violate the chronological protection conjecture and/or extraction of propulsion energy from these realms. Travel into a second veiled space-time at less than or equal to the speed of light or ever so slightly faster than light, but not in ways that would violate the chronological protection conjecture and/or extraction of propulsion energy from these realms. Travel into a third veiled space-time at less than or equal to the speed of light or ever so slightly faster than light, but

not in ways that would violate the chronological protection conjecture and/or extraction of propulsion energy from these realms.

Travel into an Ωth veiled space-time at less than or equal to the speed of light or ever so slightly faster than light, but not in ways that would violate the chronological protection conjecture and/or extraction of propulsion energy from these realms. Travel into an $(\Omega + e)$th veiled space-time at less than or equal to the speed of light or ever so slightly faster than light, but not in ways that would violate the chronological protection conjecture, where e is any finite or infinite positive integer and/or extraction of propulsion energy from these realms. Travel into a first hidden space-time at less than or equal to the speed of light or ever so slightly faster than light, but not in ways that would violate the chronological protection conjecture and/or extraction of propulsion energy from these realms. Travel into a second hidden space-time at less than or equal to the speed of light or ever so slightly faster than light, but not in ways that would violate the chronological protection conjecture and/or extraction of propulsion energy from these realms. Travel into a third hidden space-time at less than or equal to the speed of light or ever so slightly faster than light, but not in ways that would violate the chronological protection conjecture and/or extraction of propulsion energy from these realms. Travel into an Ωth hidden space-time at less than or equal to the speed of light or ever so slightly faster than light, but not in ways that would violate the chronological protection conjecture and/or extraction of propulsion energy from these realms. Travel into an $(\Omega + e)$th hidden space-time at less than or equal to the speed of light or ever so slightly faster than light, but not in ways that would violate the chronological protection conjecture, where e is any finite or infinite positive integer and/or extraction of propulsion energy from these realms. Travel into a first veiled virtual or zero-point space-time at less than or equal to the speed of light or ever so slightly faster than light, but not in ways that would violate the chronological protection conjecture and/or extraction of propulsion energy from these realms. Travel into a second veiled virtual or zero-point space-time at less than or equal to the speed of light or ever so slightly faster than light, but not in ways that would violate the chronological protection conjecture and/or extraction of propulsion energy from these realms. Travel into a third veiled virtual or zero-point space-time at less than or equal to the speed of light or ever so slightly faster than light, but not in ways that would violate the chronological protection conjecture and/or extraction of propulsion energy from these realms. Travel into an Ωth veiled virtual or zero-point space-time at less than or equal to the speed of light or ever so slightly faster than light, but not in ways that would violate the chronological protection conjecture and/or extraction of propulsion energy from these realms. Travel into an $(\Omega + e)$th virtual or zero-point veiled space-time at less than or equal to the speed of light or ever so slightly faster than light, but not in ways that would violate the chronological protection conjecture, where e is any finite or infinite positive integer and/or extraction of propulsion energy from these realms. Travel into a first hidden virtual or zero-point space-time at less than or equal to the speed of light or ever so slightly faster than light, but not in ways that would violate the chronological protection conjecture and/or extraction of propulsion energy from these realms. Travel into a second hidden virtual or zero-point space-time at less than or equal to the speed of light or ever so slightly faster than light, but not in ways that would violate the chronological protection conjecture and/or extraction of propulsion energy from these realms. Travel into a third hidden virtual or zero-point space-time at less than or equal to the speed of light or ever so slightly faster than light, but not in ways that would violate the chronological protection conjecture and/or extraction of propulsion energy from these realms. Travel into an Ωth hidden virtual or zero-point space-time at less than or equal to the speed of light or ever so slightly faster than light, but not in ways that would violate the chronological protection conjecture and/or extraction of propulsion energy from these realms. Travel into an $(\Omega + e)$th hidden virtual or zero-point space-time at less than or equal to the speed of light or ever so slightly faster than light, but not in ways that would violate the chronological protection conjecture, where e is any finite or infinite positive integer and/or extraction of propulsion energy from these realms. Travel into a first veiled hidden variables pseudo space-time at less than or equal to the speed of light or ever so slightly faster than light, but not in ways that would violate the chronological protection conjecture and/or extraction of propulsion energy from these realms. Travel into a second veiled hidden variables pseudo space-time at less than or equal to the speed of light or ever so slightly faster than light, but not in ways that would violate the chronological protection conjecture and/or extraction of propulsion energy from these realms. Travel into a third veiled hidden variables pseudo space-time at less than or equal to the speed of light or ever so slightly faster than light, but not in ways that would violate the chronological protection conjecture and/or extraction of propulsion energy from these realms. Travel into an Ωth veiled hidden variables pseudo space-time at less than or equal to the speed of light or ever so slightly faster than light, but not in ways that would violate the chronological protection conjecture and/or extraction of propulsion energy from these realms. Travel into an $(\Omega + e)$th veiled hidden variables pseudo space-time at less than or equal to the speed of light or ever so slightly faster than light, but not in ways that would violate the chronological protection conjecture, where e is any finite or infinite positive integer and/or extraction of propulsion energy from these realms. Travel into a first hidden variables pseudo space-time at less than or equal to the speed of light or ever so slightly faster than light, but not in ways that would violate the chronological protection conjecture and/or extraction of propulsion energy from these realms. Travel into a second hidden variables pseudo space-time at less than or equal to the speed of light or ever so slightly faster than light, but not in ways that would violate the chronological protection conjecture and/or extraction of propulsion energy from these realms. Travel into a third hidden variables pseudo space-time at less than or equal to the speed of light or ever so slightly faster than light, but not in ways that would violate the chronological protection conjecture and/or extraction of propulsion energy from these realms. Travel into an Ωth hidden variables pseudo space-time at less than or equal to the speed of light or ever so slightly faster than light, but not in ways that would violate the chronological protection conjecture and/or extraction of propulsion energy from these realms. Travel into an $(\Omega + e)$th hidden variables pseudo space-time at less than or equal to the speed of light or ever so slightly faster than light, but not in ways that would violate the chronological protection conjecture, where e is any finite or infinite positive integer and/or extraction of propulsion energy from these realms. Travel into a first veiled hidden variables pseudo virtual or zero-point space-time at less than or equal to the speed of light or ever so slightly faster than light, but not in ways that would violate the chronological protection conjecture and/or extraction of propulsion energy from these realms. Travel into a second veiled hidden variables pseudo virtual or zero-point space-time at less than or equal to the speed of light or ever so slightly faster than light, but not in ways that would violate the chronological protection conjecture and/or extraction of propulsion energy from these realms. Travel into a third veiled hidden variables pseudo virtual or zero-point space-time at less than or equal to the speed of light or ever so slightly faster than light, but not in ways that would violate the chronological protection conjecture and/or extraction of propulsion energy from these realms. Travel into an Ωth veiled hidden variables pseudo virtual or zero-point space-time at less than or equal to the speed of light or ever so slightly faster than light, but not in ways that would violate the chronological protection conjecture and/or extraction of propulsion energy from these realms. Travel into an $(\Omega + e)$th hidden variables pseudo virtual or zero-point veiled space-time at less than or equal to the speed of light or ever so slightly faster than light, but not in ways that would violate the chronological protection conjecture, where e is any finite or infinite positive integer and/or extraction of propulsion energy from these realms. Travel into a first hidden variables pseudo virtual or zero-point space-time at less than or equal to the speed of light or ever so slightly faster than light, but not in ways that would violate the chronological protection conjecture and/or extraction of propulsion energy from these realms. Travel into a second hidden variables pseudo virtual or zero-point space-time at less than or equal to the speed of light or ever so slightly faster than light, but not in ways that would violate the chronological protection conjecture and/or extraction of propulsion energy from these realms. Travel into a third hidden variables pseudo virtual or zero-point space-time at less than or equal to the speed of light or ever so slightly faster than light, but not in ways that would violate the chronological protection conjecture and/or extraction of propulsion energy from these realms. Travel into an Ωth hidden variables pseudo virtual or zero-point space-time at less than or equal to the speed of light or ever so slightly faster than light, but not in ways that would violate the chronological protection conjecture and/or extraction of propulsion energy from these realms. Travel into an $(\Omega + e)$th hidden variables pseudo virtual or zero-point space-time at less than or equal to the speed of light or ever so slightly faster than light, but not in ways that would violate the chronological protection conjecture, where e is any finite or infinite positive integer and/or extraction of propulsion energy from these realms. Travel into a first veiled space-time, but not in the hyperspatial or parallel history sense at less than or equal to the speed of light or ever so slightly faster than light, but not in ways that would violate the chronological protection conjecture and/or extraction of propulsion energy from these realms. Travel into a second veiled space-time but not in the hyperspatial nor parallel history sense at less than or equal to the speed of light or ever so slightly faster than light, but not in ways that would violate the chronological protection conjecture and/or extraction of propulsion energy from these realms. Travel into a third veiled space-time, but not in the hyperspatial or parallel history sense at less than or equal to the speed of light or ever so slightly faster than light, but not in ways that would violate the chronological protection conjecture and/or extraction of propulsion energy from these realms. Travel into an Ωth veiled space-time, but not in the hyperspatial or parallel history sense at less than or equal to the speed of light or ever so slightly faster than light, but not in ways that would violate the chronological protection conjecture and/or extraction of propulsion energy from these realms. Travel into an $(\Omega + e)$th veiled space-time, but not in the hyperspatial or parallel history sense at less than or equal to the speed of light or ever so slightly faster than light, but not in ways that would violate the chronological protection conjecture, where e is any finite or infinite positive integer and/or extraction of propulsion energy from these realms. Travel into a first hidden space-time, but not in the hyperspatial or parallel history sense at less than or equal to the speed of light or ever so slightly faster than light, but not in ways that would violate the chronological protection conjecture and/or extraction of propulsion energy from these realms. Travel into a second hidden space-time, but not in the hyperspatial or parallel history sense at less than or equal to the speed of light or ever so slightly faster than light, but not in ways that would violate the chronological protection conjecture and/or extraction of propulsion energy from these realms. Travel into a third hidden space-time, but not in the hyperspatial or parallel history sense at less than or equal to the speed of light or ever so slightly faster than light, but not in ways that would violate the chronological protection conjecture and/or extraction of propulsion energy from these realms. Travel into an Ωth hidden space-time, but not in the hyperspatial or parallel history sense at less than or equal to the speed of light or ever so slightly faster than light, but not in ways that would violate the chronological protection conjecture and/or extraction of propulsion energy from these realms. Travel into an $(\Omega + e)$th hidden space-time, but not in the hyperspatial or parallel history sense at less than or equal to the speed of light or ever so slightly faster than light, but not in ways that would violate the chronological protection conjecture, where e is any finite or infinite positive integer and/or extraction of propulsion energy from these realms. Travel into a first veiled virtual or zero-point space-time, but not in the hyperspatial or parallel history sense at less than or equal to the speed of light or ever so slightly faster than light, but not in ways that would violate the chronological protection conjecture and/or extraction of propulsion energy from these realms. Travel into a second veiled virtual or zero-point space-time, but not in the hyperspatial or parallel history sense at less than or equal to the speed of light or ever so slightly faster than light, but not in ways that would violate the chronological protection conjecture and/or extraction of propulsion energy from these realms. Travel into a third veiled virtual or zero-point space-time, but not in the hyperspatial or parallel history sense at less than or equal to the speed of light or ever so slightly faster than light, but not in ways that would violate the chronological protection conjecture and/or extraction of propulsion energy from these realms. Travel into an Ωth veiled virtual or zero-point space-time, but not in the hyperspatial or parallel history sense at less than or equal to the speed of light or ever so slightly faster than light, but not in ways that would violate the chronological protection conjecture and/or extraction of propulsion energy from these realms. Travel into an $(\Omega + e)$th virtual or zero-point veiled space-time, but not in the hyperspatial or parallel history sense at less than or equal to the speed of light or ever so slightly faster than light, but not in ways that would violate the chronological protection conjecture, where e is any finite or infinite positive integer and/or extraction of propulsion energy from these realms. Travel into a first hidden virtual or zero-point space-time, but not in the hyperspatial or parallel history sense at less than or equal to the speed of light or ever so slightly faster than light, but not in ways that would violate the chronological protection conjecture and/or extraction of propulsion energy from these realms. Travel into a second hidden virtual or zero-point space-time, but not in the hyperspatial or parallel history sense at less than or equal to the speed of light or ever so slightly faster than light, but not in ways that would violate the chronological protection conjecture and/or extraction of propulsion energy from these realms. Travel into a third hidden virtual or zero-point space-time, but not in the hyperspatial or parallel history sense at less than or equal to the speed of light or ever so slightly faster than light, but not in ways that would violate the chronological protection conjecture and/or extraction of propulsion energy from these realms. Travel into an Ωth hidden virtual or zero-point space-time, but not in the hyperspatial or parallel history sense at less than or equal to the speed of light or ever so slightly faster than light, but not in ways that would violate the chronological protection conjecture and/or extraction of propulsion energy from these realms. Travel into an $(\Omega + e)$th hidden virtual or zero-point space-time, but not in the hyperspatial or parallel history sense at less than or equal to the speed of light or ever so slightly faster than light, but not in ways that would violate the chronological protection conjecture, where e is any finite or infinite positive integer and/or extraction of propulsion energy from these realms. Travel into a first veiled hidden variables pseudo space-time, but not in the hyperspatial or parallel history sense at less than or equal to the speed of light or ever so slightly faster than light, but not in ways that would violate the chronological protection conjecture and/or extraction of propulsion energy from these realms. Travel into a second veiled hidden variables pseudo space-time, but not in the hyperspatial or parallel history sense at less than or equal to the speed of light or ever so slightly faster than light, but not in ways that would violate the chronological protection conjecture and/or extraction of propulsion energy from these realms. Travel into a third veiled hidden variables pseudo space-time, but not in the hyperspatial or parallel history sense at less than or equal to the speed of light or ever so slightly faster than light, but not in ways that would violate the chronological protection conjecture and/or extraction of propulsion energy from these realms. Travel into an Ωth veiled hidden variables pseudo space-time, but not in the hyperspatial or parallel history sense at less than or equal to the speed of light or ever so slightly faster than light, but not in ways that would violate the chronological protection conjecture and/or extraction of propulsion energy from these realms. Travel into an $(\Omega + e)$th veiled hidden variables pseudo space-time, but not in the hyperspatial or parallel history sense at less than or equal to the speed of light or ever so slightly faster than light, but not in ways that would violate the chronological protection conjecture, where e is any finite or infinite positive integer and/or extraction of propulsion energy from these realms. Travel into a first hidden variables pseudo space-time, but not in the hyperspatial or parallel history sense at less than or equal to the speed of light or ever so slightly faster than light, but not in ways that would violate the chronological protection conjecture and/or extraction of propulsion energy from these realms. Travel into a second hidden variables pseudo space-time, but not in the hyperspatial or parallel history sense at less than or equal to the speed of light or ever so slightly faster than light, but not in ways that would violate the chronological protection conjecture and/or extraction of propulsion energy from these realms. Travel into a third hidden variables pseudo space-time, but not in the hyperspatial or parallel history sense at less than or equal to the speed of light or ever so slightly faster than light, but not in ways that would violate the chronological protection conjecture and/or extraction of propulsion energy from these realms. Travel into an Ωth hidden variables pseudo space-time, but not in the hyperspatial or parallel history sense at less than or equal to the speed of light or ever so slightly faster than light, but not in ways that would violate the chronological protection conjecture and/or extraction of propulsion energy from these realms. Travel into an $(\Omega + e)$th hidden variables pseudo space-time, but not in the hyperspatial or parallel history sense at less than or equal to the speed of light or ever so slightly faster than light, but not in ways that would violate the chronological protection conjecture, where e is any finite or infinite positive integer and/or extraction of propulsion energy from these realms. Travel into a first veiled hidden variables pseudo virtual or zero-point space-time, but not in the hyperspatial or parallel history sense at less than or equal to the speed of light or ever so slightly faster than light, but not in ways that would violate the chronological protection conjecture and/or extraction of propulsion energy from these realms. Travel into a second veiled hidden variables pseudo virtual or zero-point space-time, but not in the hyperspatial or parallel history sense at less than or equal to the speed of light or ever so slightly faster than light, but not in ways that would violate the chronological protection conjecture and/or extraction of propulsion energy from these realms. Travel into a third veiled hidden variables pseudo virtual or zero-point space-time, but not in the hyperspatial or parallel history sense at less than or equal to the speed of light or ever so slightly faster than light, but not in ways that would violate the chronological protection conjecture and/or extraction of propulsion energy from these realms. Travel into an Ωth veiled hidden variables pseudo virtual or zero-point space-time, but not in the hyperspatial or parallel history sense at less than or equal to the speed of light or ever so slightly faster than light, but not in ways that would violate the chronological protection conjecture and/or extraction of propulsion energy from these realms. Travel into an (Ω + e)th hidden variables pseudo virtual or zero-point veiled space-time, but not in the hyperspatial or parallel history sense at less than or equal to the speed of light or ever so slightly faster than light, but not in ways that would violate the chronological protection conjecture, where e is any finite or infinite positive integer and/or extraction of propulsion energy from these realms. Travel into a first hidden variables pseudo virtual or zero-point space-time, but not in the hyperspatial or parallel history sense at less than or equal to the speed of light or ever so slightly faster than light, but not in ways that would violate the chronological protection conjecture and/or extraction of propulsion energy from these realms. Travel into a second hidden variables pseudo virtual or zero-point space-time, but not in the hyperspatial or parallel history sense at less than or equal to the speed of light or ever so slightly faster than light, but not in ways that would violate the chronological protection conjecture and/or extraction of propulsion energy from these realms. Travel into a third hidden variables pseudo virtual or zero-point space-time, but not in the hyperspatial or parallel history sense at less than or equal to the speed of light or ever so slightly faster than light, but not in ways that would violate the chronological protection conjecture and/or extraction of propulsion energy from these realms. Travel into an Ωth hidden variables pseudo virtual or zero-point space-time, but not in the hyperspatial or parallel history sense at less than or equal to the speed of light or ever so slightly faster than light, but not in ways that would violate the chronological protection conjecture and/or extraction of propulsion energy from these realms. Travel into an (Ω + e)th hidden variables pseudo virtual or zero-point space-time, but not in the hyperspatial or parallel history sense at less than or equal to the speed of light or ever so slightly faster than light, but not in ways that would violate the chronological protection conjecture, where e is any finite or infinite positive integer and/or extraction of propulsion energy from these realms.)]

Here we consider other forms of journeys. Accordingly, we provide the following list of items as a self-descriptive set of the presented types of travel:

Travel ever deeper into the accidentality of the physical cosmos and into levels of prime mattergy at analogues of the speed of light.

Travel into quantum hidden variables of a first level at analogues of the speed of light.

Travel into the quantum hidden variables of a second level at analogues of the speed of light.

Travel into the quantum hidden variables of a third level at analogues of the speed of light and so on and so on until we reach Ωth-level quantum hidden variables at analogues of the speed of light and then continue onward from there.

Travel ever deeper into the accidentality of the physical cosmos and into levels of prime mattergy at analogues of velocities vanishingly small fractions of the speed of light faster than the speed of light, but where chronological protection conjecture is secure in manifestations.

Travel into quantum hidden variables of a first level at analogues of velocities vanishingly small fractions of the speed of light faster than the speed of light, but where the chronological protection conjecture is secure in manifestations.

Travel into the quantum hidden variables of a second level at analogues of velocities vanishingly small fractions of the speed of light faster than the speed of light, but where the chronological protection conjecture is secure in manifestations.

Travel into the quantum hidden variables of a third level at analogues of velocities vanishingly small fractions of the speed of light faster than the speed of light, but where the chronological protection conjecture is secure in manifestations.

Travel into the quantum hidden variables of a Ωth level at analogues of velocities vanishingly small fractions of the speed of light faster than the speed of light, but where the chronological protection conjecture is secure in manifestations.

We may affix the following operator to formulas for gamma and related parameters to denote the above-listed mechanisms as operative in the conjectured forms of light-speed and ever-so-slightly-faster-than-light travel:

> [w(Travel ever deeper into the accidentality of the physical cosmos and into levels of prime mattergy at analogues of the speed of light. Travel into quantum hidden variables of a first level at analogues of the speed of light. Travel into the quantum hidden variables of a second level at analogues of the speed of light. Travel into the quantum hidden variables of a third level at analogues of the speed of light and so on and so on until we reach an Ωth-level quantum hidden variables at analogues of the speed of light, and then

continue onward from there. Travel ever deeper into the accidentality of the physical cosmos and into levels of prime mattery at analogues of velocities vanishingly small fractions of the speed of light faster than the speed of light, but where chronological protection conjecture is secure in manifestations. Travel into quantum hidden variables of a first level at analogues of velocities vanishingly small fractions of the speed of light faster than the speed of light, but where the chronological protection conjecture is secure in manifestations. Travel into the quantum hidden variables of a second level at analogues of velocities vanishingly small fractions of the speed of light faster than the speed of light, but where the chronological protection conjecture is secure in manifestations. Travel into the quantum hidden variables of a third level at analogues of velocities vanishingly small fractions of the speed of light faster than the speed of light, but where the chronological protection conjecture is secure in manifestations. Travel into the quantum hidden variables of a Ωth level at analogues of velocities vanishingly small fractions of the speed of light faster than the speed of light, but where the chronological protection conjecture is secure in manifestations.)]

ESSAY 31) Light-Speed Limits and Analogues Thereof for Increasingly Infinite Lorentz Factor Capable Spacecraft

Now travel at the speed of light is commonly considered associated with infinite Lorentz factors.

Accordingly, I suggest that once a spacecraft attains a suitably infinite Lorentz factor in light-speed travel, the spacecraft would run out of temporal room to travel forward in time, as well as experience such an extreme relativistic background aberration such that the spacecraft leaves the universe or larger realm of travel to enter another broader realm of travel.

However, I believe that if a spacecraft had attained light-speed in a suitably infinite Lorentz factor and then popped into another realm, and then back into the realm of origin, the effective velocity of spacecraft displacement in the realm of origin could not be greater than the speed of light in spirit.

A spacecraft that has attained a suitably infinite Lorentz factor in a first broader realm of travel, I believe, may likewise pop into a higher or second alternative realm of travel.

For a spacecraft that would thus reach a velocity of light in a first larger realm of travel having popped into a second higher realm of travel and then back into the first higher realm of travel, I believe that the velocity of displacement in the first higher realm of travel cannot be greater than the velocity of light in the first higher realm of travel. Even in cases where the spacecraft had originated in light-speed travel in the realm of origin and then popped into the first higher realm and then into the second higher realm, then directly back into the realm of origin or first back into the first higher realm then back into the realm of origin, I believe the distance of displacement in the realm of origin cannot exceed the velocity of light in the realm of origin.

We can likewise consider scenarios of still higher levels of realms, with the analogous velocity limits being equal to the respective velocities of light.

We can formulate such light-speed limits as follows:

$$[\Sigma(j = 1; j = \infty\uparrow):[[d[r[(x,1), (x, 2), ..., (x,n)]]/dt]j]] \le \text{(realm specific value of c)}$$

We may also consider derivatives with respect to other parameters such as gamma, acceleration in the ship-frame, and ship kinetic energy as follows:

$$[\Sigma(j = 1; j = \infty\uparrow):[[d[r[(x,1), (x, 2), ..., (x,n)]]/d\gamma]j]]$$

$$[\Sigma(j = 1; j = \infty\uparrow):[[d[r[(x,1), (x, 2), ..., (x,n)]]/d \text{ ship-frame acceleration}]j]]$$

$$[\Sigma(j = 1; j = \infty\uparrow):[[d[r[(x,1), (x, 2), ..., (x,n)]]/d \text{ K.E}]j]]$$

And others.

We can likewise take higher-order derivatives of the above function with respect to any variable used as a first variable of derivation. Some of these higher-order derivatives will be equal to zero, and many others will require several orders of differentiation to return a value of zero. However, some infinite order derivatives may return nonzero values.

Integration of the above derivatives can, in a sense, provide measures of distances traveled and analogues thereof.

For example, we can take the following integrals:

$$\int [\Sigma(j = 1; j = \infty\uparrow):[[d[r[(x,1), (x, 2), ..., (x,n)]]/dt]j]] \text{ dtship}$$

$$\int [\Sigma(j = 1; j = \infty\uparrow):[[d[r[(x,1), (x, 2), ..., (x,n)]]/dt]j]] \text{ d}\gamma$$

$$\int [\Sigma(j = 1; j = \infty\uparrow):[[d[r[(x,1), (x, 2), ..., (x,n)]]/dt]j]] \text{ d acceleration ship}$$

$$\int [\Sigma(j = 1; j = \infty\uparrow):[[d[r[(x,1), (x, 2), ..., (x,n)]]/dt]j]] \text{ dt KE}$$

$$\int [\Sigma(j = 1; j = \infty\uparrow):[[d[r[(x,1), (x, 2), ..., (x,n)]]/d \gamma]j]] \text{ dtship}$$

$$\int [\Sigma(j = 1; j = \infty\uparrow):[[d[r[(x,1), (x, 2), ..., (x,n)]]/d \gamma]j]] \text{ d}\gamma$$

∫ [Σ(j = 1; j = ∞↑):[[d[r[(x,1), (x, 2), …, (x,n)]]/d γ]j]] d acceleration ship

∫ [Σ(j = 1; j = ∞↑):[[d[r[(x,1), (x, 2), …, (x,n)]]/d γ]j]] d KE

∫ [Σ(j = 1; j = ∞↑):[[d[r[(x,1), (x, 2), …, (x,n)]]/d acceleration]j]] dtship

∫ [Σ(j = 1; j = ∞↑):[[d[r[(x,1), (x, 2), …, (x,n)]]/d acceleration]j]] dγ

∫ [Σ(j = 1; j = ∞↑):[[d[r[(x,1), (x, 2), …, (x,n)]]/d acceleration]j]] d acceleration ship

∫ [Σ(j = 1; j = ∞↑):[[d[r[(x,1), (x, 2), …, (x,n)]]/d acceleration]j]] d KE

∫ [Σ(j = 1; j = ∞↑):[[d[r[(x,1), (x, 2), …, (x,n)]]/d KE]j]] dtship

∫ [Σ(j = 1; j = ∞↑):[[d[r[(x,1), (x, 2), …, (x,n)]]/d KE]j]] dγ

∫ [Σ(j = 1; j = ∞↑):[[d[r[(x,1), (x, 2), …, (x,n)]]/d KE]j]] d acceleration ship

∫ [Σ(j = 1; j = ∞↑):[[d[r[(x,1), (x, 2), …, (x,n)]]/d KE]j]] d KE

We can take higher-order derivatives accordingly.

We may affix the following operator to formulas for gamma and related parameters to denote the above range of possibilities in derivatives and integrals:

{w{{{[Σ(j = 1; j = ∞↑):[[d[r[(x,1), (x, 2), …, (x,n)]]/dt]j]] ≤ (realm specific value of c)};{[Σ(j = 1; j = ∞↑):[[d[r[(x,1), (x, 2), …, (x,n)]]/dγ]j]]};{[Σ(j = 1; j = ∞↑):[[d[r[(x,1), (x, 2), …, (x,n)]]/d ship-frame acceleration]j]]};{[Σ(j = 1; j = ∞↑):[[d[r[(x,1), (x, 2), …, (x,n)]]/d K.E]j]]};{∫ [Σ(j = 1; j = ∞↑):[[d[r[(x,1), (x, 2), …, (x,n)]]/dt]j]] dtship}; {∫ [Σ(j = 1; j = ∞↑):[[d[r[(x,1), (x, 2), …, (x,n)]]/dt]j]] dγ}; {∫ [Σ(j = 1; j = ∞↑):[[d[r[(x,1), (x, 2), …, (x,n)]]/dt]j]] d acceleration ship}; {∫ [Σ(j = 1; j = ∞↑):[[d[r[(x,1), (x, 2), …, (x,n)]]/dt]j]] d KE};{∫ [Σ(j = 1; j = ∞↑):[[d[r[(x,1), (x, 2), …, (x,n)]]/d γ]j]] dtship}; {∫ [Σ(j = 1; j = ∞↑):[[d[r[(x,1), (x, 2), …, (x,n)]]/d γ]j]] dγ}; {∫ [Σ(j = 1; j = ∞↑):[[d[r[(x,1), (x, 2), …, (x,n)]]/d γ]j]] d acceleration ship}; {∫ [Σ(j = 1; j = ∞↑):[[d[r[(x,1), (x, 2), …, (x,n)]]/d γ]j]] d KE};{∫ [Σ(j = 1; j = ∞↑):[[d[r[(x,1), (x, 2), …, (x,n)]]/d acceleration]j]] dtship}; {∫ [Σ(j = 1; j = ∞↑):[[d[r[(x,1), (x, 2), …, (x,n)]]/d acceleration]j]] dγ}; {∫ [Σ(j = 1; j = ∞↑):[[d[r[(x,1), (x, 2), …, (x,n)]]/d acceleration]j]] d acceleration ship}; {∫ [Σ(j = 1; j = ∞↑):[[d[r[(x,1), (x, 2), …, (x,n)]]/d acceleration]j]] d KE};{∫ [Σ(j = 1; j = ∞↑):[[d[r[(x,1), (x, 2), …, (x,n)]]/d KE]j]] dtship}; {∫ [Σ(j = 1; j = ∞↑):[[d[r[(x,1), (x, 2), …, (x,n)]]/d KE]j]] dγ}; {∫ [Σ(j = 1; j = ∞↑):[[d[r[(x,1), (x, 2), …, (x,n)]]/d KE]j]] d acceleration ship}; {∫ [Σ(j = 1; j = ∞↑):[[d[r[(x,1), (x, 2), …, (x,n)]]/d

KE]j]] d KE }} as well as arbitrarily higher-order derivatives and integrals with regard to respective or other variables}}

Each of the following roster elements can be the basis of expressions of differentiated functions for power, energy, velocity, momentum, gamma and the like by differentiation in the following manners:

d f/dt,1 d(d f/dt,1)/dt,2 d(d f/dt,2)/dt,1 d[d(d f/dt,1)/dt,2]/dt,3 d[d(d f/dt,1)/dt,3]/dt,2

d[d(d f/dt,2)/dt,1]/dt,3 d[d(d f/dt,2)/dt,3]/dt,1 d[d(d f/dt,3)/dt,1]/dt,2 d[d(d f/dt,3)/dt,2]/dt,1

d f/dT,1 d(d f/dT,1)/dT,2 d(d f/dT,2)/dT,1 d[d(d f/dT,1)/dT,2]/dT,3 d[d(d f/dT,1)/dT,3]/dT,2

d[d(d f/dT,2)/dT,1]/dT,3 d[d(d f/dT,2)/dT,3]/dT,1 d[d(d f/dT,3)/dT,1]/dT,2 d[d(d f/dT,3)/dT,2]/dT,1

T in each case is a timelike blend of various sorts of multiple time dimensions that appropriately and usefully enables derivation in one iteration.

We can go further to consider the following:

d f/dꞀ,1 d(d f/dꞀ,1)/dꞀ,2 d(d f/dꞀ,2)/dꞀ,1 d[d(d f/dꞀ,1)/dꞀ,2]/dꞀ,3 d[d(d f/dꞀ,1)/dꞀ,3]`/dꞀ,2

d[d(d f/dꞀ,2)/dꞀ,1]/dꞀ,3 d[d(d f/dꞀ,2)/dꞀ,3]/dꞀ,1 d[d(d f/dꞀ,3)/dꞀ,1]/dꞀ,2 d[d(d f/dꞀ,3)/dꞀ,2]/dꞀ,1

Ꞇ in each case is a timelike blend of timelike blends of various sorts of multiple time dimensions that appropriately and usefully enables derivation in one iteration.

By the same token, we can also integrate functions for power, energy, velocity, momentum, and gamma and the like by respect to multiple time dimensions as follows:

∫ f(dt,1) ∫∫d f(dt,1)(dt,2) ∫∫d f(dt,2)(dt,1) ∭ f(dt,1)(dt,2)(dt,3) ∭ f(dt,1)(dt,3)(dt,2)

∭ f(dt,2)(dt,1)(dt,3) ∭ f(dt,2)(dt,3)(dt,1) ∭ f(dt,3)(dt,1)(dt,2) ∭ f(dt,3)(dt,2)(dt,1)

∫ f(dT,1) ∬ f(dT,1) (dT,2) ∬ f(dT,2) ((dT,1)) ∭ f(dT,1) (dT,2) (dT,3) ∭ f(dT,1) (dT,3) (dT,2)

∭ f(dT,2) (dT,1) (dT,3) ∭ f(dT,2) (dT,3) (dT,1) ∭ f(dT,3) (dT,1) (dT,2) ∭ f(dT,3) (dT,2) (dT,1)

We can go further to consider the following:

$$\int f(d\mathsf{T},1) \quad \int \int f(d\mathsf{T},1)(d\mathsf{T},2) \quad \int \int f(d\mathsf{T},2)(d\mathsf{T},1) \quad \int \int \int f(d\mathsf{T},1)(d\mathsf{T},2)(d\mathsf{T},3) \quad \int \int \int f(d\mathsf{T},1)(d\mathsf{T},3)(d\mathsf{T},2)$$

$$\int \int \int f(d\mathsf{T},2)(d\mathsf{T},1)(d\mathsf{T},3) \quad \int \int \int f(d\mathsf{T},2)(d\mathsf{T},3)(d\mathsf{T},1) \quad \int \int \int f(d\mathsf{T},3)(d\mathsf{T},1)(d\mathsf{T},2) \quad \int \int \int f(d\mathsf{T},3)(d\mathsf{T},2)(d\mathsf{T},1)$$

We can continue to yet higher-order derivatives and integrals. Suffice it to say that the general patterns continue for higher numbers of dimensions.

We can also consider higher-order blends of time and thus are not limited to the blends defined by T and T. We chose these two symbols for their resemblance to the letter t.

As for hyperspatial applications, we consider the following hyperspaces:

Such hyperspaces can be N-Time-M-Space, where N is an integer greater than or equal to 3 and M is any counting number.

Alternatively, N can be any rational number equal to 3 or greater and M is any rational number greater than or equal to 1.

As another set of scenarios, N can be any irrational number equal to 3 or greater and M is any irrational number greater than or equal to 1.

Alternatively, N can be any rational number greater than 0 and M is any rational number greater than 0.

As another set of scenarios, N can be any irrational number greater than 0 and M is any irrational number greater zero.

The N-Space-M-Time may optionally be flat, positively curved, negatively curved, positively curved and torsioned at one or more scales in arbitrary patterns including but not limited to fractals, negatively curved and torsioned at one or more scales in arbitrary patterns including, but not limited to, fractals, positively super-curved, negatively super-curved, positively super-curved and torsioned at one or more scales in arbitrary patterns including, but not limited to, fractals, negatively super-curved and torsioned at one or more scales in arbitrary patterns including, but not limited to, fractals, positively super-curved and super-torsioned at one or more scales in arbitrary patterns including, but not limited to, fractals, negatively super-curved and super-torsioned at one or more scales in arbitrary patterns including, but not limited to, fractals, positively curved and positively torsioned at one or more scales in arbitrary patterns including, but not limited to, fractals, negatively curved and positively torsioned at one or more scales in arbitrary patterns including, but not limited to, fractals, positively super-curved, negatively super-curved, positively super-curved and positively torsioned at one or more scales in arbitrary patterns including, but not limited to, fractals, negatively super-curved and positively torsioned at one or more scales in arbitrary patterns including, but not limited to, fractals, positively super-curved and positively super-torsioned at one or more scales in arbitrary patterns including, but not limited to, fractals, negatively super-curved and positively super-torsioned at one or more scales in arbitrary patterns including, but not limited to, fractals, positively curved and negatively torsioned at one or more scales in arbitrary patterns including, but not limited to, fractals, negatively curved and negatively torsioned at one or more scales in arbitrary patterns including, but not limited to, fractals, positively super-curved, negatively super-curved, positively super-curved and negatively torsioned at one or more scales in arbitrary patterns including, but not limited to, fractals, negatively super-curved and negatively torsioned at one or more scales in arbitrary patterns including, but not limited to, fractals, positively super-curved and negatively super-torsioned at one or more scales in arbitrary patterns including, but not limited to, fractals, negatively super-curved and negatively super-torsioned at one or more scales in arbitrary patterns including, but not limited to, fractals, positively super-...-super-curved, negatively super-...-super-curved, positively super-...-super-curved and torsioned at one or more scales in arbitrary patterns including, but not limited to, fractals, negatively super-...-super-curved and torsioned at one or more scales in arbitrary patterns including, but not limited to, fractals, positively super-...-super-curved and super-...-super-torsioned at one or more scales in arbitrary patterns including, but not limited to, fractals, negatively super-...-super-curved and super-...-super-torsioned at one or more scales in arbitrary patterns including, but not limited to, fractals, positively curved and positively torsioned at one or more scales in arbitrary patterns including, but not limited to, fractals, negatively curved and positively torsioned at one or more scales in arbitrary patterns including, but not limited to, fractals, positively super-...-super-curved, negatively super-...-super-curved, positively super-...-super-curved and positively torsioned at one or more scales in arbitrary patterns including, but not limited to, fractals, negatively super-...-super-curved and positively torsioned at one or more scales in arbitrary patterns including, but not limited to, fractals, positively super-...-super-curved and positively super-...-super-torsioned at one or more scales in arbitrary patterns including, but not limited to, fractals, negatively super-...-super-curved and positively super-...-super-torsioned at one or more scales in arbitrary patterns including, but not limited to, fractals, positively curved and negatively torsioned at one or more scales in arbitrary patterns including, but not limited to, fractals, negatively curved and negatively torsioned at one or more scales in arbitrary patterns including, but not limited to, fractals, positively super-...-super-curved, negatively super-...-super-curved, positively super-...-super-curved and negatively torsioned at one or more scales in arbitrary patterns including, but not limited to, fractals, negatively super-...-super-curved and negatively torsioned at one or more scales in arbitrary patterns including, but not limited to, fractals, positively super-...-super-curved and negatively super-...-super-torsioned at one or more scales in arbitrary patterns including, but not limited to, fractals, negatively super-...-super-curved and negatively super-...-super-torsioned at one or more scales in arbitrary patterns including, but not limited to, fractals.

All these formulations assume distance traveled through space irrespective of additional recessionary velocities that may accrue in expanding universes or higher realms including, but not limited to, hyperspaces. Recessional velocities can exceed c with respect to a location of origin, but such velocities do not accrue without space-time expansion.

So quite amazing travel itineraries are possible with infinite distances traversable in finite ship-time as well as infinite future time travel in finite ship-time.

Such travel entails travel in turn along eternal highway paths involving road segments of one light-speed after another.

At each light-speed in each associated realm, the crew can uncover new phenomena and an ever-growing sense of the meaning of light-speed from an ever-more-holistic approach.

Since the true speed of light in our universe is the same in all reference frames and assures the integrity of the chronological protection conjecture, we can intuit a major theme of Mother Nature, and this is that the speed of light is an all-encompassing parameter that ensures the integrity of the accidental aspects of Mother Nature, perhaps even down to the accidental aspect of any truly spiritual and immortal human soul.

As we explore how and why the speed of light has its own particular value in each realm, we will learn more and more about the speed of light, both from the philosophical and the mathematically scientific standpoint, which will enable untold future technological applications.

Back in the days, when I was a teenager, I attended a private school. The school psychologist was a consecrated Catholic religious brother with dark hair and a dark beard, who used to let me ride in his fancy Ford Thunderbird. The car had a black exterior and interior.

Well, at about the same time, the sitcom *The Jeffersons* was popular and the show's theme song had a refrain that went like, "Well we're movin' on up. / (Movin' on up). / To the east side."

Even back then I was interested in interstellar travel concepts.

To make a long story short, I associated with the school psychologist and rode in his Thunderbird with my internalized mantra of "Movin' on up," to the future, at near light-speed. Thus I became more hooked on special relativistic space travel and time dilation. The fire of my imagination for near-light-speed travel was lit just as assuredly as the black Ford Thunderbird resembled the black cosmic void. I knew then the ramifications of infinite time dilation and infinite forward time travel and distances through space were mathematically plausible for light-speed impulse travel.

So if you have the courage to delve farther into this book, or even study only select portions thereof, you will likely, if you have not already, also become intrigued with movin' on up into the future with special relativity. As we now have a space travel industry, we have set before us the seas of infinity.

In a way, we are all movin' on up into the future special relativistically. Such a future is unbounded in terms of forward temporal progression and the eternal duration of reality.

Even for spacecraft having attained light-speed in the least infinite Lorentz factor, there is no mathematical reason why the light-speed spacecraft cannot continue having increasing infinite Lorentz factors.

Essentially, we will always be movin' on up into the future.

Stay tuned! I have even more whimsically far-out concepts to share.

Now we can consider traveling at a first level ever so slightly faster than light.

How or what would such travel manifest?

Perhaps, just as light-speed travel results in infinite travel into the future in any finite period ship-frame, perhaps a first level of ever-so-slightly faster-than-light travel manifests in infinite travel into a preordinate coordinate relative to the future. Here, said pre-ordinate coordinate is to the future as the future is to the present.

We can intuit travel at a second level of faster-than-light travel as manifested in infinite travel into a second preordinate coordinate relative to the future or a first preordinate coordinate relative to the first preordinate coordinate relative to the future.

We can intuit travel at a third level of faster-than-light travel as manifested in infinite travel into a third preordinate coordinate relative to the future or a first preordinate coordinate relative to the second preordinate coordinate relative to the future or a second preordinate coordinate relative to the first preordinate coordinate relative to the future.

We can consider an unending series of levels of ever-so-slightly faster-than-light travel.

Here, we are not merely conjecturing travel into a hyper-future, or a future beyond a future beyond a future, and so on; nor are we conjecturing travel into a hyper-eternity, or an eternity after an eternity and so on. Instead, we are considering a kind of other than futures or other than eternities, but in ways that aren't subjugated to higher levels of future or to higher levels of eternity, but instead are pre-ordinate to levels of future or levels of eternities.

We may affix the following operator to the formula for gamma and related parameters to denote the above range of possibilities in derivatives and integrals:

{w{{{{[Σ(j = 1; j = ∞↑):[[d[r[(x,1), (x, 2), …, (x,n)]]/d coordinate preordinate to future t]j]] ≥ (realm specific value of c)};{{[Σ(j = 1; j = ∞↑):[[d[r[(x,1), (x, 2), …, (x,n)]]/dγ]j]]} returned from kth level of ever-so-slightly superluminal travel};{{[Σ(j = 1; j = ∞↑):[[d[r[(x,1), (x, 2), …, (x,n)]]/d ship-frame acceleration]j]]} returned from kth level of ever-so-slightly superluminal travel};{{[Σ(j = 1; j = ∞↑):[[d[r[(x,1), (x, 2), …, (x,n)]]/d K.E]j]]} returned from kth level of ever-so-slightly superluminal travel};{{∫ [Σ(j = 1; j = ∞↑):[[d[r[(x,1), (x, 2), …, (x,n)]]/d coordinate preordinate to future t]j]] dtship} returned from kth level of ever-so-slightly superluminal travel};{{∫ [Σ(j = 1; j = ∞↑):[[d[r[(x,1), (x, 2), …, (x,n)]]/d coordinate preordinate to future t]j]] dγ} returned from kth level of ever-so-slightly superluminal travel}; {{∫ [Σ(j = 1; j = ∞↑):[[d[r[(x,1), (x, 2), …, (x,n)]]/d coordinate preordinate to future t]j]] d acceleration ship} returned from kth level of ever-so-slightly superluminal travel}; {{∫ [Σ(j = 1; j = ∞↑):[[d[r[(x,1), (x, 2), …, (x,n)]]/d coordinate preordinate to future t]j]] d KE} returned from kth level of ever-so-slightly superluminal travel};{{∫ [Σ(j = 1; j = ∞↑):[[d[r[(x,1), (x, 2), …, (x,n)]]/d γ]j]] dtship} returned from kth level of ever-so-slightly superluminal travel}; {{∫ [Σ(j = 1; j = ∞↑):[[d[r[(x,1), (x, 2), …, (x,n)]]/d γ]j]] dγ} returned from kth level of ever-so-slightly superluminal travel}; {{∫ [Σ(j = 1; j = ∞↑):[[d[r[(x,1), (x, 2), …, (x,n)]]/d γ]j]] d acceleration ship} returned from kth level of ever-so-slightly superluminal travel}; {{∫ [Σ(j = 1; j = ∞↑):[[d[r[(x,1), (x, 2), …, (x,n)]]/d γ]j]] d KE} returned from kth level of ever-so-slightly superluminal travel};{{∫ [Σ(j = 1; j = ∞↑):[[d[r[(x,1), (x, 2), …, (x,n)]]/d acceleration]j]] dtship} returned from kth level of ever-so-slightly superluminal travel}; {{∫ [Σ(j = 1; j = ∞↑):[[d[r[(x,1), (x, 2), …, (x,n)]]/d acceleration]j]] dγ} returned from kth level of ever-so-slightly superluminal travel}; {{∫ [Σ(j = 1; j = ∞↑):[[d[r[(x,1), (x, 2), …, (x,n)]]/d acceleration]j]] d acceleration ship} returned from kth level of ever-so-slightly superluminal travel}; {{∫ [Σ(j = 1; j = ∞↑):[[d[r[(x,1), (x, 2), …, (x,n)]]/d acceleration]j]] d KE} returned from kth level of ever-so-slightly superluminal travel};{{∫ [Σ(j = 1; j = ∞↑):[[d[r[(x,1), (x, 2), …, (x,n)]]/d KE]j]] dtship} returned from kth level of ever-so-slightly superluminal travel}; {{∫ [Σ(j = 1; j = ∞↑):[[d[r[(x,1), (x, 2), …, (x,n)]]/d KE]j]] dγ} returned from kth level of ever-so-slightly superluminal travel}; {{∫ [Σ(j = 1; j = ∞↑):[[d[r[(x,1), (x, 2), ..., (x,n)]]/d KE]j]] d acceleration ship} returned from kth level of ever-so-slightly superluminal travel}; {{∫ [Σ(j = 1; j = ∞↑):[[d[r[(x,1), (x, 2), ..., (x,n)]]/d KE] j]] d KE} returned from kth level of ever-so-slightly superluminal travel}} as well as arbitrarily higher-order derivatives and integrals with regard to respective or other variables}} all, where *k* is any positive integer, finite or infinite}

ESSAY 32) The Many Ways of Light-Speed Travel

Now we know of three main ways for light-speed travel.

First, there is translational or inertial travel such as for a photon in free space.

Second, there is photon tunneling.

Third, there is quantum teleportation.

All three of these modes may be supplanted by other forms of light-speed travel such as enabled by warp drive and wormhole travel.

So we know of five plausible light-speed travel concepts.

Each of these light-speeds we may refer to as a flavor of light-speed.

So, for the above five methods, we have five distinct flavors of light-speed.

We may also intuit hybrids of two or more of the above flavors of light-speed.

Might there be other flavors of light-speed?

I would hazard a guess that there may be many more flavors that science does not yet have the language to conjecture about or to conduct any validating empirical studies. In a universe of vast scales and a greater cosmos, the range of phenomena, dimensionalities, and so on may support an infinite number of flavors of light-speed.

Now we all have generally been aware that the speed of light in a perfect vacuum is a universal constant.

We are also aware of the concept of electromagnetic waves and photons, and now gravitational waves.

However, most of us are not aware of the concept of yelm, which would be the raw energy material of initialized big bang composition before the first symmetry-breaking event caused the electro-weak-strong force to break off from gravitation at about the first Planck time unit after the start of the big bang. The Planck time unit is $T_P = \{[h/(2\pi)]\ G/(c^5)\}^{1/2}.=$ Here, h, G, and c are the Planck constant, gravitational constant, and speed of light, respectively, in MKS units. The Planck time is [5.39 x (10 EXP – 44)] second or 0.0000000000000000000000000000000000000000000539 second. This is a very short time indeed.

A fascinating prospect would Include duplicating yelm conditions to produce a sail that would be pushed forward by yelm. Here, the yelm sail would be reflective of the yelm and thus have a refractive index of zero. Negative yelm refractive index sails would be pulled on by incident yelm from in front of the sails.

Regarding types of photons, we have linearly polarized, transversely polarized, right circular polarized, left circular polarized, right elliptically polarized, left elliptically polarized, and obliquely polarized. However, we also have super-chiral photons that have topologies of twisting or quark screw features, where the classical number of twists may range from greater than 1 to infinities. Due to the limits of the Planck length, the number of twists is likely commensurately limited to one twist per Planck length unit.

The smallest spatial interval that theoretically has meaning is the Planck length unit at

$$\{[h/(2\pi)]\ G/(c^3)\}^{1/2} \approx [1.6 \times (10\ EXP - 35)]\ \text{meters}$$

Perhaps different topological forms of photons travel at ever so slightly different speeds but where the speeds trivially differ in terms of being so small in difference such that special relativity remains upheld.

Light-sails powered or propelled by the beginning of the process of creation of archetypal light where said archetypal light travels at the speed of light or perhaps ever so slightly faster than light.

The ever so slightly superluminal archetypal light may be a stand-in in, explanation, or underwriter for the maelstrom of chaos that existed before creation, or at the point of creation where the laws of creation were just coming into existence, thus implying an initial chaotic state of creation and cause-and-effect uncertainty or lack of definition, thus voiding the chronological protection conjecture at the moments of the commencement of creation and just afterward as the laws of physics were being developed and thus uncertain.

There may be many finite, if not infinite, numbers of species of yelms and archetypal light.

We may reasonably affix the following operator to formulas for gamma and related parameters to denote the number of flavors of light-speed and the rate of acquisition of light-speed flavors and first and higher orders of derivatives and integrals thereof.

Thus, we intuit that the large set of light-speed flavors may be serially acquired.

{w{{{[Σ(j = 1; j = ∞↑):[[d[r[(x,1), (x, 2), ..., (x,n)]]/dt]j]] ≤ (travel mode specific flavor of c)};{[Σ(j = 1; j = ∞↑):[[d[r[(x,1), (x, 2), ..., (x,n)]]/dγ]j]]};{[Σ(j = 1; j = ∞↑):[[d[r[(x,1), (x, 2), ..., (x,n)]]/d ship-frame acceleration]j]]};{[Σ(j = 1; j = ∞↑):[[d[r[(x,1), (x, 2), ..., (x,n)]]/d K.E]j]]};{∫ [Σ(j = 1; j = ∞↑):[[d[r[(x,1), (x, 2), ..., (x,n)]]/dt]j]] dtship}; {∫ [Σ(j = 1; j = ∞↑):[[d[r[(x,1), (x, 2), ..., (x,n)]]/dt]j]] dγ}; {∫ [Σ(j = 1; j = ∞↑):[[d[r[(x,1), (x, 2), ..., (x,n)]]/dt]j]] d acceleration ship}; {∫ [Σ(j = 1; j = ∞↑):[[d[r[(x,1), (x, 2), ..., (x,n)]]/dt]j]] d KE};{∫ [Σ(j = 1; j = ∞↑):[[d[r[(x,1), (x, 2), ..., (x,n)]]/d γ]j]] dtship}; {∫ [Σ(j = 1; j = ∞↑):[[d[r[(x,1), (x, 2), ..., (x,n)]]/d γ]j]] dγ}; {∫ [Σ(j = 1; j = ∞↑):[[d[r[(x,1), (x, 2), ..., (x,n)]]/d γ]j]] d acceleration ship}; {∫ [Σ(j = 1; j = ∞↑):[[d[r[(x,1), (x, 2), ..., (x,n)]]/d γ]j]] d KE};{∫ [Σ(j = 1; j = ∞↑):[[d[r[(x,1), (x, 2), ..., (x,n)]]/d acceleration]j]] dtship}; {∫ [Σ(j = 1; j = ∞↑):[[d[r[(x,1), (x, 2), ..., (x,n)]]/d acceleration]j]] dγ}; {∫ [Σ(j = 1; j = ∞↑):[[d[r[(x,1), (x, 2), ..., (x,n)]]/d acceleration]j]] d acceleration ship}; {∫ [Σ(j = 1; j = ∞↑):[[d[r[(x,1), (x, 2), ..., (x,n)]]/d acceleration]j]] d KE};{∫ [Σ(j = 1; j = ∞↑):[[d[r[(x,1), (x, 2), ..., (x,n)]]/d KE]j]] dtship}; {∫ [Σ(j = 1; j = ∞↑):[[d[r[(x,1), (x, 2), ..., (x,n)]]/d KE]j]] dγ}; {∫ [Σ(j = 1; j = ∞↑):[[d[r[(x,1), (x, 2), ..., (x,n)]]/d KE]j]] d acceleration ship}; {∫ [Σ(j = 1; j = ∞↑):[[d[r[(x,1), (x, 2), ..., (x,n)]]/d KE]j]] d KE }} as well as arbitrarily higher-order derivatives and integrals with regard to respective or other variables}}

We may reasonably affix the following operator to formulas for gamma and related parameters to denote the number of wave topology specific values of light-speed and the rate of acquisition of topological wave-form specific value of *c* and first and higher-orders of derivatives and integrals thereof, as well as spacecraft propulsion by one or more topological photon wave-forms.

{w{{{[Σ(j = 1; j = ∞↑):[[d[r[(x,1), (x, 2), ..., (x,n)]]/dt]j]] ≤ (topological wave-form specific value of c and where a spacecraft is propelled by one or more topological forms of light)};{[Σ(j = 1; j = ∞↑):[[d[r[(x,1), (x, 2), ..., (x,n)]]/dγ]j]]};{[Σ(j = 1; j = ∞↑):[[d[r[(x,1), (x, 2), ..., (x,n)]]/d ship-frame acceleration]j]]};{[Σ(j = 1; j = ∞↑):[[d[r[(x,1), (x, 2), ..., (x,n)]]/d K.E]j]]};{∫ [Σ(j = 1; j = ∞↑):[[d[r[(x,1), (x, 2), ..., (x,n)]]/dt]j]] dtship}; {∫ [Σ(j = 1; j = ∞↑):[[d[r[(x,1), (x, 2), ..., (x,n)]]/dt]j]] dγ}; {∫ [Σ(j = 1; j = ∞↑):[[d[r[(x,1), (x, 2), ..., (x,n)]]/dt]j]] d acceleration ship}; {∫ [Σ(j = 1; j = ∞↑):[[d[r[(x,1), (x, 2), ..., (x,n)]]/dt]j]] d KE};{∫ [Σ(j = 1; j = ∞↑):[[d[r[(x,1), (x, 2), ..., (x,n)]]/d γ]j]] dtship}; {∫ [Σ(j = 1; j = ∞↑):[[d[r[(x,1), (x, 2), ..., (x,n)]]/d γ]j]] dγ}; {∫ [Σ(j = 1; j =

∞↑):[[d[r[(x,1), (x, 2), …, (x,n)]]/d γ]j]] d acceleration ship}; {∫ [Σ(j = 1; j = ∞↑):[[d[r[(x,1), (x, 2), …, (x,n)]]/d γ]j]] d KE};{∫ [Σ(j = 1; j = ∞↑):[[d[r[(x,1), (x, 2), …, (x,n)]]/d acceleration]j]] dtship}; {∫ [Σ(j = 1; j = ∞↑):[[d[r[(x,1), (x, 2), …, (x,n)]]/d acceleration]j]] dγ}; {∫ [Σ(j = 1; j = ∞↑):[[d[r[(x,1), (x, 2), …, (x,n)]]/d acceleration]j]] d acceleration ship}; {∫ [Σ(j = 1; j = ∞↑):[[d[r[(x,1), (x, 2), …, (x,n)]]/d acceleration]j]] d KE};{∫ [Σ(j = 1; j = ∞↑):[[d[r[(x,1), (x, 2), …, (x,n)]]/d KE]j]] dtship}; {∫ [Σ(j = 1; j = ∞↑):[[d[r[(x,1), (x, 2), …, (x,n)]]/d KE]j]] dγ}; {∫ [Σ(j = 1; j = ∞↑):[[d[r[(x,1), (x, 2), …, (x,n)]]/d KE]j]] d acceleration ship}; {∫ [Σ(j = 1; j = ∞↑):[[d[r[(x,1), (x, 2), …, (x,n)]]/d KE]j]] d KE }} as well as arbitrarily higher-order derivatives and integrals with regard to respective or other variables}}

We may reasonably affix the following operator to formulas for gamma and related parameters to denote the number of yelms and the rate of acquisition of yelm specific values of ever so slightly greater than light-speed and first and higher orders of derivatives and integrals thereof as well as spacecraft propulsion by one or more yelms:

{w{{{[Σ(j = 1; j = ∞↑):[[d[r[(x,1), (x, 2), …, (x,n)]]/dt]j]] ≤ (Yelm light specific values of ever so slightly greater than light-speed, whereby a spacecraft is propelled by any one or multiplicity of yelms)};{[Σ(j = 1; j = ∞↑):[[d[r[(x,1), (x, 2), …, (x,n)]]/dγ]j]]};{[Σ(j = 1; j = ∞↑):[[d[r[(x,1), (x, 2), …, (x,n)]]/d ship-frame acceleration]j]]};{[Σ(j = 1; j = ∞↑):[[d[r[(x,1), (x, 2), …, (x,n)]]/d K.E]j]]};{∫ [Σ(j = 1; j = ∞↑):[[d[r[(x,1), (x, 2), …, (x,n)]]/dt]j]] dtship}; {∫ [Σ(j = 1; j = ∞↑):[[d[r[(x,1), (x, 2), …, (x,n)]]/dt]j]] dγ}; {∫ [Σ(j = 1; j = ∞↑):[[d[r[(x,1), (x, 2), …, (x,n)]]/dt]j]] d acceleration ship}; {∫ [Σ(j = 1; j = ∞↑):[[d[r[(x,1), (x, 2), …, (x,n)]]/dt]j]] d KE};{∫ [Σ(j = 1; j = ∞↑):[[d[r[(x,1), (x, 2), …, (x,n)]]/d γ]j]] dtship}; {∫ [Σ(j = 1; j = ∞↑):[[d[r[(x,1), (x, 2), …, (x,n)]]/d γ]j]] dγ}; {∫ [Σ(j = 1; j = ∞↑):[[d[r[(x,1), (x, 2), …, (x,n)]]/d γ]j]] d acceleration ship}; {∫ [Σ(j = 1; j = ∞↑):[[d[r[(x,1), (x, 2), …, (x,n)]]/d γ]j]] d KE};{∫ [Σ(j = 1; j = ∞↑):[[d[r[(x,1), (x, 2), …, (x,n)]]/d acceleration]j]] dtship}; {∫ [Σ(j = 1; j = ∞↑):[[d[r[(x,1), (x, 2), …, (x,n)]]/d acceleration]j]] dγ}; {∫ [Σ(j = 1; j = ∞↑):[[d[r[(x,1), (x, 2), …, (x,n)]]/d acceleration]j]] d acceleration ship}; {∫ [Σ(j = 1; j = ∞↑):[[d[r[(x,1), (x, 2), …, (x,n)]]/d acceleration]j]] d KE};{∫ [Σ(j = 1; j = ∞↑):[[d[r[(x,1), (x, 2), …, (x,n)]]/d KE]j]] dtship}; {∫ [Σ(j = 1; j = ∞↑):[[d[r[(x,1), (x, 2), …, (x,n)]]/d KE]j]] dγ}; {∫ [Σ(j = 1; j = ∞↑):[[d[r[(x,1), (x, 2), …, (x,n)]]/d KE]j]] d acceleration ship}; {∫ [Σ(j = 1; j = ∞↑):[[d[r[(x,1), (x, 2), …, (x,n)]]/d KE]j]] d KE }} as well as arbitrarily higher-order derivatives and integrals with regard to respective or other variables}}

We may reasonably affix the following operator to formulas for gamma and related parameters to denote the number of archetypal lights and the rate of acquisition of archetypal lights specific values of ever so slightly greater than light-speed and first and higher orders of derivatives and integrals thereof, as well as options for spacecraft propulsion by archetypal lights:

{w{{{[Σ(j = 1; j = ∞↑):[[d[r[(x,1), (x, 2), ..., (x,n)]]/dt]j]] ≤ (Archetypal light specific values of ever so slightly greater than light speed, whereby a spacecraft is propelled by any one or multiplicity of archetypal lights)};{[Σ(j = 1; j = ∞↑):[[d[r[(x,1), (x, 2), ..., (x,n)]]/dγ]j]]};{[Σ(j = 1; j = ∞↑):[[d[r[(x,1), (x, 2), ..., (x,n)]]/d ship-frame acceleration]j]]};{[Σ(j = 1; j = ∞↑):[[d[r[(x,1), (x, 2), ..., (x,n)]]/d K.E]j]]};{∫ [Σ(j = 1; j = ∞↑):[[d[r[(x,1), (x, 2), ..., (x,n)]]/dt]j]] dtship}; {∫ [Σ(j = 1; j = ∞↑):[[d[r[(x,1), (x, 2), ..., (x,n)]]/dt]j]] dγ}; {∫ [Σ(j = 1; j = ∞↑):[[d[r[(x,1), (x, 2), ..., (x,n)]]/dt]j]] d acceleration ship}; {∫ [Σ(j = 1; j = ∞↑):[[d[r[(x,1), (x, 2), ..., (x,n)]]/dt]j]] d KE};{∫ [Σ(j = 1; j = ∞↑):[[d[r[(x,1), (x, 2), ..., (x,n)]]/d γ]j]] dtship}; {∫ [Σ(j = 1; j = ∞↑):[[d[r[(x,1), (x, 2), ..., (x,n)]]/d γ]j]] dγ}; {∫ [Σ(j = 1; j = ∞↑):[[d[r[(x,1), (x, 2), ..., (x,n)]]/d γ]j]] d acceleration ship}; {∫ [Σ(j = 1; j = ∞↑):[[d[r[(x,1), (x, 2), ..., (x,n)]]/d γ]j]] d KE};{∫ [Σ(j = 1; j = ∞↑):[[d[r[(x,1), (x, 2), ..., (x,n)]]/d acceleration]j]] dtship}; {∫ [Σ(j = 1; j = ∞peed where by a spacecraft is propelled by any one or mutiple [Σ(j = 1; j = ∞↑):[[d[r[(x,1), (x, 2), ..., (x,n)]]/d acceleration]j]] d acceleration ship}; {∫ [Σ(j = 1; j = ∞↑):[[d[r[(x,1), (x, 2), ..., (x,n)]]/d acceleration]j]] d KE};{∫ [Σ(j = 1; j = ∞↑):[[d[r[(x,1), (x, 2), ..., (x,n)]]/d KE]j]] dtship}; {∫ [Σ(j = 1; j = ∞↑):[[d[r[(x,1), (x, 2), ..., (x,n)]]/d KE]j]] dγ}; {∫ [Σ(j = 1; j = ∞↑):[[d[r[(x,1), (x, 2), ..., (x,n)]]/d KE]j]] d acceleration ship}; {∫ [Σ(j = 1; j = ∞↑):[[d[r[(x,1), (x, 2), ..., (x,n)]]/d KE]j]] d KE }} as well as arbitrarily higher-order derivatives and integrals with regard to respective or other variables}}

Note that spacecraft propulsion by yelms and archetypal lights conceptually is an intermediary between truly direct divine supernatural intervention and ordinary visible accidental manifestations of light-speed and impulse travel forms that are the result of completely natural processes. So a technological civilization able to manifest propulsion by yelms and archetypal lights will necessarily be highly evolved and responsible.

In actuality, we may affix the following compound operator as a composition of the four previously presented operators to formulas for gamma and related parameters to denote the respective propulsion scenarios:

{{w{{{[Σ(j = 1; j = ∞↑):[[d[r[(x,1), (x, 2), ..., (x,n)]]/dt]j]] ≤ (travel mode specific flavor of c)};{[Σ(j = 1; j = ∞↑):[[d[r[(x,1), (x, 2), ..., (x,n)]]/dγ]j]]};{[Σ(j = 1; j = ∞↑):[[d[r[(x,1), (x, 2), ..., (x,n)]]/d ship-frame acceleration]j]]};{[Σ(j = 1; j = ∞↑):[[d[r[(x,1), (x, 2), ..., (x,n)]]/d K.E]j]]};{∫ [Σ(j = 1; j = ∞↑):[[d[r[(x,1), (x, 2), ..., (x,n)]]/dt]j]] dtship}; {∫ [Σ(j = 1; j = ∞↑):[[d[r[(x,1), (x, 2), ..., (x,n)]]/dt]j]] dγ}; {∫ [Σ(j = 1; j = ∞↑):[[d[r[(x,1), (x, 2), ..., (x,n)]]/dt]j]] d acceleration ship}; {∫ [Σ(j = 1; j = ∞↑):[[d[r[(x,1), (x, 2), ..., (x,n)]]/dt]j]] d KE};{∫ [Σ(j = 1; j =

∞↑):[[d[r[(x,1), (x, 2), …, (x,n)]]/d γ]j]] dtship}; {∫ [Σ(j = 1; j = ∞↑):[[d[r[(x,1), (x, 2), …, (x,n)]]/d γ]j]] dγ}; {∫ [Σ(j = 1; j = ∞↑):[[d[r[(x,1), (x, 2), …, (x,n)]]/d γ]j]] d acceleration ship}; {∫ [Σ(j = 1; j = ∞↑):[[d[r[(x,1), (x, 2), …, (x,n)]]/d γ]j]] d KE};{∫ [Σ(j = 1; j = ∞↑):[[d[r[(x,1), (x, 2), …, (x,n)]]/d acceleration]j]] dtship}; {∫ [Σ(j = 1; j = ∞↑):[[d[r[(x,1), (x, 2), …, (x,n)]]/d acceleration]j]] dγ}; {∫ [Σ(j = 1; j = ∞↑):[[d[r[(x,1), (x, 2), …, (x,n)]]/d acceleration]j]] d acceleration ship}; {∫ [Σ(j = 1; j = ∞↑):[[d[r[(x,1), (x, 2), …, (x,n)]]/d acceleration]j]] d KE};{∫ [Σ(j = 1; j = ∞↑):[[d[r[(x,1), (x, 2), …, (x,n)]]/d KE]j]] dtship}; {∫ [Σ(j = 1; j = ∞↑):[[d[r[(x,1), (x, 2), …, (x,n)]]/d KE]j]] dγ}; {∫ [Σ(j = 1; j = ∞↑):[[d[r[(x,1), (x, 2), …, (x,n)]]/d KE]j]] d acceleration ship}; {∫ [Σ(j = 1; j = ∞↑):[[d[r[(x,1), (x, 2), …, (x,n)]]/d KE]j]] d KE }} as well as arbitrarily higher-order derivatives and integrals with regard to respective or other variables}}:{w{{{[Σ(j = 1; j = ∞↑):[[d[r[(x,1), (x, 2), …, (x,n)]]/dt]j]] ≤ (topological wave-form specific value of c and where a spacecraft is propelled by one or more topological forms of light)};{[Σ(j = 1; j = ∞↑):[[d[r[(x,1), (x, 2), …, (x,n)]]/dγ]j]]};{[Σ(j = 1; j = ∞↑):[[d[r[(x,1), (x, 2), …, (x,n)]]/d ship-frame acceleration]j]]};{[Σ(j = 1; j = ∞↑):[[d[r[(x,1), (x, 2), …, (x,n)]]/d K.E]j]]};{∫ [Σ(j = 1; j = ∞↑):[[d[r[(x,1), (x, 2), …, (x,n)]]/dt]j]] dtship}; {∫ [Σ(j = 1; j = ∞↑):[[d[r[(x,1), (x, 2), …, (x,n)]]/dt]j]] dγ}; {∫ [Σ(j = 1; j = ∞↑):[[d[r[(x,1), (x, 2), …, (x,n)]]/dt]j]] d acceleration ship}; {∫ [Σ(j = 1; j = ∞↑):[[d[r[(x,1), (x, 2), …, (x,n)]]/dt]j]] d KE};{∫ [Σ(j = 1; j = ∞↑):[[d[r[(x,1), (x, 2), …, (x,n)]]/d γ]j]] dtship}; {∫ [Σ(j = 1; j = ∞↑):[[d[r[(x,1), (x, 2), …, (x,n)]]/d γ]j]] dγ}; {∫ [Σ(j = 1; j = ∞↑):[[d[r[(x,1), (x, 2), …, (x,n)]]/d γ]j]] d acceleration ship}; {∫ [Σ(j = 1; j = ∞↑):[[d[r[(x,1), (x, 2), …, (x,n)]]/d γ]j]] d KE};{∫ [Σ(j = 1; j = ∞↑):[[d[r[(x,1), (x, 2), …, (x,n)]]/d acceleration]j]] dtship}; {∫ [Σ(j = 1; j = ∞↑):[[d[r[(x,1), (x, 2), …, (x,n)]]/d acceleration]j]] dγ}; {∫ [Σ(j = 1; j = ∞↑):[[d[r[(x,1), (x, 2), …, (x,n)]]/d acceleration]j]] d acceleration ship}; {∫ [Σ(j = 1; j = ∞↑):[[d[r[(x,1), (x, 2), …, (x,n)]]/d acceleration]j]] d KE};{∫ [Σ(j = 1; j = ∞↑):[[d[r[(x,1), (x, 2), …, (x,n)]]/d KE]j]] dtship}; {∫ [Σ(j = 1; j = ∞↑):[[d[r[(x,1), (x, 2), …, (x,n)]]/d KE]j]] dγ}; {∫ [Σ(j = 1; j = ∞↑):[[d[r[(x,1), (x, 2), …, (x,n)]]/d KE]j]] d acceleration ship}; {∫ [Σ(j = 1; j = ∞↑):[[d[r[(x,1), (x, 2), …, (x,n)]]/d KE]j]] d KE }} as well as arbitrarily higher-order derivatives and integrals with regard to respective or other variables}}:{w{{{[Σ(j = 1; j = ∞↑):[[d[r[(x,1), (x, 2), …, (x,n)]]/dt]j]] ≤ (Yelm light specific values of ever so slightly greater than light-speed, where by a spacecraft is propelled by any one or multiplicity of yelms)};{[Σ(j = 1; j = ∞↑):[[d[r[(x,1), (x, 2), …, (x,n)]]/dγ]j]]};{[Σ(j = 1; j = ∞↑):[[d[r[(x,1), (x, 2), …, (x,n)]]/d ship-frame acceleration]j]]};{[Σ(j = 1; j = ∞↑):[[d[r[(x,1), (x, 2), …, (x,n)]]/d K.E]j]]};{∫ [Σ(j = 1; j = ∞↑):[[d[r[(x,1), (x, 2), …, (x,n)]]/dt]j]] dtship}; {∫ [Σ(j = 1; j = ∞↑):[[d[r[(x,1), (x, 2), …, (x,n)]]/dt]j]] dγ}; {∫ [Σ(j = 1; j = ∞↑):[[d[r[(x,1), (x, 2), …, (x,n)]]/dt]j]] d acceleration ship}; {∫ [Σ(j = 1; j = ∞↑):[[d[r[(x,1), (x, 2), …, (x,n)]]/dt]j]] d KE};{∫ [Σ(j = 1; j = ∞where by a spacecraft is propelled by any one or mult∫ [Σ(j = 1; j = ∞↑):[[d[r[(x,1), (x, 2), …, (x,n)]]/d γ]j]] dγ}; {∫ [Σ(j = 1; j = ∞↑):[[d[r[(x,1), (x, 2), …, (x,n)]]/d γ]j]] d acceleration ship}; {∫ [Σ(j = 1; j = ∞↑):[[d[r[(x,1), (x, 2), …, (x,n)]]/d γ]j]] d KE};{∫ [Σ(j = 1; j = ∞↑):[[d[r[(x,1), (x, 2), …, (x,n)]]/d acceleration]j]] dtship}; {∫ [Σ(j = 1; j = ∞↑):[[d[r[(x,1), (x, 2), …, (x,n)]]/d acceleration]j]] dγ}; {∫ [Σ(j = 1; j = ∞↑):[[d[r[(x,1), (x, 2), …, (x,n)]]/d acceleration]j]] d acceleration ship}; {∫ [Σ(j = 1; j = ∞↑):[[d[r[(x,1), (x, 2), …, (x,n)]]/d acceleration]j]] d KE};{∫ [Σ(j = 1; j = ∞↑):[[d[r[(x,1), (x, 2), …, (x,n)]]/d KE]j]] dtship}; {∫ [Σ(j = 1; j = ∞↑):[[d[r[(x,1), (x, 2), …, (x,n)]]/d KE]j]] dγ}; {∫ [Σ(j = 1; j = ∞↑):[[d[r[(x,1), (x, 2), …, (x,n)]]/d KE]j]] d acceleration ship}; {∫ [Σ(j = 1; j = ∞↑):[[d[r[(x,1), (x, 2), …, (x,n)]]/d KE]j]] d KE }} as well as arbitrarily higher-order derivatives and integrals with regard to respective or other variables}}:{w{{{[Σ(j = 1; j = ∞↑):[[d[r[(x,1), (x, 2), …, (x,n)]]/dt]j]] ≤ (Archetypal light specific values of ever so slightly greater than light-speed, whereby a spacecraft is propelled by any one or multiplicity of archetypal lights)};{[Σ(j = 1; j = ∞↑):[[d[r[(x,1), (x, 2), …, (x,n)]]/dγ]j]]};{[Σ(j = 1; j = ∞↑):[[d[r[(x,1), (x, 2), …, (x,n)]]/d ship-frame acceleration]j]]};{[Σ(j = 1; j = ∞peed where by a spa∫ [Σ(j = 1; j = ∞↑):[[d[r[(x∫ [Σ(j = 1; j = ∞↑):[[d[r[(x,1), (x, 2), …, (x,n)]]/dt]j]] dtship}; {∫ [Σ(j = 1; j = ∞↑):[[d[r[(x,1), (x, 2), …, (x,n)]]/dt]j]] dγ}; {∫ [Σ(j = 1; j = ∞↑):[[d[r[(x,1), (x, 2), …, (x,n)]]/dt]j]] d acceleration ship}; {∫ [Σ(j = 1; j = ∞↑):[[d[r[(x,1), (x, 2), …, (x,n)]]/dt]j]] d KE};{∫ [Σ(j = 1; j = ∞↑):[[d[r[(x,1), (x, 2), …, (x,n)]]/d γ]j]] dtship}; {∫ [Σ(j = 1; j = ∞↑):[[d[r[(x,1), (x, 2), …, (x,n)]]/d γ]j]] dγ}; {∫ [Σ(j = 1; j = ∞↑):[[d[r[(x,1), (x, 2), …, (x,n)]]/d γ]j]] d acceleration ship}; {∫ [Σ(j = 1; j = ∞↑):[[d[r[(x,1), (x, 2), …, (x,n)]]/d γ]j]] d KE};{∫ [Σ(j = 1; j = ∞↑):[[d[r[(x,1), (x, 2), …, (x,n)]]/d acceleration]j]] dtship}; {∫ [Σ(j = 1; j = ∞↑):[[d[r[(x,1), (x, 2), …, (x,n)]]/d acceleration]j]] dγ}; {∫ [Σ(j = 1; j = ∞↑):[[d[r[(x,1), (x, 2), …, (x,n)]]/d acceleration]j]] d acceleration ship}; {∫ [Σ(j = 1; j = ∞↑):[[d[r[(x,1), (x, 2), …, (x,n)]]/d acceleration]j]] d KE};{∫ [Σ(j = 1; j = ∞↑):[[d[r[(x,1), (x, 2), …, (x,n)]]/d KE]j]] dtship}; {∫ [Σ(j = 1; j = ∞↑):[[d[r[(x,1), (x, 2), …, (x,n)]]/d KE]j]] dγ}; {∫ [Σ(j = 1; j = ∞↑):[[d[r[(x,1), (x, 2), …, (x,n)]]/d KE]j]] d acceleration ship}; {∫ [Σ(j = 1; j = ∞↑):[[d[r[(x,1), (x, 2), …, (x,n)]]/d KE]j]] d KE }} as well as arbitrarily higher-order derivatives and integrals with regard to respective or other variables}}}

ESSAY 33) From Ordinary Infinite to Nonmathematically Describable Infinite Spacecraft Lorentz Factors and Consequences Thereof

Now we know of concepts for the finite and the infinite.

Before continuing, we need to do a mild digression into the concept of infinities.

Note that Hyper4(a, n) is equal to a tetrated n, or a raised to the power of itself n-1 times. The latter value is symbolically written as n subscript a.

For example,

3 EXP 4 = 81, but 4 subscript 3 is approximately equal to 10 EXP (1,000,000,000,000).

Alternatively,

4 subscript 2 = 2 EXP 2 EXP 2 EXP 2 = 2 EXP [2 EXP [2 EXP 2]] = 2 EXP (2 EXP 4) = 2 EXP 16 = 65,536.

For example,

Hyper5(4, 4)is equal to 4 tetrated 4 tetrated 4 tetrated 4. This value is commonly referred to as 4 pentated 4.

Hyper 6, (4,4) is 4 pentated 4 pentated 4 pentated 4 and is also referred to as 4 hexataed 4.

Hyper 7, (4,4) is 4 hexated 4 hexated 4 hexated 4 and so on.

Aleph 0 is the infinite number of integers.

Aleph 1, according to the perhaps unprovable, and thus unfalsifiable, continuum hypothesis, is the number of real numbers that is greater than Aleph 0 by a multiplicative factor of infinity.

Aleph 2 is similarly greater than Aleph 1.

Aleph 3 is similarly greater than Aleph 2.

Aleph 4 is similarly greater than Aleph 3.

And so on

In general, Aleph n = 2 EXP [Aleph (n-1)]

The number Ω is commonly stated as the least infinite positive integer or ordinal.

Now here is a real zinger.

So we can produce the abstraction of [Hyper Aleph Ω (Aleph Ω, Aleph Ω)].

We can go to ever greater infinities.

Now, there are mathematically recognized countable and uncountable infinities. Uncountable infinities are usually considered in contexts as being greater than countable infinities although the distinction is a bit more detailed.

So can you imagine the following infinities!

[Hyper Aleph Uncountable Infinities (Aleph Uncountable Infinities, Aleph Uncountable Infinities)]

We can go on to consider infinities so large that they have no mathematical language to explain them.

Now can you imagine the following infinities!

[Hyper Aleph (infinities so large that they have no mathematical language to explain them) [[Aleph (infinities so large that they have no mathematical language to explain them)], [Aleph (infinities so large that they have no mathematical language to explain them)]]]

Perhaps a spacecraft may obtain Lorentz factor in the range of the latter infinities.

Now consider a set of mathematical objects that are another whole class distinct from and elevated with respect to the infinities as infinities are to finite numbers.

This class of mathematical objects is not simply just larger than the infinities but instead, is a class of objects so elevated above infinities and analogously related to infinities as infinities are related to finite numbers.

We have no mathematical language to further elaborate on this class of mathematical objects. Suffice it to say that this class of mathematical objects has levels just as the infinities have levels.

And so continues the awesome scope of reality and our to never end process of exploration of the cosmos.

Now infinite Lorentz factors realized in light-speed travel in theory enable a spacecraft to travel infinite numbers of light-years through space and infinite numbers of years into the future thus effectively enabling travel at infinite multiples of the speed of light.

Thus, it seems as if a sufficiently infinite Lorentz factor scenario might lead to a spacecraft running out of future eternity in a local eternity to travel forward in time. Of course, such a spacecraft may pop into a larger and broader realm with a larger and broader future eternity. As with the first eternity, a spacecraft may thus run out of future eternity room to travel forward in time for the second eternity entered upon which the spacecraft attaining still greater infinite Lorentz factors may pop into a third and larger eternity than the second eternity considered previously. The process of hopping from one eternity to the next may be ongoing.

This being said, it behooves us to consider the process of hopping from one eternity to the next greater eternity, as if occurring in the spaceship-frame in an eternity like continuum for which the proper eternity-hopping process of the spacecraft is as if a mere local embossment on the eternity-like continuum proper to the spacecraft.

What's more, just as the spacecraft would hop from one eternity to the next, the spacecraft can realize emplacement in one eternity-like continuum after another. The number of eternities and eternity-like realms of entrance may be ever-growing and unboundedly infinite.

We may affix the following operator to formulas for gamma and related parameters to denote the spacecraft hopping from one eternity after another and emplacement in one eternity-like continuum after another in a never-ending series of ever higher levels and arbitrary orders of differentials and integrals of the rate of hopping progression with respect to gamma, spacecraft kinetic energy, spacecraft acceleration, spacecraft-time, and so on:

> [[w(Increasingly infinite Lorentz factor Spacecraft hopping from one eternity after another and emplacement in one eternity-like continuum after another in a never ending series of ever higher levels)] and arbitrary orders of differentials and integrals of the rate of hopping progression with respect to gamma, spacecraft kinetic energy, spacecraft acceleration, spacecraft-time and so on].

Now we are all familiar with the Cartesian number line and how it extends from positive infinity to negative infinity. We are also aware of the concept of 1 divided by 0 being equal to infinity.

Those of us who have had advanced college math are aware of the various aleph numbers, where Aleph n = 2 EXP [Aleph (n-1)], where *n* is any finite or infinite positive integer and the least aleph number, Aleph 0, is equal to the infinite number of positive integers.

We who are schooled in advanced math also understand the distinction between countable and uncountable infinities with uncountable infinities being generally larger than countable infinities.

However, we of higher math training often overlook the concept of the ultimate infinity or the ultimately large real number. We tend to view such inquiries as childish or otherwise intellectually immature.

Well, I would like to assert the real possibility that there is an ultimate infinity and will define the ultimate infinity by the symbol ф.

We can imagine that the number of GOD's ontological transcendentals is equal to ф, and perhaps even greater as nonlogically possible.

Ontological transcendentals, as I define them, are principles such as ontological goodness, value, purpose, and the like.

As such, ontological transcendentals surpass and are pre-ordinate even to constructs such as the all-powerful, the all-knowing, the all-loving, the incorruptible, and the like.

We humans are aware of only a few ontological transcendentals so I believe we are going to have great fun in Heaven discovering more and more of them in a never-ending learning process.

Now, coming back to the concepts of light-speed and so on aspects of extreme travel, we can always hold attainment of Lorentz factors, ship-frame acceleration, and the like as great as ф, as an ideal to work toward as if an eternally ever-distant beacon, but one that gives a mysterious and subtle light that forever beckons us forward to forever continue designing and assembling spacecraft that can travel ever closer to the speed of light and capable of ever more extreme Lorentz factors and magnitudes of acceleration.

We will take this digression still further to consider values beyond the ultimate infinity value. We denote hyper-ultimate infinities by the phrase >ф.

We can consider values beyond the values beyond the ultimate infinity by the phrase >>ф.

We can consider values beyond the values beyond the values beyond the ultimate infinity by the phrase >>>ф.

We label these three terms as follows:

$$1>ф, 2>ф, \text{ and } 3>ф.$$

We can continue as $4>ф$, $5>ф$, all the way up to and beyond $(\text{Aleph } \Omega)>ф$. GOD has these many and even more ontological transcendentals.

So we have so much exploration to do about GOD in the next life, the ever-growing cosmos in the next life, the ever-growing set of experiences in the next life, and so on.

As for GOD, I would even hazard a guess that He has incomprehensibly more than $[(\text{Aleph } \Omega)>ф]>ф$ ontological transcendentals.

So do not fret about any light-speed limits. Such limits make and imply travel at ever greater finite and eventually infinite Lorentz factors. We have an eternal voyage before us in exploring the cosmos, Mother Nature, the mysterious existential depth of our souls and GOD.

Now, just about every physics theory and formula is related to or inclusive of the speed of light. As such, the speed of light is an all-important physical parameter that stitches the laws of physics and ordered causality in the universe together, and, presumably, in analogues in other realms such as multiverses, forests, biospheres, hyperspaces, and the like.

So a spacecraft traveling inertially able to attain the velocity of light in a sense becomes more related and relevant commensurately with the existential importance of the speed of light.

Attaining the speed of light in infinite Lorentz factors is like achieving a great existential awakening for which a spacecraft and crew become nonsentiently more existentially full.

However, there are some expected consequences for any living and awake crew members on a light-speed infinite Lorentz factor spacecraft. For example, the technology, culture, science, evolution of life, and so on will manifest in truly extreme ways for the light-speed crew when they arrive at a destination having experienced but even a mere Planck time unit ship-frame at the speed of light. So those who hold that divine providence will radically operate on the background civilizations, cultures, spirituality, and so on for the background personal lifeforms, and sub-personal life-forms that evolve to personal lifeforms must hold that the discontinuities experienced by the spacecraft crew from just before a light-speed mission to just after arrival at a destination will manifest as Lorentz varying scenarios. After all, while the craft traveled at the speed of light for finite ship-time intervals, the cosmic order would have drastically changed most likely additionally having undergone universal phase changes, symmetry-breaking events, vacuum state energy decays, and many other similar events perhaps an infinite number of times. So this alone assures that many results of light-speed travel at infinite Lorentz factors are not Lorentz invariant.

Now the speed of light according to Special Relativity is the same in all reference frames, presumably even with respect to light-speed reference frames.

Accordingly, a spacecraft traveling at the speed of light will experience a beam of light from an internal laser also traveling at the speed of light. As ironic as this sounds, this would be true.

So we are left with a conjecture that the speed of light is multiply so or that the velocity of light is multiply existentially present with itself.

Multiply existential self-presence of the speed of light would imply perhaps that a spacecraft traveling at the speed of light in a vacuum while obtaining increasingly infinite Lorentz factors would, in a sense, seem to become increasingly multiply present with itself.

Additionally, the light-speed spacecraft may become locked into the speed of light in one infinite Lorentz factor to another greater infinite Lorentz factor and so on. Thus, a spacecraft experiencing nonzero ship-frame acceleration may become perpetually locked in acceleration or Lorentz factor increases and the acceleration and Lorentz factors may in a sense continue to the runway to ever greater values in an asymptotic like manner, except that the predecessor Lorentz factors of the spacecraft before said asymptotic-like runaways would already be infinite. This being said, the spacecraft would thus achieve one qualitatively functional greater infinite Lorentz factor after another.

In a sense, the spacecraft would experience increasingly rapid asymptotic gains in qualitatively different and greater quantitatively functionally infinite increases in Lorentz factors.

We may affix the following operator to formulas for gamma and related parameters to denote the relevant mechanisms of the rates of the following:

1) light-speed spacecraft experiencing potential divine providence radically operating on the background civilizations, cultures, spirituality, and so on for the background personal life-forms and sub-personal life-forms that evolve to personal life-forms; 2) the extent of the spacecraft becoming increasingly multiply present with itself; 3) the spacecraft experiencing increasingly rapid asymptotic gains in qualitatively different and greater quantitatively functionally infinite increases in Lorentz factors; as well as single and higher-order derivatives and integrals of these three itemized rates with respect to ship-time, gamma, acceleration, kinetic energy, and other parameters.

> [w(Relevant mechanisms of the rates of 1) light-speed spacecraft experiencing potential divine providence radically operating on the background civilizations, cultures, spirituality, and so on for the background personal lifeforms and sub-personal lifeforms that evolve to personal lifeforms; 2) the extent of the spacecraft becoming increasingly multiply present with itself; 3) the spacecraft experiencing increasingly rapid asymptotic gains in qualitatively different and greater quantitatively functionally infinite increases in Lorentz factors; as well as single and higher-order derivatives and integrals of these three itemized rates with respect to ship-time, gamma, acceleration, kinetic energy, and other parameters)]

Now we can imagine a light-speed spacecraft running out of future room to travel forward in time as a result of time dilation. Thus, we can intuit that such a light-speed spacecraft having attained a sufficiently infinite Lorentz factor might pop out of a universe of travel to enter in a broader realm such as a multiverse, a forest, a biosphere, and the like or a hyperspace.

Obtaining still a greater infinite Lorentz factor, a spacecraft may pop out of a first greater realm of travel and into a second and greater realm of travel.

Obtaining still yet a greater infinite Lorentz factor, a spacecraft may pop out of a second greater realm of travel and into a third and greater realm of travel.

We can continue likewise to ever greater realms.

However, suppose a spacecraft could make an impossible jump to an incredibly great infinite Lorentz factor to such an extent that the spacecraft leaps out of the entire cosmic order, indeed, out of the entire created order and into some kind of realm or state beyond creation. Here, we are not considering supernatural realms or the life of grace but, instead, something much more abstract and perhaps even more sublime.

Here, we conjecture that there are real realms that are neither GOD nor creation but which, in a sense, depend on GOD to be manifested or to have been manifested. What such realms would be is a question alluding to some real, non-divine, and non-created reality or multiple realities.

There may be more than one such non-created non-divine realities, perhaps in numbers so large that the number cannot be defined.

Another option would be that the spacecraft having attained sufficiently huge infinite Lorentz factors may pop out of the created order to enter negation. Perhaps there is more than one level of negation. There may even be an undefinable infinite number of negations or levels of negation.

Just as the number of objects, realms, and potential realms of negation that do not properly exist cannot be defined, so cannot be defined the number of non-created and non-divine levels of reality.

The prospects of travel into such realms are so extreme in the conjecture that we will have all eternity to explore these realms.

This brings us to another conjecture. Presumably, there is only one GOD, which is what I believe as a practicing Catholic. However, the number of GODs otherwise, but which do not exist, would be undefinable. To say that the number of such nonexisting GODs is infinite does not grasp the untold indefinability.

Similarly, the number of non-GOD objects that do not exist is also undefinable and is so large that infinite values cannot be assigned to such nonexistent, non-divine numbers of non-GOD objects.

So prospects for sufficiently infinite Lorentz factor spacecraft overshooting the cosmos and into non-GOD objects or realms are enormous, so much so that the number of such objects or realms cannot be defined.

We should, however, not view the above speculations as a "no show" for and of GOD in any way that diminishes Him. After all, GOD is existence Himself in the most pure and possible ultimate way. GOD as pure existence definitely dwarfs all nonexisting realities to an undefinable extent.

Thus, to the extent that GOD dwarfs all nonexisting realities to an undefinable extent, GOD is likewise commensurately larger than the number of non-existing objects and non-exiting realities.

Now, if GOD is undefinably larger than the number of nonexisting objects and nonexisting realities, which is in turn undefinably larger than the number of existing objects and realities that are not GOD, how large, then, must be GOD and how extreme must His existence be. After all, GOD is existence in its very self.

Now ontological transcendentals are properties such as value, purpose, ontological goodness, worth, and so on. We created persons who recognize a relatively few ontological transcendentals. However, GOD being pure existence to such an extent that He dwarfs the number of nonexistent realities by a factor that is undefinable, GOD must likewise have a commensurately undefinable number of ontological transcendentals and must really be inherently each of these ontological transcendentals.

So we have all future eternity to explore reality, especially the reality of GOD, but also the good cosmos He made and continues to make.

GOD as pure existence, absolutely simple is pre-ordinate over His Three Divine Persons although each Divine Person is truly GOD.

As pure existence with preordination, we can say GOD is more existence than pure existence. We will refer to this attribute as more-existence-than-pure-existence-1.

We may also say that GOD is more existence than more-existence-than-pure-existence-1 or more-existence-than-pure-existence-2.

We may also say that GOD is more existence than more-existence-than-pure-existence-2 or more-existence-than-pure-existence-3.

We can continue the series in numbers of serial terms that are beyond definition.

We may affix the following operator to formulas for gamma and related parameters to define the number of and scope of scenarios and realms potentially accessible and outside of the created order but which are not GOD or any aspects of GOD and how GOD is undefinably greater than these scenarios and realms as pure existence Himself and as denoted by the terms more-existence-than-pure-existence-k, where *k* ranges from 1 to undefinable integers, along with His commensurately greater number of ontological transcendentals for us to explore.

> [w(Undefinable number of scopes, scenarios, and realms potentially accessible and outside of the created order but which are not GOD nor any aspects of GOD and how GOD is undefinably greater than these scenarios and realms as pure existence Himself as denoted by the terms, more-existence-than-pure-existence-k, where k ranges from 1 to undefinable integers, along with His commensurately greater number of ontological transcendentals for us to explore)]

Now the speed of light is perhaps the most important physical constant.

For example, the speed of light as perhaps and hopefully an inviolable speed limit ensures that effects cannot happen before their causes. So the inviolability of the speed of light ensures that physical creation and, perhaps by coupling, spiritual creations, are not corrupted by causal chaos.

Second, the speed of light features so prominently in the laws of physics that it ensures the conservation of energy and the mass-energy equivalence.

Third, as a result of time dilation, the speed of light obtained by massive spacecraft in some conjectural scenarios enables infinite distance of travel, effectively infinite multiples of the speed of light in travel velocities, and infinite travel into the future in one Planck time unit ship-frame.

Fourth, the inviolability of the speed of light ensures, to a reasonable degree, that extraterrestrial civilizations traveling between stars and galaxies have sufficiently spiritually matured such that they are not a direct threat to other civilizations by armed aggression. Overly aggressive species would likely have caused themselves to become extinct by armed conflict and intraivilization warfare before, they master interstellar and intergalactic travel.

Now the speed of light is the same in all reference frames according to special relativity. Thus, a spacecraft having attained the velocity of light may have crew members communicating acoustically and visually within their spacecraft even via electronic or photonic signals that travel at the speed of light in the spacecraft reference frame.

So we may surmise that the speed of light is actually a spectrum of velocities having a vanishingly small finite, or perhaps even infinitesimal width.

Thus the velocity of light may be as if comprised of an infinite number of metaphorical tiles that form the tapestry on which the order of cause and effect and existential integrity of the accidental aspects of created reality is maintained.

The tiles of light-speed may as if form their own tapestry, which may be akin to an analogue of space-times or higher dimensional continuums.

All this being said, it is possible that spacecraft having attained the speed of light in any of an infinite number of infinite Lorentz factors may be converted from a mass-energy-space-time composition to a composition of the speeds of light or tiles of light-speed.

Additionally, a light-speed spacecraft may travel in a continuum composed of light-speed tiles as analogues of subluminal spacecraft traveling among the realms of mass-energy-space-times.

Just as there may be other universes, multiverses, forests, biospheres, and the like and hyperspaces of different dimensionalities, and then so in an infinite number of varieties, so might there also exist an infinite number of continuums fabricated of light-speed tiles.

Additionally, since there would be a spectral width of the velocity of light, there may exist infinite numbers of species of light-speed tiles. Thus, the varieties of different and distinct light-speed tile tapestries or continuums may be of infinite numbers.

We may affix the following operator to formulas for gamma and related parameters to denote the above light-speed tiling and tapestry scenarios.

> [w(The speed of light as perhaps and hopefully an inviolable speed limit assuring that: 1) Effects cannot happen before their causes so preserving the order of physical creation, and perhaps of coupled spiritual creations from causal chaos; 2) Ensuring the conservation of energy and the mass-energy equivalence; 3) Enabling infinite distance of travel, effectively infinite multiples of the speed of light, and infinite travel into the future in one Planck time unit ship-frame; and 4) Ensuring, to a reasonable degree, that extraterrestrial civilizations traveling between stars and galaxies have sufficiently spiritually matured such that they are not a direct threat to other civilizations by armed aggression)]:[w(Enabling light-speed spacecraft crew members and the craft internal thermodynamic processes proceed as normal due to the universal speed of light in spirit)]: [w(The speed of light as a spectrum of velocities having a vanishingly small finite or perhaps even infinitesimal width)]:[w(The velocity of light as comprised of an infinite number

of metaphorical tiles that form the tapestry on which the order of cause and effect and existential integrity of the accidental aspects of created reality is thus maintained in forming their own tapestry which may be akin to an analogue of space-time or higher dimensional continuums)]:[w(The infinite number of speeds of light in a vanishingly small width velocity spectrum enabling infinite numbers of differing light-speed tile tapestries thus forming an infinite number of varieties of analogues of universes, multiverses, forests, biospheres, and the like, and any infinite number of different analogues of hyperspaces or hyper-space-times)]

Now we are all aware of numbers, quantities, and the like. However, how many of us have pondered quantities that are nonnumerical but at the same time not qualitative? Most likely, we rarely consider such constructs, and when they start to come to the conscious awareness of folks, they are quickly filtered out.

Well, I would like to suggest that spacecraft having attained the velocity of light in the least requisite infinite Lorentz factors may continue to accelerate to develop nonnumeric increases in velocity while the numerical velocity of the craft remains at the speed of light.

Such nonnumeric increases in spacecraft velocity may at first be associated with slight to great increases in infinite Lorentz factors. As the infinite Lorentz factors become so infinite that nature would have trouble defining still greater infinite Lorentz factors, the already-huge infinite Lorentz factors would experience nonnumerical but quantitative increases.

Accordingly, the first level of nonnumerical quantitative increases we label as nonnumerical-Lorentz-factor-increases-1.

After the spacecraft achieved sufficiently great nonnumerical-Lorentz-factor-increases-1, the spacecraft would go on to attain nonnumerical-Lorentz-factor-increases-2 which are to nonnumerical-Lorentz-factor-increases-1, as nonnumerical-Lorentz-factor-increases-1 are to numerical infinite Lorentz factors.

After the spacecraft achieved sufficiently great nonnumerical-Lorentz-factor-increases-2, the spacecraft would go on to attain nonnumerical-Lorentz-factor-increases-3 which are to nonnumerical-Lorentz-factor-increases-2, as nonnumerical-Lorentz-factor-increases-2 which are to nonnumerical-Lorentz-factor-increases-1, as nonnumerical-Lorentz-factor-increases-1 are to numerical infinite Lorentz factors.

We can consider the above repeating patterns carried on perpetually to ever greater levels.

Accordingly, nonnumerical increases in the spacecraft velocity of light could follow analogous patterns.

For example, the first level of nonnumerical quantitative increases in the light-speed of the craft we label as nonnumerical-light-speed-increases-1.

After the spacecraft achieved sufficiently great nonnumerical-light-speed-increases-1, the spacecraft would go on to attain nonnumerical-light-speed-increases-2 which are to nonnumerical-light-speed-increases-1, as nonnumerical-light-speed-increases-1 are to the numerical speed of light.

After the spacecraft achieved sufficiently great nonnumerical-light-speed-increases-2, the spacecraft would go on to attain non-numerical-light-speed-increases-3 which are to nonnumerical-light-speed-increases-2, as nonnumerical-light-speed-increases-2 which are to nonnumerical-light-speed-increases-1, as nonnumerical-light-speed-increases-1 are to the numerical speed of light.

We can continue the above serial pattern perpetually.

The above speculations are far afield from what is normally considered even very theoretical; but if we want to affirm the universal velocity of light, the above mathematical constructs seem a new reasonable first-order calculus approach.

Commensurate with the above nonnumerical schemata are the exotic realms of spacecraft travel as well as exotic, perhaps very bizarre, kinematic, topological, thermodynamic, and informational consequences of such nonnumerical-Lorentz-factor-increases-k and nonnumerical-light-speed-increases-k. Here, k can be any positive integer finite or infinite.

So achieving numerical light-speed in small numerically infinite Lorentz factors is only a first stepping-stone down an eternal path of future opportunities to go ever faster and achieve ever more extreme Lorentz factors.

However, before we accomplish the first step, eons of smaller foot shifts are likely required, for which we will learn how to achieve ever greater finite Lorentz factors. This in itself will be a long road, but the eventual, perhaps virtually eternally distant, beacon, of obtaining the first true speed of light in the least infinite Lorentz factor can serve as a mysterious, ever-distant light we can look forward to arriving at. All the while, we can contemplate what lies beyond the beacon and be assured of a never-ending sense of the mystical and whimsically far-out.

We may affix the following operator to formulas for gamma and related parameters to denote the above nonnumerical increases in spacecraft speed and Lorentz factors, the exotic realms of spacecraft travel, as well as exotic, perhaps very bizarre, kinematic, topological, thermodynamic, and informational consequences of such nonnumerical-Lorentz-factor-increases-k and nonnumerical-light-speed-increases-k. We also consider the rates of progression or increases in these nonnumerical values with respect to ship-time, ship-frame acceleration, nonnumerical-Lorentz-factor-increases-k and nonnumerical-light-speed-increases-k and single and higher-order derivatives and integrals thereof.

> [w(Non-numerical increases in spacecraft speed and Lorentz factors, the exotic realms of spacecraft travel as well as exotic, perhaps very bazaar, kinematic, topological, thermodynamic, and informational consequences of such non-numerical-Lorentz-factor-increases-k and non-numerical-light-speed-increases-k and the rates of progression or increases in these non-numerical values with respect to ship-time, ship-frame acceleration, non-numerical-Lorentz-factor-increases-k and non-numerical-light-speed-increases-k and single and higher-order derivatives and integrals thereof and where k can be any finite or infinite counting number)]

Now the Old Testament has reference to the waters from which creations are drawn, with an emphasis on the Heavens.

These waters, no doubt, are metaphors and not actually the compound we refer to as water.

However, if we nonetheless take a more or less literal interpretation of the reference to waters, we seem to be implying that something non-GOD existed before or pre-ordinately of creation, most especially the Heavens.

Could Sacred Scripture thereby be contradicting contemporary Catholic theology, which holds that GOD is the creator of all that is seen and unseen? Theological and doctrinal theorists would hold this to be the case. However, I say, "Not so fast!"

Accordingly, we refer to GOD as pure existence itself.

Since GOD is pure existence, then all things that exist only do so because GOD as existence underwrites the existence of all that is seen and unseen.

Thus, the mere fact that GOD is existence in a loose interpretation allows non-GOD realities pre-ordinate to creations and the Heavens to exist simply by the reality that GOD exists and is existence itself.

So perhaps the waters above the highest Heavens as such, although perhaps scientifically and cosmologically best referred to as nothing, actually exist as preordinate elements of non-GOD reality, which are also not created in the ordinary sense and meaning of the word.

Now imagine if we could somehow access and harness these metaphorical waters above the highest Heavens as a means to power spacecraft, as well as providing spacecraft shielding, and even materials, out of which spacecraft may be fabricated!

Additionally, imagine if we could ply the depths of the vast oceans of these waters above the highest Heavens!

Perhaps there are multiple species and even multiple levels of water above the highest Heavens. Thus, we may be able to travel among the species of such waters for which the number of species and levels of species may be arbitrarily infinite.

We might access said waters by light-speed inertial travel at sufficiently great infinite Lorentz factors commensurate with popping out of our universe or a broader realm of travel.

We may affix the following operator to formulas for gamma and related parameters to denote the water travel concepts and their resulting manifestations for travel:

> [w(Inertial travel at c in sufficiently ∞ Lorentz factors commensurate with popping out of our universe or broader realm of travel to enter and travel among any of k species or h levels thereof of waters above the highest Heavens where said waters are a manifestation of GOD being pure existence where said pure existence of GOD being real thus underwrites the existence of said pre-creational waters: k, h, = 1, 2, 3, …)]

Now, it has been said that the velocity of light is a very special velocity. I have stated this often in my writing on the subject of the speed of light as over the past year as has my academic advisor back in the days as an undergrad physics student at George Mason University.

A relatively well-known but rather basic concept of the speed of light is such that time for massive objects traveling at the speed of light slows to a stop relative to the background.

Another fascinating aspect of, say, a light-speed spacecraft is such that the background will appear infinitely aberrated, and it is as if the entire universe of travel becomes located directly in front of the spacecraft while contracted to an angular diameter of a point.

Another fascinating aspect of light-speed travel is that the length of the craft along the direction of travel will appear to have been infinitely contracted to appear as a sheet of zero thickness along the direction of travel.

Even if faster-than-light travel can occur, the speed of light remains a special speed.

For example, faster-than-light impulse travel results in backward time travel. Thus, the boundary of less-than-light-speed travel and faster-than-light travel would seem to be a big, fat, dumb, happy medium, in much the same way that the distinction between past, present, and future, as well as the distinction between cause and effect, as well as the order of cause and effect, is eliminated.

However, it appears as if the speed of light is actually a spectrum of velocities of vanishingly small finite, or perhaps even an infinitesimal, width.

Accordingly, an object traveling at the velocity of light in the least transfinite real number Lorentz factor such as a spacecraft would experience a flashlight beam or an electronic signal transmitted within the spacecraft in any direction as still traveling at the speed of light in the spacecraft translational reference frame.

So eventually, as the already-infinite Lorentz factor of a light-speed impulse spacecraft continues to increase, even to levels of uncountable infinities, we may expect that the order of cause and effect, the distinction between past, present, and future, and other timelike distinctions become blurred, uncertain, and upon still greater infinite Lorentz factors become completely indistinguishable to yield one super-causal state and one super-temporal state.

Such a state would be ontologically simple and have no causal sub-composition and no timelike distinctions between past, present, or future.

As the spacecraft Lorentz factor climbed well into the mathematically conceptualized uncountable infinities, at some point, the super-causal state will contract into an existential point with no features, as will the super-temporal state. We will refer to these states as super-super-causal-states and super-super-temporal-states.

Now, as for the precedent super-causal-states and super-temporal-states, we will refer to these as super-causal-states-1 and super-temporal-states-1, while referring to super-super-causal-states and super-super-temporal-states and super-causal-states-2 and super-temporal-states-2.

We can continue the march to still greater uncountably infinite Lorentz factors for which the associated states contract to an existential subpoint to become super simple. We refer to these states as super-causal-states-3 and super-temporal-states-3.

We can continue the march to still greater uncountably infinite Lorentz factors for which the associated states contract to an existential sub-subpoint to become super-super-simple. We refer to these states as super-simple-2, super-causal-states-4, and super-temporal-states-4.

We can continue the march to still greater uncountably infinite Lorentz factors for which the associated states contract to an existential sub-sub-subpoint to become super-super-super-simple. We refer to these states as super-simple-3, super-causal-states-5, and super-temporal-states-5.

We can consider the following series:

$$\{\Sigma(k = 1;\ k = \infty\uparrow):[w[\text{super-simple-}k,\ \text{super-causal-states-}(k + 2)\ \text{and}\ \text{super-temporal-states-}(k + 2)]]\}$$

These are really freaky concepts indeed but are likely just the finite tip of an infinite iceberg floating in eternal abstract space of light-speed.

We may affix the following simple operator to formulas for gamma and related parameters to denote the above conjectures and associated ramifications:

$$\{\Sigma(k = 1;\ k = \infty\uparrow):[w[\text{super-simple-}k,\ \text{super-causal-states-}(k + 2)\ \text{and}\ \text{super-temporal-states-}(k + 2)]]\}$$

The notation, "$\infty\uparrow$", indicates ever open infinite ordinals, thus with no set limits.

Now, assuming the velocity of light is a vanishingly small spectrum of velocities, we can intuit that each element in the spectrum is as if a horizontal line attained in a qualitatively distinct but infinite Lorentz factor.

Thus, the velocity of light may be as if a set of blinds such as a set of venetian blinds with an unspecified infinite number of slats while having a height or vertical length that is vanishingly finitely or infinitesimally small.

However, each slat may have an associated infinite number of Lorentz factors. Thus, instead of a set of monolithic slats, the speed of light may more properly be represented as a two-dimensional array of squares or rectangles.

In yet another correction, the number of meanings of the speed of light and types of the speed of light may likewise be infinite.

For example, there can be the speed of light associated with impulse travel such as that of light in a vacuum.

Then there can be the speed of light associated with a finitely massive object with infinite kinetic energy.

Then there can be the speed of light associated with quantum signal transmission via a light-speed classical feedback loop to close the teleportation process via two entangled particles.

Then there is the speed of light associated with electric currents along a conductor or superconductor.

Then there is the speed of light as the escape velocity of a particle at the event horizon surface of a black hole.

Then there is the speed of light as the maximum possible rotation rate of a black hole.

So each two-dimensional array actually has perhaps an infinite number of arrays placed behind it to manifest a three-dimensional block.

Tachyonic theories of particles that travel faster than light, if they exist, should never be able to slow down to the speed of light. Accordingly, it would take an infinite amount of energy to slow any finite mass tachyonic body to the speed of light. Here, the sign of the tachyonic masses is negative.

So most truly, the speed of light is as if abstractly represented as a luminary placed in the center of a blocklike matrix, each composed of fluorescent materials, in the form of a row of closely spaced two-dimensional arrays of separated slats.

When we look directly into the three-dimensional matrix, we spot the light source as a luminary surrounded by fluorescing slats, but where the light of fluorescence becomes dimmer as one looks away from the central luminary.

We can also consider scenarios for which a spacecraft's velocity has an associated infinite Lorentz factor but whose velocity, although not greater than that of light, is nonetheless, outside of the velocity of light.

The number of velocities outside the velocity of light, but with associated infinite Lorentz factors, may itself be infinite.

What these outside velocities would be is anyone's guess, but more than likely, a complete, or evolved approximate theory for such will require the development of new mathematics to explain.

We may affix the following operator to formulas for gamma and related parameters to denote the three-dimensional discrete block packing pattern of light-speeds and the infinite number of velocities outside the speed of light but each is associated with an infinite Lorentz factor:

[w(Three dimensional discrete block packing pattern of light-speeds and the infinite number of velocities outside the speed of light but each associated with an infinite Lorentz factor)]

We consider the notion of N-Space-M-Time extended into an ordinary 4-D space-time. Accordingly, said N-Space-M-Time would be different from a parallel space-time. Additionally, the N-Space-M-Time would be different from space-time tags, appendages, extensions, and the like.

We consider the notion of N-Space-M-Time enmeshed with an ordinary 4-D space-time. Accordingly, said N-Space-M-Time would be different from a parallel space-time. Additionally, the N-Space-M-Time would be different from space-time tags, appendages, extensions, and the like.

Now it is plausible that there may exist an infinite number of parallel histories according to the so-called many-worlds interpretation of quantum mechanics and various backward time travel conjectures.

Additionally, theories of parallel universes abound in different interpretations.

However, what is lacking is the notion of parallel realms that are neither historical nor universes.

Accordingly, these realms should be referred to as parallel abstract realms so as to provide a label until the heuristics and lexicography of physics and philosophy become suitably advanced to provide the necessary explanations and theoretical framework.

One such abstract realm might at first ostensibly be the realm of consciousness. However, dualistic theories of human nature are already familiar with such constructs. So what I am conjecturing are realms that are completely distinct from those posited by dualistic consciousness theories.

So these parallel abstract realms are really, really, far-out and whimsical. Perhaps these exotic realms can be accessed by light-speed travel and associated infinite spacecraft Lorentz factors.

Accordingly, a spacecraft having attained an inertial velocity of the speed of light might experience so great of time dilation that it no longer has future room in a given universe or broader realm to travel forward in time. Thus, the spacecraft itinerary would overflow the future eternity of a given universe or broader realm of travel and pop into additional related parallel realms.

Other prospects include mildly relativistic spacecraft, such as, for example, nuclear fusion– or nuclear fission–powered worldship that is designed to simply plod along at, say, 10 percent to 20 percent of the speed of light until the next universal phase change, symmetry-breaking event, inflationary epoch, or until universal energy well emplacement decays or tunnels through in potential well barrier.

Such mildly relativistic spacecraft velocities might just work, provided the entire spacecraft is cooled to sufficiently close to absolute zero.

Accordingly, the craft would experience such internally slowed processes that the craft would experience pseudo–time dilation or the complete effective halt to its internal thermodynamic processes.

Actually, attaining a temperature of literally absolute zero would cause the craft to become internally motionless.

However, absolute zero may be difficult to attain because the Heisenberg uncertainty principle would ordinarily keep the internal composition of the craft at a finite, although very low, temperature.

For such spacecraft carrying cryogenically asleep humans, the eternal placement of the craft may be implied by the natural processes that human persons have continuity in conscious states, at least under ordinary circumstances.

So a fusion-powered craft that has an interior completely shielded from its exterior and cryogenically at its lowest possible energy state may as if drift into the infinitely distant eternal limits of a given universe or broader realm of travel. Here, the human or other personal life-forms within the spacecraft may, by virtue of the naturalness of conscious continuity, force the craft to manifest post-eternal future travel, and thus travel outside the aspects of a universe or broader realm of travel.

Ideally, such craft would travel into one or more parallel abstract realms.

The mechanisms of human or other life-form consciousness may thus act in a loose sense as analogues of the Pauli exclusion principle by which fermions resist being placed in the same energy level in the same location of space by what is referred to as degeneracy pressure.

The number of parallel abstract realms may be infinite, perhaps even uncountably infinite.

Taking two methane molecules and placing them base against base, and then pushing each extended atom toward the middle of the bipartite system to the extent that the otherwise extended atoms bond, the stored strain energy in these molecules can thus be very high.

Now, a black hole's radius may be larger than it would take light to travel the distance of the black hole radius for which the space-time curvature of the black hole is taken into account.

The extension of the black hole is in first order as such is due to the extension of the radius of the black hole as follows:

$$R_{\text{Black-Hole}} = [(x_{1a} - x_{1b})^2 + (x_{2a} - x_{2b})^2 + (x_{3a} - x_{3b})^2 + (t_{1a} - t_{1b})^2]^{1/2}$$

Taking space-time curvature into account, the extension of the black hole is in first order as such is due to the extension of the radius of the black hole as follows:

$$R_{\text{Black-Hole}} = (\text{Space-Time Curvature Pattern})[(x_{1a} - x_{1b})^2 + (x_{2a} - x_{2b})^2 + (x_{3a} - x_{3b})^2 + (t_{1a} - t_{1b})^2]^{1/2}$$

The extension of the black hole is in better first order as such is due to the extension of the radius of the black hole as follows:

$$R_{\text{Black-Hole}} = [(x_{1a} - x_{1b})^2 + (x_{2a} - x_{2b})^2 + (x_{3a} - x_{3b})^2 + (t_{1a,\text{forward}} - t_{1b,\text{forward}})^2 + (t_{1a,\text{backward}} - t_{1b,\text{backward}})^2]^{1/2}$$

Taking space-time curvature into account, the extension of the black hole is in first order as such is due to the extension of the radius of the black hole as follows:

$$R_{\text{Black-Hole}} = (\text{Space-Time Curvature Pattern})[(x_{1a} - x_{1b})^2 + (x_{2a} - x_{2b})^2 + (x_{3a} - x_{3b})^2 + (t_{1a,\text{forward}} - t_{1b,\text{forward}})^2 + (t_{1a,\text{backward}} - t_{1b,\text{backward}})^2]^{1/2}$$

Here, t is time, and x is spatial distance. Accordingly, we use the Pythagorean distance formula.

One consequence of a black hole radius being slightly larger than the time it would take light to travel the curved radius is such that the overshoot of the radius may imply nonlocality and be related to prospects for velocities being ever so slightly greater than light

Now the backward temporal and forward temporal extensions of a black hole may be as small as the Planck time, thus corresponding to an equivalent spatial overextension being equal to the Planck length unit.

In cases where the temporal extensions of black holes would include backward and forward temporal extensions, and where these temporal coordinates would have a spacelike aspect, the dimensionality of black holes would include extensions into a bipartite fourth spatial dimension.

Here, the fourth spatial dimension is bipartite to account for the bidirectional temporal extension.

The fourth spatial dimension would be bipartite in manifestation in the sense that the backward and forward temporal extension would be a medial condition between topologies of no temporal spacelike extension and outright spatial extension as having five coordinates—three for conventional spatial extension and one each for backward temporal extension and forward temporal extension.

Now, the backward and forward temporal extension interval may itself require second order adjustments as follows:

$R_{Black-Hole}$ = (Space-Time Curvature Pattern)[$(x_{1a} - x_{1b})^2 + (x_{2a} - x_{2b})^2 + (x_{3a} - x_{3b})^2 +$ [[[[$t_{1a,forward}$ + [$t_{1a,forward}$ /[(Space-Time Curvature Pattern)[$(x_{1a} - x_{1b})^2 + (x_{2a} - x_{2b})^2 + (x_{3a} - x_{3b})^2 + (t_{1a,forward} - t_{1b,forward})^2 + (t_{1a,backward} - t_{1b,backward})^2]^{1/2}$]] + [[$t_{1a,forward}$ /[(Space-Time Curvature Pattern)[$(x_{1a} - x_{1b})^2 + (x_{2a} - x_{2b})^2 + (x_{3a} - x_{3b})^2 + (t_{1a,forward} - t_{1b,forward})^2 + (t_{1a,backward} - t_{1b,backward})^2]^{1/2}$]] / [(Space-Time Curvature Pattern)[$(x_{1a} - x_{1b})^2 + (x_{2a} - x_{2b})^2 + (x_{3a} - x_{3b})^2 + (t_{1a,forward} - t_{1b,forward})^2 + (t_{1a,backward} - t_{1b,backward})^2]^{1/2}$]] + [[[$t_{1a,forward}$ /[(Space-Time Curvature Pattern)[$(x_{1a} - x_{1b})^2 + (x_{2a} - x_{2b})^2 + (x_{3a} - x_{3b})^2 + (t_{1a,forward} - t_{1b,forward})^2 + (t_{1a,backward} - t_{1b,backward})^2]^{1/2}$]]/ [(Space-Time Curvature Pattern)[$(x_{1a} - x_{1b})^2 + (x_{2a} - x_{2b})^2 + (x_{3a} - x_{3b})^2 + (t_{1a,forward} - t_{1b,forward})^2 + (t_{1a,backward} - t_{1b,backward})^2]^{1/2}$]] / [(Space-Time Curvature Pattern)[$(x_{1a} - x_{1b})^2 + (x_{2a} - x_{2b})^2 + (x_{3a} - x_{3b})^2 + (t_{1a,forward} - t_{1b,forward})^2 + (t_{1a,backward} - t_{1b,backward})^2]^{1/2}$]] + ...]

-

[[[$t_{1b,forward}$ + [$t_{1b,forward}$ /[(Space-Time Curvature Pattern)[$(x_{1a} - x_{1b})^2 + (x_{2a} - x_{2b})^2 + (x_{3a} - x_{3b})^2 + (t_{1a,forward} - t_{1b,forward})^2 + (t_{1a,backward} - t_{1b,backward})^2]^{1/2}$]] + [[$t_{1b,forward}$ /[(Space-Time Curvature Pattern)[$(x_{1a} - x_{1b})^2 + (x_{2a} - x_{2b})^2 + (x_{3a} - x_{3b})^2 + (t_{1a,forward} - t_{1b,forward})^2 + (t_{1a,backward} - t_{1b,backward})^2]^{1/2}$]] / [(Space-Time Curvature Pattern)[$(x_{1a} - x_{1b})^2 + (x_{2a} - x_{2b})^2 + (x_{3a} - x_{3b})^2 + (t_{1a,forward} - t_{1b,forward})^2 + (t_{1a,backward} - t_{1b,backward})^2]^{1/2}$]] + [[[$t_{1b,forward}$ /[(Space-Time Curvature Pattern)[$(x_{1a} - x_{1b})^2 + (x_{2a} - x_{2b})^2 + (x_{3a} - x_{3b})^2 + (t_{1a,forward} - t_{1b,forward})^2 + (t_{1a,backward} - t_{1b,backward})^2]^{1/2}$]]/ [(Space-Time Curvature Pattern)[$(x_{1a} - x_{1b})^2 + (x_{2a} - x_{2b})^2 + (x_{3a} - x_{3b})^2 + (t_{1a,forward} - t_{1b,forward})^2 + (t_{1a,backward} - t_{1b,backward})^2]^{1/2}$]] / [(Space-Time Curvature Pattern)[$(x_{1a} - x_{1b})^2 + (x_{2a} - x_{2b})^2 + (x_{3a} - x_{3b})^2 + (t_{1a,forward} - t_{1b,forward})^2 + (t_{1a,backward} - t_{1b,backward})^2]^{1/2}$]] + ...]]^2

+

[[[[$t_{1a,backward}$ + [$t_{1a,backward}$ /[(Space-Time Curvature Pattern)[$(x_{1a} - x_{1b})^2 + (x_{2a} - x_{2b})^2 + (x_{3a} - x_{3b})^2 + (t_{1a,forward} - t_{1b,forward})^2 + (t_{1a,backward} - t_{1b,backward})^2]^{1/2}$]] + [[$t_{1a,backward}$ /[(Space-Time Curvature Pattern)[$(x_{1a} - x_{1b})^2 + (x_{2a} - x_{2b})^2 + (x_{3a} - x_{3b})^2 + (t_{1a,forward} - t_{1b,forward})^2 + (t_{1a,backward} - t_{1b,backward})^2]^{1/2}$]] / [(Space-Time Curvature Pattern)[$(x_{1a} - x_{1b})^2 + (x_{2a} - x_{2b})^2 + (x_{3a} - x_{3b})^2 +$

$(t_{1a,forward} - t_{1b,forward})^2 + (t_{1a,backward} - t_{1b,backward})^2]^{1/2}]] + [[[t_{1a,backward} /[(Space-Time Curvature Pattern)[(x_{1a} - x_{1b})^2 + (x_{2a} - x_{2b})^2 + (x_{3a} - x_{3b})^2 + (t_{1a,forward} - t_{1b,forward})^2 + (t_{1a,backward} - t_{1b,backward})^2]^{1/2}]]/ [(Space-Time Curvature Pattern)[(x_{1a} - x_{1b})^2 + (x_{2a} - x_{2b})^2 + (x_{3a} - x_{3b})^2 + (t_{1a,forward} - t_{1b,forward})^2 + (t_{1a,backward} - t_{1b,backward})^2]^{1/2}]] / [(Space-Time Curvature Pattern)[(x_{1a} - x_{1b})^2 + (x_{2a} - x_{2b})^2 + (x_{3a} - x_{3b})^2 + (t_{1a,forward} - t_{1b,forward})^2 + (t_{1a,backward} - t_{1b,backward})^2]^{1/2}]] + ...]$

-

$[[[t_{1b,backward} + [t_{1b,backward} /[(Space-Time Curvature Pattern)[(x_{1a} - x_{1b})^2 + (x_{2a} - x_{2b})^2 + (x_{3a} - x_{3b})^2 + (t_{1a,forward} - t_{1b,forward})^2 + (t_{1a,backward} - t_{1b,backward})^2]^{1/2}]] + [[t_{1b,backward} /[(Space-Time Curvature Pattern)[(x_{1a} - x_{1b})^2 + (x_{2a} - x_{2b})^2 + (x_{3a} - x_{3b})^2 + (t_{1a,forward} - t_{1b,forward})^2 + (t_{1a,backward} - t_{1b,backward})^2]^{1/2}]] / [(Space-Time Curvature Pattern)[(x_{1a} - x_{1b})^2 + (x_{2a} - x_{2b})^2 + (x_{3a} - x_{3b})^2 + (t_{1a,forward} - t_{1b,forward})^2 + (t_{1a,backward} - t_{1b,backward})^2]^{1/2}]] + [[[t_{1b,backward} /[(Space-Time Curvature Pattern)[(x_{1a} - x_{1b})^2 + (x_{2a} - x_{2b})^2 + (x_{3a} - x_{3b})^2 + (t_{1a,forward} - t_{1b,forward})^2 + (t_{1a,backward} - t_{1b,backward})^2]^{1/2}]]/ [(Space-Time Curvature Pattern)[(x_{1a} - x_{1b})^2 + (x_{2a} - x_{2b})^2 + (x_{3a} - x_{3b})^2 + (t_{1a,forward} - t_{1b,forward})^2 + (t_{1a,backward} - t_{1b,backward})^2]^{1/2}]] / [(Space-Time Curvature Pattern)[(x_{1a} - x_{1b})^2 + (x_{2a} - x_{2b})^2 + (x_{3a} - x_{3b})^2 + (t_{1a,forward} - t_{1b,forward})^2 + (t_{1a,backward} - t_{1b,backward})^2]^{1/2}]] + ...]]^2]^{1/2}$

ESSAY 34) Instantaneous and Ascending Levels of Super-Instantaneous Spacecraft Travel and Manifestations Thereof

Herein, we consider travel velocities that are near-instantaneous, instantaneous, faster-than-instantaneous, or faster-than-instantaneous-1, faster than faster than instantaneous or faster-than-instantaneous-2, faster than faster than faster than instantaneous or faster-than-instantaneous-3, and so on to faster-than-instantaneous-Ω, and so on to faster-than-instantaneous-(Aleph 1), and so on to faster-than-instantaneous-(Aleph 2), and so on to faster-than-instantaneous-(Aleph 3), and so on to faster-than-instantaneous-(Aleph Ω), and eternally onward from there. Here, we are talking background reference frame velocities, although we may consider light-speed ship-time reference frame velocities effectively so based on ever-greater infinite Lorentz factors. For variously and increasingly great infinite spacecraft Lorentz factors, the effective travel through space and forward in time becomes instantaneous relative to ship-time.

For backward time traveling instantaneous velocity craft, and faster-than-instantaneous-k craft, the craft might overshoot the history initiation to travel into the pre-creational era, even in cases where a Divine origin of creation is a valid concept.

For faster-than-instantaneous-1 craft, the time of transport to a finitely distant location is [[0/(faster-than-instantaneous-1velocity)](distance traveled)] seconds.

For faster-than-instantaneous-2 craft, the time of transport to a finitely distant location is [[0/(faster-than-instantaneous-2 velocity)](distance traveled)] seconds.

For faster-than-instantaneous-3 craft, the time of transport to a finitely distant location is [[0/(faster-than-instantaneous-3 velocity)](distance traveled)] seconds.

For faster-than-instantaneous-(Aleph 0) craft, the time of transport to a finitely distant location is [[0/(faster-than-instantaneous-(Aleph 0) velocity)](distance traveled)] seconds.

For faster-than-instantaneous-(Aleph 1) craft, the time of transport to a finitely distant location is [[0/(faster-than-instantaneous-(Aleph 1) velocity)](distance traveled)] seconds.

For faster-than-instantaneous-(Aleph 2) craft, the time of transport to a finitely distant location is [[0/(faster-than-instantaneous-(Aleph 2) velocity)](distance traveled)] seconds.

For faster-than-instantaneous-(Aleph 3) craft, the time of transport to a finitely distant location is [[0/(faster-than-instantaneous-(Aleph 3) velocity)](distance traveled)] seconds.

For faster-than-instantaneous-(Aleph Ω) craft, the time of transport to a finitely distant location is [[0/(faster-than-instantaneous-(Aleph Ω) velocity)](distance traveled)] seconds.

And so on.

These formulas also apply to infinite travel distances as well.

For time difference of arrival between a faster-than-instantaneous-2 craft, and a faster-than-instantaneous-1 craft, over a finitely distant location, we have: [[0/(faster-than-instantaneous-2 velocity)](distance traveled)] - [[0/(faster-than-instantaneous-1velocity)](distance traveled)].

For time difference of arrival between a faster-than-instantaneous-3 craft, and a faster-than-instantaneous-1 craft, over a finitely distant location, we have: [[0/(faster-than-instantaneous-3 velocity)](distance traveled)] - [[0/(faster-than-instantaneous-1velocity)](distance traveled)].

For time difference of arrival between a faster-than-instantaneous-3 craft, and a faster-than-instantaneous-2 craft, over a finitely distant location, we have: [[0/(faster-than-instantaneous-3 velocity)](distance traveled)] - [[0/(faster-than-instantaneous-2 velocity)](distance traveled)].

For time difference of arrival between a faster-than-instantaneous-(Aleph 1) craft, and a faster-than-instantaneous-(Aleph 0) craft, over a finitely distant location, we have: [[0/[faster-than-instantaneous-(Aleph 1) velocity]](distance traveled)] - [[0/[faster-than-instantaneous-(Aleph 0) velocity]](distance traveled)].

For time difference of arrival between a faster-than-instantaneous-(Aleph 2) craft, and a faster-than-instantaneous-(Aleph 0) craft, over a finitely distant location, we have: [[0/[faster-than-instantaneous-(Aleph 2) velocity]](distance traveled)] - [[0/[faster-than-instantaneous-(Aleph 0) velocity]](distance traveled)].

For time difference of arrival between a faster-than-instantaneous-(Aleph 2) craft, and a faster-than-instantaneous-(Aleph 1) craft, over a finitely distant location, we have: [[0/[faster-than-instantaneous-(Aleph 2) velocity]](distance traveled)] - [[0/[faster-than-instantaneous-(Aleph 1) velocity]](distance traveled)].

For time difference of arrival between a faster-than-instantaneous-(Aleph 3) craft, and a faster-than-instantaneous-(Aleph 0) craft, over a finitely distant location, we have: [[0/[faster-than-instantaneous-(Aleph 3) velocity]](distance traveled)] - [[0/[faster-than-instantaneous-(Aleph 0) velocity]](distance traveled)].

For time difference of arrival between a faster-than-instantaneous-(Aleph 3) craft, and a faster-than-instantaneous-(Aleph 1) craft, over a finitely distant location, we have: [[0/[faster-than-instantaneous-(Aleph 3) velocity]](distance traveled)] - [[0/[faster-than-instantaneous-(Aleph 1) velocity]](distance traveled)].

For time difference of arrival between a faster-than-instantaneous-(Aleph 3) craft, and a faster-than-instantaneous-(Aleph 2) craft, over a finitely distant location, we have: [[0/[faster-than-instantaneous-(Aleph 3) velocity]](distance traveled)] - [[0/[faster-than-instantaneous-(Aleph 2) velocity]](distance traveled)].

For time difference of arrival between a faster-than-instantaneous-[Aleph (Ω + 1)] craft, and a faster-than-instantaneous-(Aleph Ω) craft, over a finitely distant location, we have: [[0/[faster-than-instantaneous-[Aleph (Ω + 1)] velocity]](distance traveled)] - [[0/[faster-than-instantaneous-(Aleph Ω) velocity]](distance traveled)].

For time difference of arrival between a faster-than-instantaneous-[Aleph (Ω + 2)] craft, and a faster-than-instantaneous-[Aleph (Ω + 1)] craft, over a finitely distant location, we have: [[0/[faster-than-instantaneous-[Aleph (Ω + 2)] velocity]](distance traveled)] - [[0/[faster-than-instantaneous-[Aleph (Ω + 1)] velocity]](distance traveled)].

For time difference of arrival between a faster-than-instantaneous-[Aleph (Ω + 2)] craft, and a faster-than-instantaneous-[Aleph Ω] craft, over a finitely distant location, we have: [[0/[faster-than-instantaneous-[Aleph (Ω + 2)] velocity]](distance traveled)] - [[0/[faster-than-instantaneous-[Aleph Ω] velocity]](distance traveled)].

For time difference of arrival between a faster-than-instantaneous-[Aleph (Ω + 3)] craft, and a faster-than-instantaneous-[Aleph (Ω + 1)] craft, over a finitely distant location, we have: [[0/[faster-than-instantaneous-[Aleph (Ω + 3)] velocity]](distance traveled)] - [[0/[faster-than-instantaneous-[Aleph (Ω + 1)] velocity]](distance traveled)].

For time difference of arrival between a faster-than-instantaneous-[Aleph (Ω + 3)] craft, and a faster-than-instantaneous-[Aleph (Ω + 2)] craft, over a finitely distant location, we have: [[0/[faster-than-instantaneous-[Aleph (Ω + 3)] velocity]](distance traveled)] - [[0/[faster-than-instantaneous-[Aleph (Ω + 2)] velocity]](distance traveled)].

We could cover an infinite number of examples, but by now, the reader should see the pattern even if they do not agree with the conclusions.

Now, at least in our universe, theoretical paradigms hold that the smallest unit of time that has any meaning is the Planck time unit, which is finite.

The Planck time is $\{[h/(2\pi)]\ G/(c\ EXP\ 5)]\}\ EXP\ (1/2)$ or about $[5.39 \times (10\ EXP - 44)]$ second.

The smallest time unit in other universes topologically coupled to our universe is likely also finitely quantized.

So the above formulas for transit time assume continuous and super-continuous space-times of travel instead of finitely quantized space-time.

So if the smallest time unit is of finite length in duration, then it behooves us to conjecture on travel times, which are so small that these travel times are sub-times in dimensionality, which loosely translate into being pieces of a Planck time unit but not fractions of a Planck time unit.

Thus, these sub-time intervals would be a fractional parcel of the one time dimension we know of.

Perhaps another approach is to consider the Planck time unit as a composition of infinitely many sub-time objects. These sub-time objects would not be intervals of time but instead serve as building blocks that once properly arranged would holistically as a set generate one Planck time unit.

Instantaneous and faster-than-instantaneous spacecraft in various levels of faster-than-instantaneous travel velocities would travel, for lack of a better phrase, on or under a sub-time object or a finite number of sub-time objects.

Travel at sufficiently faster-than-instantaneous velocities may likewise enable transits, for lack of a better phrase, on or under a sub-sub-time object, which would be loosely related to a sub-time object as a sub-time object is to one Planck time unit.

Travel at still greater sufficiently faster-than-instantaneous velocities may likewise enable transits, for lack of a better phrase, on or under a sub-sub-sub-time object which would be loosely related to a sub-sub-time object as a sub-sub-time object would be to a sub-time object as a sub–time unit is to one Planck time unit.

We refer to sub-time, sub-sub-time, and sub-sub-sub-time as sub-1-time, sub-2-time, and sub-3-time, respectively.

We can go still farther to consider sub-4-time, sub-5-time, sub-6-time, …, sub-Ω-time, …, sub-(Aleph 1)-time, …, sub-(Aleph 2)-time, …, sub-(Aleph 3)-time, …, sub-(Aleph Ω)-time, and the like.

All these travel velocities and time may effectively have analogues in light-speed impulse travel at various associated infinite values of Lorentz factors. Here the sub-k-times would be in the spacecraft reference frames.

So, above are some really cool concepts that I am just getting started working on.

ESSAY 35) Traveling Less Than C but Faster Than the Practical Speed of Light in Space and Why Light-Speed Limits Strictly Imposed by Nature Work Out for All the Better

Consider a photon traveling in space at a velocity precisely equal to c or $[1/(\mu_0\epsilon_0)]^{1/2}$ a distance equal to one cosmic light-cone radius or 13.77 billion light-years. This works out to be $[1.377 \times 10^{10}](10^{16})\{1/[1.6 \times (10^{-35})]\}$ Planck length units= 8.521×10^{60} Planck length units.

So a spacecraft traveling $\{[8.521 \times (10^{60})]-e\}$ Planck length units in 1.377×10^{10} years, where e is less than one will have velocity that is cosmically locally indistinguishable from the speed of light.

The latter boundary value condition may most naturally define the transition of the spacecraft into travel in the first level of space-time foam.

Gamma = $\{1/\{1- [(v/c)^2]\}\}^{1/2}$.

Gamma^2 = $\{1/\{1-[(v/c)^2]\}\}$

$1/(\text{Gamma}^2)$ = $\{1-[(v/c)^2]\}$

$[1/(\text{Gamma}^2)] -1 =-[(v/c)^2]$

$(1 - [1/(\text{Gamma}^2)]\}^{1/2}$ = v/c

So a Lorentz factor of 5 corresponds to about 0.98 c.

A Lorentz factor of 50 corresponds to about 0.9998 c.

A Lorentz factor of 500 corresponds to about 0.999998 c.

Now we are considering travel at $\{c - \{\{1/[[8.521 \times (10^{60})] + e]\} c\}\}$.

However, consider the above patterns Lorentz factors versus velocities.

For example, 5^{-2} = 0.025.

While 50^{-2} = 0.00025

While 500^{-2} = 0.0000025.

So in the first order, the number of decimal places for each digitally pentilated Lorentz factor of the examples provided above to the right of the decimal point is twice that of the associated velocity up to and including the digit *8*.

This tells us that the lower cutoff value of the Lorentz factor of a spacecraft that would travel less than *c* mathematically, but which would have actual velocity cosmically indistinguishable from light-speed, would be about $[8.521 \times (10^{60})]^{1/2}$ = $[2.91907 \times (10^{30})]$.

So a principally chemically fueled spacecraft or a spacecraft with chemical energy initiated–background ionic, magnetic, electric, and/or electromagnetic energy uptake may, in principle, attain a finite Lorentz factor that is large enough to enable velocities indistinguishable from the speed of light over cosmically local travel intervals.

The cutoff spacecraft Lorentz factor would be as follows:

$$\{1/\{[1 - [(v/c)^2]]^{1/2}\}\} = \{1/\{[1 - [(\{c - \{\{1/[[8.521 \times (10^{60})] + e]\} c\}\}/c)^2]]^{1/2}\}\}.$$

Perhaps the much greater probability of a photon interacting with a gas atom one meter to the side of the photon's classical travel path, such as would be the norm in deep intergalactic space for every cubic meter of photon propagation, can enhance the refractive index of intergalactic space to a value of $1 + \epsilon$ where ϵ is the inverse of a suitable sub-ensemble number.

Since the universe is suffused with CMBR and starlight, even though photons obey Bose-Einstein statistics, they may nonetheless undergo ever so slight nonlinear superpositions when photons cross paths of travel at the same time. Perhaps such nonlinear effects can induce a greater-than-nonzero but very small increase in the refractive index of deep space. Additional factors that may influence the refractive index of deep space include the following:

- Dark matter that may interact ever so slightly with the photonic background to provide a very small greater-than-zero increase in the refractive index of deep space. Such a mechanism for interaction may include electroweak mixing parameters, electrostrong mixing parameters, electroweak-strong mixing parameter, and any supersymmetric analogues. The presence of more or less static magnetic and electric fields that may induce a greater-than-zero component to the refractive index of deep space by nonlinear electromagnetic superposition effects.

- The presence of gravitational fields throughout deep space that may provide a greater-than-zero component to the refractive index of deep space by nonlinear energy densifying for which the sum of the gravitational and electromagnetic energy in a given volume of space is somehow greater than the two energy species separately added.

- The presence of dark energy throughout deep space may provide a greater-than-zero component to the refractive index of deep space by nonlinear energy densifying for which the sum of the dark energy and electromagnetic energy in a given volume of space is somehow greater than the two energy species separately added.

Because deep space may have a slightly greater-than-one variable and varying electromagnetic refractive index, perhaps the true cosmic speed limit as actually occurs in nature is equal to that for gravitation and gravitational waves.

Since the electromagnetic refractive index is currently defined as $n = C/v$, where C is the velocity of light in an absolute vacuum, perhaps the electromagnetic refractive index should alternatively be defined as $N_{em} = [C + \epsilon_{atomic\text{-}gas} + \epsilon_{em\text{-}non\text{-}linear} + \epsilon_{em\text{-}dark\text{-}matter} + \epsilon_{static\ e\text{-}\&\text{-}m\text{-}fields} + \epsilon_{grav\text{-}non\text{-}linear} + \epsilon_{dark\text{-}energy\text{-}non\text{-}linear}]/v$, where C is the average velocity of light in the vacuum of space.

Alternatively, the electromagnetic refractive index might be defined as $N_{em} = [C - \epsilon_{atomic-gas} - \epsilon_{em-non-linear} - \epsilon_{em-dark-matter} - \epsilon_{static\ e-\&-m-fields} - \epsilon_{grav-non-linear} - \epsilon_{dark-energy-non-linear}]/v$, where C is the maximum possible velocity of light in a vacuum.

The true refractive electromagnetic index might be defined as $N_{em} = [C_{grav} - \epsilon_{atomic-gas} - \epsilon_{em-non-linear} - \epsilon_{em-dark-matter} - \epsilon_{static\ e-\&-m-fields} - \epsilon_{grav-non-linear} - \epsilon_{dark-energy-non-linear}]/v$, where C_{grav} is the velocity of gravitation in vacuum.

Since gravitation is the weakest force and involves the propagation of topological deformities in space-time itself, perhaps gravitational waves are impeded less as a result of the atomic gas, electromagnetic gas, cold dark matter, and dark energy in deep intragalactic and deep extragalactic space. Regardless, since gravitational energy is real, perhaps the above-delineated mechanisms for bosonic slowing are also operative in gravitational waves.

Thus, the gravitational refractive index could be defined as $N_{grav} = C_{grav-max}/v$, where $C_{grav-max}$ is the maximum possible velocity of gravitational radiation in vacuum, or alternatively be defined as $n = [C_{grav} + \epsilon_{atomic-gas} - \epsilon_{em-non-linear} - \epsilon_{em-dark-matter} - \epsilon_{static\ e-\&-m-fields} - \epsilon_{grav-non-linear} - \epsilon_{dark-energy-non-linear}]/v$, where C_{grav} is the average velocity of gravitational radiation in the vacuum of space.

Alternatively, the gravitational refractive index might be defined as $N_{grav} = [C_{grav-max} - \epsilon_{atomic-gas} - \epsilon_{em-non-linear} - \epsilon_{em-dark-matter} - \epsilon_{static\ e-\&-m-fields} - \epsilon_{grav-non-linear} - \epsilon_{dark-energy-non-linear}]/v$, where $C_{grav-max}$ is the maximum possible velocity of gravitation in vacuum.

Inertial bodies traveling as sub-ensemble gamma factors might experience a non–Lorentz component increase or decrease in gamma factor. Alternatively, such bodies might travel faster than the currently accepted velocity of light and thereby experience backward time travel in ironic contrast to forward time travel. Another possible consequence is that the bodies might have a compounded superluminal travel, resulting in the bodies already traveling slightly in excess of the accepted velocity, arriving slightly earlier at a cosmic distant destination than if they traveled at or below the accepted value of C.

Exotic discontinuities in the propagation path might occur as we achieve sub-ensemble gamma factors or discreet jumps in along the orientation of a spacecraft velocity vector.

In considering the concept of a variable speed of light, we open prospects of more rigorously probing the maximum theoretical speed limit and the maximum speed limit as commonly and routinely permitted by nature. For example, research and development aimed at attaining ever higher gamma factors, at some point of time in the future, may enable humans to travel faster than light does in interstellar, and even intergalactic, space. Such practical superluminal travel may reveal nothing new in the way of novel physics, or it may lead to the manifestation of exotic space-time topology states, as well as fascinating and novel mass-energy transport physics.

How such effective and practical superluminal travel could be employed for enhanced space-time travel is anyone's guess. For all we know, achieving this may open up numerous useful modes of space-time travel or a Pandora's box of dangers. One beneficial result could be the entrance into a new continuum or topological realm that exists at or just above the practical speed of light. Another result may be slightly backward time travel. The time travel might not be backward in an absolute sense but instead may manifest as an earlier arrival date in the future at the destination point than would otherwise result. It is conceivable that although the superluminal craft travels forward in time, the extent of forward time travel may be muted, thus adding an additional component to the superluminal velocity of the spacecraft.

This additional superluminal component to spacecraft velocity may provide for a second order lessening of the extent of forward time travel. This second-order lessening may in turn enable a third order component to the superluminal velocity of the spacecraft. The superluminal effect would result from the time interval over which the spacecraft travels from its point of origin to its destination being less than that for a conventional light-speed particle such as a photon.

The third-order velocity component may lead to a third-order lessening of the extent of forward time travel, thus enabling a fourth-order component to the superluminal velocity of the spacecraft.

Likewise, we may consider the orders of the lessening of the extent of forward time travel and that the components of the spacecraft's superluminal velocity might be defined as follows:

M_{ftt} = mth, m = 1,2,3, …

M_{ssv} = (m + 1)th, m = 1,2,3, …

Thus, the compounding effect of the first-order and higher-order modes or terms may progress to an increasingly nonlinear extent as the spacecraft kinetic energy or gamma factor increases. An interesting question is at what rate would the effects of the higher-order terms increase the spacecraft velocity and slow forward time travel? Perhaps the effects would produce a runway effect to either an absolute finite limiting value for superluminal travel and/or slow forward time travel. The limiting values, if finite, may either be very small to minuscule compared to the practical velocity of light or the spacecraft's superluminal velocity and the conjectured first-order forward time travel slowing. Alternatively, the conjectured mechanism may enable modest but significant increases in the velocity of the spacecraft relative to the practical velocity of light and/or a modest but significant reduction in forward time travel. Another possibility is that this effect may enable extreme increases in the velocity of the spacecraft relative to the practical velocity of light and/or extreme reductions in forward time travel. An interesting result would be a compounding of the effect to the extent that the spacecraft actually travels backward in time.

Another possibility may be that such a superluminal spacecraft experiences an infinite runaway velocity, reducing the degree of forward time travel to zero, or the craft may experience backward time travel at an infinite rate.

Below, V_{ssc} and C_p are the spacecraft velocity and the practical natural speed of light and the following limits describe these scenarios:

Lim V_{ssc} = C_p + e: where e is the inverse of a sub-ensemble number.

Σ M_{ftt} ➔ Σ M_{ftt} (1, ↑)

Lim V_{ssc} = C_p + e: where e is the inverse of a sub-ensemble number.

Σ M_{ssv} ➔ Σ M_{ssv} (1, ↑)

Lim V_{ssc} = C_p + [C_p/ ensemble]

Σ M_{ftt} ➔ Σ M_{ftt} (1, ↑)

Lim V_{ssc} = C_p + [C_p/ ensemble]

Σ M_{ssv} ➔ Σ M_{ssv} (1, ↑)

Lim V_{ssc} = C_p + [C_p/ infinity scrapper]

Σ M_{ftt} ➔ Σ M_{ftt} (1, ↑)

Lim V_{ssc} = C_p + [C_p/ infinity scrapper]

Σ M_{ssv} ➔ Σ M_{ssv} (1, ↑)

Lim V_{ssc} = C_p + {C_p/ [f(Ω) or f(Alephj 0) or f(Aleph 1) or f(Aleph 2) or, …, or f(Aleph i),or …]}: i = 3, 4, 5, ,,,

Σ M_{ftt} ➔ Σ M_{ftt} (1, ↑)

Lim V_{ssc} = C_p + {C_p/ [f(Ω) or f(Alephj 0) or f(Aleph 1) or f(Aleph 2) or, …, or f(Aleph i),or …]}: i = 3, 4, 5, ,,,

Σ M_{ssv} ➔ Σ M_{ssv} (1, ↑)

$\text{Lim } V_{ssc} = C_p + [C_p\ g]$, where g is modest.

$\Sigma\ M_{ftt} \rightarrow \Sigma\ M_{ftt}\ (1, \uparrow)$

$\text{Lim } V_{ssc} = C_p + [C_p\ g]$, where g is modest.

$\Sigma\ M_{ssv} \rightarrow \Sigma\ M_{ssv}\ (1, \uparrow)$

$\text{Lim } V_{ssc} = C_p + [C_p\ f]$, where f is sub-ensemble but very large.

$\Sigma\ M_{ftt} \rightarrow \Sigma\ M_{ftt}\ (1, \uparrow)$

$\text{Lim } V_{ssc} = C_p + [C_p\ f]$, where f is sub-ensemble but very large.

$\Sigma\ M_{ssv} \rightarrow \Sigma\ M_{ssv}\ (1, \uparrow)$

$\text{Lim } V_{ssc} = C_p + [(C_p)(\text{ensemble})]$

$\Sigma\ M_{ftt} \rightarrow \Sigma\ M_{ftt}\ (1, \uparrow)$

$\text{Lim } V_{ssc} = C_p + [(C_p)(\text{ensemble})]$

$\Sigma\ M_{ssv} \rightarrow \Sigma\ M_{ssv}\ (1, \uparrow)$

$\text{Lim } V_{ssc} = C_p + [(C_p)(\text{infinity scrapper})]$

$\Sigma\ M_{ftt} \rightarrow \Sigma\ M_{ftt}\ (1, \uparrow)$

$\text{Lim } V_{ssc} = C_p + [(C_p)(\text{infinity scrapper})]$

$\Sigma\ M_{ssv} \rightarrow \Sigma\ M_{ssv}\ (1, \uparrow)$

$\text{Lim } V_{ssc} = C_p + \{C_p\ [f(\Omega)\ \text{or}\ f(\text{Alephj } 0)\ \text{or}\ f(\text{Aleph } 1)\ \text{or}\ f(\text{Aleph } 2)\ \text{or, ..., or}\ f(\text{Aleph } i),\text{or ... }]\}$: $i = 3, 4, 5, ...$

$\Sigma\ M_{ftt} \rightarrow \Sigma\ M_{ftt}\ (1, \uparrow)$

$\text{Lim } V_{ssc} = C_p + \{C_p\ [f(\Omega)\ \text{or}\ f(\text{Alephj } 0)\ \text{or}\ f(\text{Aleph } 1)\ \text{or}\ f(\text{Aleph } 2)\ \text{or, ..., or}\ f(\text{Aleph } i),\text{or ... }]\}$: $i = 3, 4, 5, ...$

$\Sigma\ M_{ssv} \rightarrow \Sigma\ M_{ssv}\ (1, \uparrow)$

The following limits may describe the lessening of forward time travel.

$\Delta T_{sscb} = [(\text{Distance Traveled}/C] - \{e\ [(\text{Distance Traveled})/C]\}$

$\Sigma\ M_{ftt} \rightarrow \Sigma\ M_{ftt}\ (1, \uparrow)$

ΔT$_{sscb}$ = [(Distance Traveled/C] – {e [(Distance Traveled)/C]}

Σ M$_{ssv}$ ➜= Σ M$_{ssv}$ (1, ↑)

ΔT$_{sscb}$ = [(Distance Traveled/C] – {g [(Distance Traveled)/C]}

Σ M$_{ftt}$ ➜ Σ M$_{ftt}$ (1, ↑)

ΔT$_{sscb}$ = [(Distance Traveled/C] – {g [(Distance Traveled)/C]}

Σ M$_{ssv}$ ➜ Σ M$_{ssv}$ (1, ↑)

ΔT$_{sscb}$ = [(Distance Traveled/C] – {f [(Distance Traveled)/C]}

Σ M$_{ftt}$ ➜ Σ M$_{ftt}$ (1, ↑)

ΔT$_{sscb}$ = [(Distance Traveled/C] – {f [(Distance Traveled)/C]}

Σ M$_{ssv}$ ➜ Σ M$_{ssv}$ (1, ↑)

ΔT$_{sscb}$ = [(Distance Traveled/C] – {[f(Ω) or f(Alephj 0) or f(Aleph 1) or f(Aleph 2) or, …, or f(Aleph i),or …]

[(Distance Traveled)/C]}

Σ M$_{ftt}$ ➜ Σ M$_{ftt}$ (1, ↑)

ΔT$_{sscb}$ = [(Distance Traveled/C] – {[f(Ω) or f(Alephj 0) or f(Aleph 1) or f(Aleph 2) or, …, or f(Aleph i),or …]

[(Distance Traveled)/C]}

Σ M$_{ssv}$ ➜ Σ M$_{ssv}$ (1, ↑)

The following limits express the growth relations for M$_{ftt}$ and M$_{ssv}$.

Σ M$_{ftt}$ = Σ M$_{ftt}$ (1, infinity)

Distance Traveled ➜ infinity

Σ M$_{ssv}$ = Σ M$_{ssv}$ (1, infinity)

Distance Traveled ➜ infinity

Σ M$_{ftt}$ = Σ M$_{ftt}$ (1, infinity)

gamma ➜ infinity

$$\Sigma \ M_{ssv} = \Sigma \ M_{ssv} \ (1, \text{infinity})$$

$$\text{gamma} \rightarrow \text{infinity}$$

We present only some of the possibilities here. The simple fact is, we do not know whether backward time travel is supported by superluminal velocities from an empirical standpoint. From a theoretical standpoint, superluminal inertial travel would result in backward time travel according to most special relativistic conventions. To what extent or to what degree backward time travel would result from superluminal velocities remains a mystery that is further compounded in that such time travel may simply be a lessening of forward time travel or an absolute backward time travel.

We recognize that the speculations presented above may be outside the normal bounds of theoretical inquiry. Physicists generally do not accept local infinite values or real infinities that can crop up in mathematical formulations. More likely, only the sub-ensemble and ensemble boundary limits would apply unless a reified mathematical phase transition occurs for finite mass-energy depletions in interstellar and intergalactic space.

However, nature may itself play with local infinities. The real energy of a local system seems philosophically bound to conserve energy to the absolute letter of the law. Violations of real energy conservancy have never been observed. Thus, nature seems to do energy accounting with infinite precision. Other similarly precise natural accountings include electric charge conservation, quantum spin conservation, the complete lack of observed absolute backward time travel, and the like. Thus, it is understandable that theorists would be uncomfortable with the above conjectures on absolute backward time travel.

In the case for which photons in interstellar space do travel exactly at C, no more and no less, nature, in a sense, already permits infinite time dilation. Perhaps nature does this also for neutrinos, although the current experimental evidence supports a nonzero invariant mass for the three species or flavors of neutrinos. It would prove interesting if neutrino velocities could never be measured differently from C as measurement capabilities continue to improve.

Next, we consider notions such as the utterly maximum speed of quantum teleportation based on quantum mechanical entanglement and the need to send a classical feedback loop signal that travels at the speed of light.

For example, a particle may effectively enable light-speed teleportive transport even if this particle acting as a classical feedback loop has a finite Lorentz factor. Here, the finite Lorentz factor would be about the same approximately 10 EXP 30 value conjectured about previously.

As another example, a classical feedback signal may travel faster than the practical velocity of light in cases for which the signal was a specially prepared photon that would experience the index of refraction of the background medium vanishingly small fractions of unity greater than 1 but less than the effective index of refraction of the background medium experienced by almost all other photons. Perhaps such more precisely unitary refractive index values for the specially prepared photon would imbue the photon with a slightly recessed forward time travel based on the addition of a slightly backward time travel fraction component which is negative in sign relative to the otherwise resulting speed of the photon.

Such mechanism may lead to some fascinating subtle modifications of special relativity, which can manifest asymptotically exaggerated Lorentz factors of time dilation (some of which may already be infinite), asymptotically exaggerated relativistic aberration, and asymptotically exaggerated Lorentz contraction. A truly fascinating result would include the spontaneous generation of infinite kinetic energies under conditions where the kinetic energies of a spacecraft traveling at these enhanced velocities would otherwise be finite.

A fascinating result would include an (Aleph 0) order ship-time derivatives of the increase in spacecraft Lorentz factor increase, which in themselves runs away to positive infinity in an asymptotic manner. We can, in spacecraft acceleration in the ship, consider likewise analogous increases-frame, as well as kinetic energy gains and extents of relativistic aberration.

We can consider analogous scenarios, including an (Aleph 1) order ship-time derivatives of the increase in spacecraft Lorentz factor increase, which in themselves runs away to positive infinity in an asymptotic manner. We can in spacecraft acceleration in the ship consider likewise-analogous increases-frame, as well as kinetic energy gains, and extents of relativistic aberration.

We can consider analogous scenarios including an (Aleph 2) order ship-time derivatives of the increase in spacecraft Lorentz factor increase, which in themselves runs away to positive infinity in an asymptotic manner. We can in spacecraft acceleration in the ship consider likewise analogous increases-frame, as well as kinetic energy gains, and extents of relativistic aberration.

We can consider analogous scenarios, including an (Aleph 3) order ship-time derivatives of the increase in spacecraft Lorentz factor increase, which in themselves runs away to positive infinity in an asymptotic manner. We can, in spacecraft acceleration in the ship, consider likewise-analogous increases-frame, as well as kinetic energy gains, and extents of relativistic aberration.

We can go all the way to consider analogous scenarios, including an (Aleph Ω) order ship-time derivatives of the increase in spacecraft Lorentz factor, which in themselves runs away to positive infinity in an asymptotic manner. We can, in spacecraft acceleration in the ship, consider likewise analogous increases-frame, as well as kinetic energy gains, and extents of relativistic aberration. Here, Ω, is the least transfinite ordinal, a.k.a., the least infinite integer. We can march on from here to ever greater respective infinities.

In essence, we thus explore scenarios of so-called stacked asymptotes.

The kinematic, thermodynamic, and topological aspects of these stacked asymptotes as such are very profound.

The associated curve plots become incredibly skewed and distorted relative to those from ordinary asymptotes. Thus, the ship-frame Lorentz factors, relativistic aberration, kinetic energy gains, and the like will take on exotic values as will the kinematic, thermodynamic, and topological aspects of the travel of the associated spacecraft.

Now, the velocity of light in a vacuum is also calculated as the inverse square root of the product of the electrical permittivity of free space and the magnetic permeability of free space.

In formulaic jargon,

$$c = \{1/[(\epsilon_0)(\mu_0)]\} \text{ EXP } (1/2).$$

Now we can conjecture whether there are more than one velocity of light.

For example, there may be other velocities of light that manifest in other dimensions, hyperspaces, hyperspace-times, and the like.

Additionally, the one velocity of light in a vacuum that we know of may have various other manifestations in other dimensions, hyperspaces, and hyperspace-times.

Additionally, the one known velocity of light may have other meanings, parallel distinct meanings, and the like in other dimensions, hyperspaces, and hyperspace-times.

Now consider the energy formula $E = mc^2$.

Since we conjectured on other speeds of light such as in other dimensions, hyperspaces, hyperspace-times, and the like, the energy value of the associated matter may be different and/or outright distinct from the energy equivalent of matter in said other dimensions, hyperspaces, hyperspace-times, and the like.

In scenarios where the speed of light has different natural interpretations or manifestations such as in other dimensions, hyperspaces, hyperspace-times, and the like, analogues of the formula $E = mc^2$ may apply, thus indicating the possibility of analogues of energy, alternate types of energy, or different meanings of energy.

These constructs can be denoted by the following compound expressions:

For different types of energy,

$$[\Sigma(k = 1; k = \infty\uparrow): E_k = m_k c_k^2]$$

For analogues of energy,

$$[\Sigma(k = 1; k = \infty\uparrow): E_{analog,k} = m_{analog,k} c_{analog,k}^2]$$

For different meanings of energy,

$$[\Sigma(k = 1; k = \infty\uparrow): [f(E_k)] = [f(m_k)] [[f(c_k)]^2]]]$$

Electromagnetic energy is a very pure form of energy, and as a manifestation of the strongest known long-range force, it has a special status.

Because light as pure energy is so ephemeral and appears from a medical and scientific standpoint, highly related to the conscious workings of the human mind/brain, the soul may have aspects that ontologically reach down to couple with the living human body, especially the living human brain.

So the para-electromagnetic aspects of the human soul would seem to be highly ephemeral and truly spiritual, if for no other reason than that they are coupled with the human soul and related to the human soul in this life.

When we consider that the human soul has both a substance and accidents, the aspects of the human soul that existentially tower over electromagnetic and para-electromagnetic energy are even more ephemeral and elevated.

So traveling at the speed of light is thus a profound plausibility for future spacecraft in terms of the physical and metaphysical aspects of such travel and the relation between spacecraft travel at the speed of light and electromagnetic radiation itself. Such light-speed travel may even have loosely related applications for enhancing the para-electromagneticality of the soul at least in terms of the accidents of the soul.

Perhaps in some cosmic future era, all human persons that have ever existed will be brought into the light of the electromagnetic and para-electromagnetic and into a civilization of light.

Here we consider light-speed travel as opening up vast future realms for travel and associated profound and far-out opportunities.

The conceptual grounding for the above assertion is that the cosmos keeps growing in size and accrued number of events of quantum and macroscopic in scale as it ages. Thus, light-speed travelers at variously infinite Lorentz factors behold a cosmos that expands in scale and numbers of elements at an ever-increasing infinite rate relative to the spacecraft as the spacecraft Lorentz factor abounds to ever greater infinite values.

Extremely great finite Lorentz factor transport and infinite Lorentz factor transport offers means to populate the future with human technologies and human persons, as well as nonhuman space alien persons.

For example, the population of our universe in the future assuming a constant light-cone radius is defined as follows:

(# of extreme Lorentz factor travelers per cosmic light-cone volumetric element in a given Planck time unit background time)(# of Planck time units background time per Aeon)(Average # of cosmic light-cone volume units per Aeon)(# of aeons of universal age period considered)

In another formulation, said population may be defined as follows:

[(Average Lorentz factor of crewed starships)(# of relativistic starships per average ship-frame Planck time unit per cosmic light-cone)(# of average ship-frame Planck time units per Aeon)](Future universal age number of Aeons)(Time averaged number of cosmic light-cone volumes in universe over future era considered)(Average population per starship)

Now, it may be possible to send out signals from a central location in a given cosmic light-cone in many directions to signal the need to move matter or matter and energy into a common cosmic region.

Accordingly, the matter and energy buildup in the said cosmic region will greatly slow and/or reverse cosmic expansion in the said region perhaps enabling the region to host civilizations that will thrive forever.

Accordingly, these civilizations would maintain a balance between gravitational collapse and universal expansion.

These habitats may continue to grow in size and mass energy content as said habitats collect more mass and energy from outside the main habitat region.

Now, in some cases, the habitats may acquire sufficient mass and energy to form a black hole. However, electrical charge patterns installed within the habitats may result in a negation of gravitational collapse, thus maintaining a balance between gravitational collapse and electrodynamic outward-directed decompressive forces.

Other mechanisms of maintaining habitats amid eternal cosmic expansion may include the conversion of local stores of dark energy into baryonic matter. The baryonic matter produced would be distributed to cancel the expansionary effects of universal expansion.

Now, we can intuit natural infinities.

For example, we can understand that under normal mathematical scenarios, a spacecraft would need to acquire an infinite amount of kinetic energy to reach the speed of light.

We can also consider the open-ended eternal future of infinite temporal extent.

Likewise, we can conjecture that our universe may be infinite in extent.

We can all relate to notions of an infinite GOD.

Now, there are countable and uncountable infinities.

But before we go further, consider the hyper-operator notation that was designed to express huge values not otherwise expressible.

For example, note that Hyper4(a, n) is equal to a tetrated n, or a raised to the power of itself n-1 times. The latter value is symbolically written as n subscript a.

For example,

3 EXP 4 = 81, but 4 subscript 3 is approximately equal to 10 EXP (1,000,000,000,000).

Alternatively,

4 subscript 2 = 2 EXP 2 EXP 2 EXP 2 = 2 EXP [2 EXP [2 EXP 2]] = 2 EXP (2 EXP 4) = 2 EXP 16 = 65,536.

For example,

Hyper5(4, 4)is equal to 4 tetrated 4 tetrated 4 tetrated 4. This value is commonly referred to as 4 pentated 4.

Hyper 6, (4,4) is 4 pentated 4 pentated 4 pentated 4 and is also referred to as 4 hexataed 4.

Hyper 7, (4,4) is 4 hexated 4 hexated 4 hexated 4 and so on.

Aleph 0 is the infinite number of integers.

Aleph 1, according to the perhaps unprovable, and thus unfalsifiable, continuum hypothesis, is the number of real numbers which is greater than Aleph 0 by a multiplicative factor of infinity.

Aleph 2 is similarly greater than Aleph 1.

Aleph 3 is similarly greater than Aleph 2.

Aleph 4 is similarly greater than Aleph 3.

And so on.

In general, Aleph n = 2 EXP [Aleph (n-1)].

The number Ω is commonly stated as the least infinite positive integer or ordinal.

Now here is a real zinger.

So we can produce the abstraction of Hyper Aleph Ω (Aleph Ω, Aleph Ω).

Now, the above infinities as specified are mathematical infinities and are thus somewhat removed from concrete physical infinities.

The next level of infinities is what we would refer to as abstract infinities.

For example, we can intuit a GOD that has infinite goodness, infinite value, infinite worth, infinite purpose, and so on. These infinities are more of an ontological transcendental type and presumably transcend even GOD's power, intellect, and other active infinities. The number of ontological transcendentals of GOD must be unboundable, even mathematically.

However, we can also consider abstract quantitative infinities. These infinities would be greater than definable mathematical infinities.

In a sense, abstract quantitative infinities would be to mathematical infinities as mathematical infinities are to physical infinities.

We can go still farther to consider hyper-abstract-infinities, which would be to abstract quantitative infinities as abstract quantitative infinities are to mathematical infinities as mathematical infinities are to physical infinities.

We can go still further to consider hyper-hyper-abstract-infinities, which would be to hyper-abstract-infinities as hyper-abstract-infinities are to abstract quantitative infinities as abstract quantitative infinities are to mathematical infinities as mathematical infinities are to physical infinities.

We can go still further to consider hyper-hyper-hyper-abstract-infinities, which would be to hyper-hyper-abstract-infinities as hyper-hyper-abstract-infinities are to hyper-abstract-infinities as hyper-abstract-infinities are to abstract quantitative infinities as abstract quantitative infinities are to mathematical infinities as mathematical infinities are to physical infinities.

We will label hyper-abstract-infinities as hyper-1-abstract-infinities; hyper-hyper-abstract-infinities as hyper-2-abstract-infinities; and hyper-hyper-hyper-abstract-infinities as hyper-3-abstract-finities.

Well, we can continue to conjecture hyper-4-abstract-infinities, hyper-5-abstract-infinities, and so on to Hyper-[Hyper Aleph Ω (Aleph Ω, Aleph Ω)]-abstract-infinities, and continue forever onward.

Now, light-speed impulse travel provides, mathematically, at least for the prospect of (Aleph 0) Lorentz factors. As such, a craft of (Aleph 0) values of gamma would travel an (Aleph 0) light-years in one-year ship-frame through space, an (Aleph 0) years of travel into the future in one-year ship-frame, and an effective (Aleph 0) multiple of the speed of light in travel velocity.

We can boldly conjecture about light-speed craft attaining (Aleph 1) Lorentz factors and then do the simple respective math and go on from there.

How hyper-k-abstract-infinities are related to an unbounded future evolution of the fabric of reality is anyone's guess.

However, we now have the prospects of developing starships that achieve 20 percent of the speed of light or a small Lorentz factor of about 1.02, or just a little bit greater than 1.

These crafts, I am almost certain, could be designed, assembled, and crewed within the next forty years.

Next, we perhaps would aim for the big goal of craft able to attain 30 percent of the speed of light or of a Lorentz factor of 1.048, which is still only a little bit greater than 1.

Next, we perhaps would aim for the big goal of craft able to attain 40 percent of the speed of light or of a Lorentz factor of 1.091.

By some time next century, we should be able to attain 98 percent of light-speed and a Lorentz factor of about 5. Such a craft upon reaching 98 percent of the speed of light could travel almost 500 light-years in 100 years' ship-frame, travel 500 years into the future in 100 years' ship-frame, and travel effectively close to five times the speed of light.

It only gets better from here.

So if you feel bummed out by any light-speed limit, take heart. It works out all the better in the end.

ESSAY 36) The Primary Importance of Light-Speed as a Proverbial Thread That Binds Visible Reality Together

Now we of science fiction interests of the Star Trek and Star Wars movies generations can be put off by notions that faster-than-light warp drive, wormhole travel, and the like might simply be undefined by Mother Nature.

Regarding how big infinities, and thus light-speed spacecraft Lorentz factors, can be, consider the following digression that is repeated in this book.

Now, there are countable and uncountable infinities. Uncountable infinities can be much larger than countable infinities although the precise distinction is a little bit technical.

But before we go further, consider the hyper-operator notation that was designed to express huge values not otherwise expressible.

For example, note that Hyper4(a, n) is equal to a tetrated n, or a raised to the power of itself n-1 times. The latter value is symbolically written as n subscript a.

For example,

3 EXP 4 = 81, but 4 subscript 3 is approximately equal to 10 EXP (1,000,000,000,000).

Alternatively,

4 subscript 2 = 2 EXP 2 EXP 2 EXP 2 = 2 EXP [2 EXP [2 EXP 2]] = 2 EXP (2 EXP 4) = 2 EXP 16 = 65,536.

For example,

Hyper5(4, 4) is equal to 4 tetrated 4 tetrated 4 tetrated 4. This value is commonly referred to as 4 pentated 4.

Hyper 6, (4,4) is 4 pentated 4 pentated 4 pentated 4 and is also referred to as 4 hexataed 4.

Hyper 7, (4,4) is 4 hexated 4 hexated 4 hexated 4 and so on.

Aleph 0 is the infinite number of integers.

Aleph 1, according to the perhaps unprovable, and thus unfalsifiable, continuum hypothesis, is the number of real numbers which is greater than Aleph 0 by a multiplicative factor of infinity. Aleph 1 = 2 EXP (Aleph 0).

Aleph 2 is similarly greater than Aleph 1. Aleph 2 = 2 EXP (Aleph 1).

Aleph 3 is similarly greater than Aleph 2. Aleph 3 = 2 EXP (Aleph 2).

Aleph 4 is similarly greater than Aleph 3. Aleph 4 = 2 EXP (Aleph 3).

And so on.

In general, Aleph n = 2 EXP [Aleph (n-1)], where *n* is any positive integer,

The number Ω is commonly stated as the least infinite positive integer or ordinal.

Now here is a real zinger.

So we can produce the abstraction of Hyper Aleph Ω (Aleph Ω, Aleph Ω).

Here is another zinger.

We can produce the abstraction of,

> Hyper Aleph (Uncountable infinity) [Aleph (Uncountable infinity), Aleph (Uncountable infinity)]

The uncountable infinities can be as great as we desire.

Now here comes a real zinger!

Imagine an infinite number so great that it cannot be named.

Well, we can ironically produce the following construct.

We can produce the abstraction of,

> Hyper Aleph (Infinity number so great that it cannot be named) [Aleph (Infinity number so great that it cannot be named), Aleph (Infinity number so great that it cannot be named)].

Now we can imagine light-speed travel at the least infinite number. The least infinite positive integer is often denoted as Ω.

Once we are able to attain Lorentz factors of $[f(\Omega)] = n\Omega$, where n is finite but greater than 1, we will need an entirely new category of mathematics.

From there, we will forever climb the gamma factor ladder.

Even Lorentz factors of $[f(\Omega)]$ will provide us an eternity of delight. Functions of greater Aleph numbers such as [f(Aleph 1)], [f(Aleph 2)], [f(Aleph 3)], and so on as values of Lorentz factors to be achieved will delight us for eternity after eternity in a never-ending ascension.

Tantalum hafnium carbide is sort of an old wonder material that is known for its extreme melting point. The material may conceivably be incorporated into starlight and CMBR shielding for extremely relativistically blue-shifted background radiations. Other uses for the material can include monolithic or weavelike grid light sails, charged Project Medusa–type sails, and perhaps pusher plates for Project Orion–style starships.

Now, before we can design, build, and fly starships capable of infinite Lorentz factors in light-speed travel, more than likely, we will need to develop technology for a period of an infinite number of human familial generations. However, this is not a bad thing because we will have loads of whimsical joy overcoming each next Lorentz factor hurdle. It will almost be like achieving light-speed in infinite Lorentz factors will be like working toward an infinite distance goal as if a subtly dim beacon of mystery that is infinite distance.

Now, most stars are red dwarfs. However, when we first fly starships capable of a Lorentz factor of 2, the red dwarfs will appear blue, with more stars in view as you look farther toward the front of the spacecraft heading relative to perpendicular to the spacecraft heading. The red-dwarf stars in the back will look brownish red.

Eventually, we will be able to reach a Lorentz factor of 10, which will place any stars within a 500-light-year radius from Earth in range of the starship in one average working lifetime of a human.

Our next major goal will be obtaining a Lorentz factor of about 1,500. As such, the outer hull of the spacecraft would be made of tantalum hafnium carbide, which melts at about 7,610 degrees F. Accordingly, the cosmic microwave background radiation will appear to be about 7,000 degrees F, so a really refractory compound will be required to prevent the ship from being destroyed by the heat. Additionally, the outer hull will need to be polished to near-perfect reflectiveness, as well as being highly elongated and pointed much like a sewing needle.

In the fictional Star Trek series of TV episodes and movies, warp factor 10 is about as fast as the *Enterprise* can travel. This corresponds to a velocity of 1,024 times the speed of light. Well, for an impulse craft crew of a starship traveling at a Lorentz factor of 1,500, the effective travel velocity will be greater than warp factor 10. What's more, the crew will travel 1,500 years into the future for every year ship-time. Assuming a longevity of 100 or more years' ship-time for crew members, the crew members would have the opportunity to travel 150,000 years into the future.

Now, can you imagine a time, perhaps 200 years into the future from now, booking a flight on Galactic Starlines?

When everyone has boarded, space traffic control will provide flight clearance, and the pilot will radio in, setting the course for the Pleiades at G-1,500. The term G-1,500 will mean a value of gamma or Lorentz factor equal to 1,500.

Eventually, we will develop even better refractories to enable all the greater spacecraft Lorentz factors.

At some point, I believe we will figure out how to make bulk neutronium, then quarkonium, then higgsinium, the mono-higgsinium, then monopolium, and onward from there.

Now, usually, contemporary cosmologists hold that the universe, at least the one we live in, is huge but likely finite. However, there are a growing number of cosmologists open to a universe of infinite expanse. An infinite universe gives us awesome travel prospects.

Accordingly, it offers us a way to attain light-speed in inertial travel with infinite kinetic energy and infinite time dilation per the associated infinite Lorentz factors.

Even a Lorentz factor equal to Ω, the least transfinite ordinal would enable travel an infinite number of light-years through space and an infinite number of years traveled forward in time in one-year ship-time, as well as effectively travel at an infinite multiple of the speed of light.

If the universe we live in is infinite in size, we may then consider how great this infinity is and how much greater it will become as the universe continues to expand.

Now, we who graduated high school are all well aware of the infinite Cartesian number line and how it extends from positive infinity to negative infinity. We have this notion that there are infinite positive integers hanging way out along the expanse of the abstract number line. However, most folks as such have not really contemplated, at least in depth, hyperextended number lines.

Accordingly, the ticks on hyperextended number lines might not even be integers, much less any real numbers. Perhaps these ticks are best paired with what we will refer to as hyperintegers.

The degree of hyperextension of number lines may be arbitrary, as may the level of hyperintegers.

Eventually, we can consider number lines so extended that we need a new lexicography to denote them. As such, I propose the term ultra-integers. Like hyperintegers, ultra-integers may come in various levels, as can the extension of associated ultra-extended number lines.

At some point, in philosophy, especially metaphysics, we may need to develop the concept of transnumerical objects, which would be assigned to ticks on a number line or analogues thereof.

Now back to the subject of Lorentz factors and the size of our universe.

Given the above digression on number lines, we have open-ended possibilities for traveling as transfinite, then hyper-integral, then ultra-integral, and perhaps even transnumerical Lorentz factors and numbers of light-years through space.

If our universe is not large enough to support the above Lorentz factors and travel distances, then perhaps a greater multiverse, forest, biosphere, and the like may. If not, then perhaps one or more levels of hyperspaces can. Note that a forest as such is a topologically related set of multiverses, a biosphere is a topologically related set of forests, and so on. These funny-sounding names are simply used as metaphors to denote fractal-verse theory levels.

So do not fret over light-speed limits. It actually works out better in the end. And no one need be left behind. For it is plausible that world-ships may be developed at some time in the future that can carry entire civilizations to the speed of light.

All of us have likely pondered what GOD will reveal to us an entire eternity into the future. I know for sure I have.

Well, the prospect of obtaining truly light-speed in even the least infinite Lorentz factor can serve as a proverbial archetypal light that spurs us on to develop spacecraft that can travel all the more close to the speed of light in superabounding finite but ever greater Lorentz factors.

We might reflect on this beacon of hope in abstract mental and emotional imagery as if it were some mysterious, ever-so-dim, but magical light that beckons us forward. This beacon can serve as a guiding principle of hope.

Even more whimsical is the concept of obtaining ever-greater infinite Lorentz factors that may imply a never-ending series of beacons guiding us forward with the open-ended possibilities of traveling into the associated ever-greater future eternities.

So, I hope I have added some joy and mystique to your day. Just as we are all in this together to save our planet from natural and/or man-made calamities, we are likewise all together as we begin our wonderful journey down the road of eternity. The sights and attractions along the way will in themselves be grand affairs with their own mystique.

Now, the speed of light is perhaps the most important constant in all of physics.

By providing a limited velocity in the value it is, the velocity of light ensures that backward time travel is impossible in any sense that it could actually be used to change the present. Thus, Mother Nature has worked physics out so as to avoid events that would corrupt the integrity of cause and effect.

Mother Nature has evolved to develop the lowest energy state she can, and this serves as a self-protection mechanism.

However, the speed of light ensures that infinite time dilation is possible in theory, as is infinite kinetic energy, infinite momentum, infinite Lorentz contraction, infinite relativistic aberration, and travel effectively at infinite multiples of the speed of light.

Now we ponder just how great these infinities may be.

Now Aleph 0 is the infinite number of positive integers.

Aleph 1 is, according to the perhaps unprovable and unfalsifiable continuum hypothesis, equal to the number of real numbers.

Either way, Aleph 1 = 2 EXP (Aleph 0).

Aleph 2 = 2 EXP (Aleph 1).

Aleph 3 = 2 EXP (Aleph 2).

Aleph 4 = 2 EXP (Aleph 3).

And so on.

In general, [Aleph (N +1)] = 2 EXP (Aleph N), where N is any whole number.

Regarding how big infinities can be, consider the following digression that is repeated in this book.

Now, there are countable and uncountable infinities.

But before we go further, consider the hyper-operator notation that was designed to express huge values not otherwise expressible.

For example, note that Hyper4(a, n) is equal to a tetrated n, or a raised to the power of itself n-1 times. The latter value is symbolically written as n subscript a.

For example,

3 EXP 4 = 81, but 4 subscript 3 is approximately equal to 10 EXP (1,000,000,000,000).

Alternatively,

4 subscript 2 = 2 EXP 2 EXP 2 EXP 2 = 2 EXP [2 EXP [2 EXP 2]] = 2 EXP (2 EXP 4) = 2 EXP 16 = 65,536.

For example,

Hyper5(4, 4)is equal to 4 tetrated 4 tetrated 4 tetrated 4. This value is commonly referred to as 4 pentated 4.

Hyper 6, (4,4) is 4 pentated 4 pentated 4 pentated 4 and is also referred to as 4 hexataed 4.

Hyper 7, (4,4) is 4 hexated 4 hexated 4 hexated 4 and so on.

The number Ω is commonly stated as the least infinite positive integer or ordinal.

Now here is a real zinger.

So we can produce the abstraction of Hyper Aleph Ω (Aleph Ω, Aleph Ω).

We may continue the hyper-operator process in any finite or infinite number of nestings.

For example, the first higher nesting is as follows:

Hyper [Hyper Aleph Ω (Aleph Ω, Aleph Ω)] {[Hyper Aleph Ω (Aleph Ω, Aleph Ω)], [Hyper Aleph Ω (Aleph Ω, Aleph Ω)]}

The second higher nesting is as follows:

{Hyper [Hyper Aleph Ω (Aleph Ω, Aleph Ω)] {[Hyper Aleph Ω (Aleph Ω, Aleph Ω)], [Hyper Aleph Ω (Aleph Ω, Aleph Ω)]}}

The third higher nesting is as follows:

Hyper { Hyper [Hyper Aleph Ω (Aleph Ω, Aleph Ω)] {[Hyper Aleph Ω (Aleph Ω, Aleph Ω)], [Hyper Aleph Ω (Aleph Ω, Aleph Ω)]}} {{ Hyper [Hyper Aleph Ω (Aleph Ω, Aleph Ω)] {[Hyper Aleph Ω (Aleph Ω, Aleph Ω)], [Hyper Aleph Ω (Aleph Ω, Aleph Ω)]}}, { Hyper [Hyper Aleph Ω (Aleph Ω, Aleph Ω)] {[Hyper Aleph Ω (Aleph Ω, Aleph Ω)], [Hyper Aleph Ω (Aleph Ω, Aleph Ω)]}}}

Now we can consider the first higher nesting as a mere point on the Cartesian number line that is to the immediate right and adjacent to the point associated with 0.

The second higher nesting will be paired with a mere point that is located to the immediate right and adjacent to the point associated with the first higher nesting.

We can lay down points for the third higher nesting, the fourth higher nesting, and so on. For the least infinite series of nestings, we derive a line segment that is of the smallest finite length.

So naturally, we can fill the number line past the integer 1 and continue from there to the least infinite integer.

We can then continue on to develop a hyperextended number line for which the ticks eventually are associated with hyperintegers. We can consider any arbitrary levels of hyperextensions such as hyper-hyperextensions, hyper-hyper-hyperextensions, and so on.

At some point, the mathematical objects at the ticks and the number line will become qualitatively differentiated to morph into what we refer to as an ultra-extended number line and various levels of ultra-integers or whatever we choose to call these bizarre mathematical entities.

So you can now have a sense of how great a Lorentz factor just might be, provided that Mother Nature cooperates.

So nature provides us unlimited options for infinite Lorentz factors in light-speed travel.

Now, since the speed of light is at the very heart of observable physical reality in our universe, traveling ever more closely to the speed of light with ever-increasing but finite Lorentz factors is akin to diving into the ocean of observable physical reality. Once we attain the velocity of light in a first and least infinite Lorentz factor, we will be fully submerged in an ocean of unfathomably infinite depth and extent.

As we continue to increase the spacecraft's already infinite Lorentz factor, we will be setting sail for the very heart of the observable aspects of Mother Nature.

As we attain still greater infinite Lorentz factors, we will come closer and closer to the throne and majesty of Mother Nature.

This throne is of itself infinite in scale beyond fathoming and has level upon level in a never-ending ascending series of levels.

Thus, the speed of light is as if a thread that binds observable physical reality together and serves as a great protector of Mother Nature.

ESSAY 37) The Infinite Number of Coordinates In Abstract Light-Speed Space

Now here is a far-out set of ideas.

First, we need to agree that we approach light-speed from speeds less than that of light. When we reach infinite Lorentz factors, we are precisely at the speed of light.

Now for the zinger.

Instead of approaching the speed of light from below or from superluminal velocities in such a way that light-speed would be represented as the origin or point of 0 on a y-axis, suppose we were to approach the speed of light as if from speeds represented by the x-axis.

Furthermore, suppose we were to approach the speed of light as if from speeds represented by a z-axis.

We can continue the general pattern perhaps defined by an infinite dimensional coordinate system.

These ideas are not the same as higher-dimensional space-time travel such as travel in higher-dimensional coordinate systems but, instead, are concepts much more abstract where each axis, except the y-axis, is distinct from the greater-than or less-than metric.

The above obtusely abstract system plausibly has analogues, which may optionally be as follows:

> flat, positively curved, negatively curved, positively curved and torsioned at one or more scales in arbitrary patterns including, but not limited to, fractals, negatively curved and torsioned at one or more scales in arbitrary patterns including, but not limited to, fractals, positively super-curved, negatively super-curved, positively super-curved and torsioned at one or more scales in arbitrary patterns including, but not limited to, fractals, negatively super-curved and torsioned at one or more scales in arbitrary patterns including, but not limited to, fractals, positively super-curved and super-torsioned at one or more scales in arbitrary patterns including, but not limited to, fractals, negatively super-curved and super-torsioned at one or more scales in arbitrary patterns including, but not limited to, fractals,

positively curved and positively torsioned at one or more scales in arbitrary patterns including, but not limited to, fractals, negatively curved and positively torsioned at one or more scales in arbitrary patterns including, but not limited to, fractals, positively super-curved, negatively super-curved, positively super-curved and positively torsioned at one or more scales in arbitrary patterns including, but not limited to, fractals, negatively super-curved and positively torsioned at one or more scales in arbitrary patterns including, but not limited to, fractals, positively super-curved and positively super-torsioned at one or more scales in arbitrary patterns including, but not limited to, fractals, negatively super-curved and positively super-torsioned at one or more scales in arbitrary patterns including, but not limited to, fractals, positively curved and negatively torsioned at one or more scales in arbitrary patterns including, but not limited to, fractals, negatively curved and negatively torsioned at one or more scales in arbitrary patterns including, but not limited to, fractals, positively super-curved, negatively super-curved, positively super-curved and negatively torsioned at one or more scales in arbitrary patterns including, but not limited to, fractals, negatively super-curved and negatively torsioned at one or more scales in arbitrary patterns including, but not limited to, fractals, positively super-curved and negatively super-torsioned at one or more scales in arbitrary patterns including, but not limited to, fractals, negatively super-curved and negatively super-torsioned at one or more scales in arbitrary patterns including, but not limited to, fractals, positively super-...-super-curved, negatively super-...-super-curved, positively super-...-super-curved and torsioned at one or more scales in arbitrary patterns including, but not limited to, fractals, negatively super-...-super-curved and torsioned at one or more scales in arbitrary patterns including, but not limited to, fractals, positively super-...-super-curved and super-...-super-torsioned at one or more scales in arbitrary patterns including but not limited to fractals, negatively super-...-super-curved and super-...-super-torsioned at one or more scales in arbitrary patterns including, but not limited to fractals, positively curved and positively torsioned at one or more scales in arbitrary patterns including, but not limited to, fractals, negatively curved and positively torsioned at one or more scales in arbitrary patterns including, but not limited to, fractals, positively super-...-super-curved, negatively super-...-super-curved, positively super-...-super-curved and positively torsioned at one or more scales in arbitrary patterns including, but not limited to, fractals, negatively super-...-super-curved and positively torsioned at one or more scales in arbitrary patterns including, but not limited to, fractals, positively super-...-super-curved and positively super-...-super-torsioned at one or more scales in arbitrary patterns including, but not limited to, fractals, negatively super-...-super-curved and positively super-...-super-torsioned at one or more scales in arbitrary patterns including, but not limited to fractals, positively curved and negatively torsioned at one or more scales in arbitrary patterns including, but not limited to, fractals, negatively curved and negatively torsioned at one or more scales in arbitrary patterns including, but not limited to, fractals, positively super-...-super-curved, negatively super-...-super-curved, positively super-...-super-curved and negatively torsioned at one or more scales in arbitrary patterns including, but not limited to, fractals, negatively super-...-super-curved and negatively torsioned at one or more scales in arbitrary patterns including, but not limited to, fractals, positively super-...-super-curved and negatively super-...-super-torsioned at one or more scales in arbitrary patterns including, but not limited to, fractals, negatively super-...-super-curved and negatively super-...-super-torsioned at one or more scales in arbitrary patterns including, but not limited to, fractals.

ESSAY 38) Infinite Spacecraft Lorentz Factors and the Relative Rotation of an Entire Universe to a Point Directly in Front of the Spacecraft

Now, traveling at light-speed in a way results in essentially an entire universe, or at least a light-cone of travel to be rotated to a location in front of you, as well as being contracted in spatial extent in front of you by a multiple of infinity. Thus, the cosmic light-cone of travel, and even finite portions of the universe of travel, become relativistically aberrated and infinitely Lorentz contracted for spacecraft traveling at light-speed and associated infinite Lorentz factors. Additionally, a spacecraft ostensibly would run out of future room to travel forward in time in a given universe of travel as the spacecraft reaches future eternal limits of time for the given universe of travel. So perhaps a light-speed impulse spacecraft would leave the universe of travel and enter a larger realm, perhaps one that is infinitely larger than the universe of origin, such as a multiverse or a hyperspace.

Such a light-speed spacecraft might travel down a portal of tiny, perhaps infinitesimally angular, diameter in the spacecraft reference frame.

Accordingly, the portal may be of infinite length yet appear infinitesimal in length relative to the spacecraft. Any such portals of travel would be distinct from wormholes but may have relevance for understanding wormhole travel in related topology, kinematics, and thermodynamics.

As a spacecraft slips into such a portal, it may, in a sense, become automatically cloaked from the universe of travel and any finite larger realms of serial entrance. So the spacecraft, in a sense, would have its presence completely separated from the space-time of travel until it enters its quasi-permanent realm of entrance.

In cases where the serially entered realms were of a smaller infinity of extent than the spacecraft Lorentz factors, the spacecraft would plausibly remain hidden within and cloaked by the portal of travel.

In a sense, the spacecraft might permanently reside in the portal of travel until it attains its first omega point. As such, the omega point would be as if divinely enabled, perhaps as part of GOD's grand plan for the cosmos.

Note that the time the spacecraft spends in a portal as such may be infinite both in the background reference frames and also in the spacecraft reference frames. Moreover, the interior of the portal may become as if a realm all to itself, especially for spacecraft having a size equal to a cosmic light-cone or greater. Additionally, there likely would be realms or offshoots from the portal as the spacecraft travels within it such that the spacecraft might stop by one or more of these realms.

Now, just about every physics theory and formula is related or inclusive of the speed of light. As such, the speed of light is an all-important physical parameter that stitches the laws of physics and ordered causality in the universe together, and presumably, in analogues in other realms such as multiverses, forests, biospheres, hyperspaces, and the like.

So a spacecraft traveling inertially and able to attain the velocity of light, in a sense, becomes more related and relevant commensurately with the existential importance of the speed of light.

Attaining the speed of light in infinite Lorentz factors is like achieving a great existential awakening for which a spacecraft and crew become nonsentiently more existentially full.

However, there are some expected consequences for any living and awake crew members on a light-speed infinite Lorentz factor spacecraft. For example, the technology, culture, science, evolution of life, and so on will manifest in truly extreme ways for the light-speed crew when they arrive at a destination having experienced but even a mere Planck Time unit ship-frame at the speed of light. So those who hold that divine providence will radically operate on the background civilizations, cultures, spirituality, and so on for the background personal lifeforms, and sub-personal life-forms that evolve to personal life-forms, must hold that the discontinuities experienced by the spacecraft crew from just before a light-speed mission to just after arrival at a destination will manifest as Lorentz varying scenarios. After all, while the craft traveled at the speed of light for finite ship-time intervals, the cosmic order would have drastically changed, most likely additionally having undergone universal phase changes, symmetry-breaking events, vacuum-state energy decays, and many other similar events perhaps an infinite number of times. So this alone ensures that many results of light-speed travel at infinite Lorentz factors are not Lorentz invariant.

Now the speed of light, according to special relativity, is the same in all reference frames, presumably even with respect to light-speed reference frames.

Accordingly, a spacecraft traveling at the speed of light will experience a beam of light from an internal laser also traveling at the speed of light. As ironic as this sounds, this would be true.

So we are left with a conjecture that the speed of light is multiply so or that the velocity of light is multiply existentially present with itself.

Multiply existential self-presence of the speed of light would imply perhaps that a spacecraft traveling at the speed of light in a vacuum while obtaining increasingly infinite Lorentz factors would, in a sense, seem to become increasingly multiply present with itself.

Additionally, the light-speed spacecraft may become locked into the speed of light in one infinite Lorentz factor to another greater infinite Lorentz factor and so on. Thus, a spacecraft experiencing nonzero ship-frame acceleration may become perpetually locked in acceleration, or Lorentz factor increases, and the acceleration and Lorentz factors may, in a sense, continue to runway to ever greater values in an asymptotic like manner, except that the predecessor Lorentz factors of the spacecraft before said asymptotic-like runaways would already be infinite. This being said, the spacecraft would thus achieve one qualitatively functional greater infinite Lorentz factor after another.

In a sense, the spacecraft would experience increasingly rapid asymptotic gains in qualitatively different and greater quantitatively functionally infinite increases in Lorentz factors.

So a craft might as if entering the speed of light for which the speed of light might serve as if a tower that can be forever climbed upward by a spacecraft having perpetually increasing infinite Lorentz factors. Even though the width of the spectrum in the velocity of light would be vanishingly finite or infinitesimal in width, for a light-speed craft, the interior of the velocity of light may be as if a tower of unlimited height.

Now, as we pointed out earlier, the velocity of light shows up almost ubiquitously in theoretical physics. Thus, as a constant, the velocity of light is very rich, and then so, commensurate with its ubiquitousity in the formulas of theoretical physics.

So perhaps the velocity of light is not simply a tower of infinite intra-height, but also an N-dimensional solid, where *N* may itself be very large, perhaps even infinite. The light-speed solid may have infra-extensity of infinite scale.

Other manifestations of reified light-speed abstraction may present themselves.

Accordingly, the lower surface of a substantially one-dimensional light-speed tower may have a planar face commensurate with the speed of light being the same in all reference frames. This boundary condition may be analogous to a bottom cap on a graduated liquid-containing cylinder.

So the actual velocity of light may manifest as a hyper-cubic solid, which may be approached from a limited subset of external edges, vertices, planes, and the like. The other sub-ordinate features of the hyper-cubic solid would remain unapproachable from the outside but would manifest as an infinite series of steps for craft traveling inside of the speed of light. Thus, we obtain a view of the speed of light in its hypercubic form as if a vast ocean, of unfathomably infinitely many dimensions and dimensional extensions, as well as in numbers of sub-ordinate intra-light-speed topological objects.

Approaching any boundary of said sub-ordinate intra-light-speed topological objects may result in exotic transport effects by virtue of the existential and mathematical stresses and strains in the velocity of light at these locations. These stresses and strains would be analogues of extremely curved or abrupt transitions in space-time due to extreme curvature or distortion.

So travel at the velocity of light perhaps has an unlimited number of surprises for us once we learn how to do such.

ESSAY 39) The Calculus of Inter-Eternity Travel at Infinite Lorentz Factors

We can consider the following calculi:

d(Number of eternities passed thru)/d(gamma)

d(Number of eternities passed thru)/d(ship-frame acceleration)

d(Number of eternities passed thru)/d(ship-frame-time)

d(Number of eternities passed thru)/d(kinetic energy)

d(…(d(Number of eternities passed thru)/d(gamma))/…)/d(gamma)

d(…(d(Number of eternities passed thru)/d(ship-frame acceleration))/…)/d(ship-frame acceleration)

d(…(d(Number of eternities passed thru)/d(ship-frame-time))/…)/d(ship-frame-time)

d(…(d(Number of eternities passed thru)/d(kinetic energy))/…)/d(kinetic energy)

∫(Number of eternities passed thru)d(gamma)

∫(Number of eternities passed thru)d(ship-frame acceleration)

∫(Number of eternities passed thru)d(ship-frame-time)

∫(Number of eternities passed thru)d(kinetic energy)

∫…∫(Number of eternities passed thru)d(gamma)… d(gamma)

∫…∫(Number of eternities passed thru)d(ship-frame acceleration)…d(ship-frame acceleration)

∫…∫(Number of eternities passed thru)d(ship-frame-time)…d(ship-frame-time)

∫…∫(Number of eternities passed thru)d(kinetic energy)… d(kinetic energy)

d(Number of big bang cycles passed thru)/d(gamma)

d(Number of big bang cycles passed thru)/d(ship-frame acceleration)

d(Number of big bang cycles passed thru)/d(ship-frame-time)

d(Number of big bang cycles passed thru)/d(kinetic energy)

d(…(d(Number of big bang cycles passed thru)/d(gamma))/…)/d(gamma)

d(…(d(Number of big bang cycles passed thru)/d(ship-frame acceleration))/…)/d(ship-frame acceleration)

d(…(d(Number of big bang cycles passed thru)/d(ship-time))/…)/d(ship-time)

d(…(d(Number of big bang cycles passed thru)/d(kinetic energy))/…)/d(kinetic energy)

∫(Number of big bang cycles passed thru)d(gamma)

∫(Number of big bang cycles passed thru)d(ship-frame acceleration)

∫(Number of big bang cycles passed thru)d(ship-frame-time)

∫(Number of big bang cycles passed thru)d(kinetic energy)

∫…∫(Number of big bang cycles passed thru)d(gamma)… d(gamma)

∫…∫(Number of big bang cycles passed thru) d(ship-frame acceleration)…d(ship-frame acceleration)

∫…∫(Number of big bang cycles passed thru) d(ship-time)…d(ship-time)

∫…∫(Number of big bang cycles passed thru) d(kinetic energy)… d(kinetic energy)

d(Number of multiverses passed thru)/d(gamma)

d(Number of multiverses passed thru)/d(ship-frame acceleration)

d(Number of multiverses passed thru)/d(ship-frame-time)

d(Number of multiverses passed thru)/d(kinetic energy)

d(…(d(Number of multiverses passed thru)/d(gamma))/…)/d(gamma)

d(…(d(Number of multiverses passed thru)/d(ship-frame acceleration))/…)/d(ship-frame acceleration)

d(…(d(Number of multiverses passed thru)/d(ship-time))/…)/d(ship-time)

d(…(d(Number of multiverses passed thru)/d(kinetic energy))/…)/d(kinetic energy)

∫(Number of multiverses passed thru)d(gamma)

∫(Number of multiverses passed thru)d(ship-frame acceleration)

∫(Number of multiverses passed thru)d(ship-frame-time)

∫(Number of multiverses passed thru)d(kinetic energy)

∫…∫(Number of multiverses passed thru)d(gamma)… d(gamma)

∫…∫(Number of multiverses passed thru) d(ship-frame acceleration)…d(ship-frame acceleration)

∫…∫(Number of multiverses passed thru) d(ship-time)…d(ship-time)

∫…∫(Number of multiverses passed thru) d(kinetic energy)… d(kinetic energy)

d(Number of forests passed thru)/d(gamma)

d(Number of forests passed thru)/d(ship-frame acceleration)

d(Number of forests passed thru)/d(ship-frame-time)

d(Number of forests passed thru)/d(kinetic energy)

d(…(d(Number of forests passed thru)/d(gamma))/…)/d(gamma)

d(…(d(Number of forests passed thru)/d(ship-frame acceleration))/…)/d(ship-frame acceleration)

d(…(d(Number of forests passed thru)/d(ship-time))/…)/d(ship-time)

d(…(d(Number of forests passed thru)/d(kinetic energy))/…)/d(kinetic energy)

∫(Number of forests passed thru)d(gamma)

∫(Number of forests passed thru)d(ship-frame acceleration)

∫(Number of forests passed thru)d(ship-frame-time)

∫(Number of forests passed thru)d(kinetic energy)

∫…∫(Number of forests passed thru)d(gamma)… d(gamma)

∫…∫(Number of forests passed thru) d(ship-frame acceleration)…d(ship-frame acceleration)

∫…∫(Number of forests passed thru) d(ship-time)…d(ship-time)

∫…∫(Number of forests passed thru) d(kinetic energy)… d(kinetic energy)

We can run similar calculi for biospheres, solar systems, and the like for higher levels of fractal-verse theories as well as for hyperspaces.

Each of the following roster elements can be the basis of expressions of differentiated functions for power, energy, velocity, momentum, gamma, and the like by differentiation in the following manners. The above functions for realms passed through or hyperspaces passed through can likewise be temporally differentiated or integrated.

D f/dt,1 d(d f/dt,1)/dt,2 d(d f/dt,2)/dt,1 d[d(d f/dt,1)/dt,2]/dt,3 d[d(d f/dt,1)/dt,3]/dt,2

d[d(d f/dt,2)/dt,1]/dt,3 d[d(d f/dt,2)/dt,3]/dt,1 d[d(d f/dt,3)/dt,1]/dt,2 d[d(d f/dt,3)/dt,2]/dt,1 d f/dT,1 d(d f/dT,1)/dT,2 d(d f/dT,2)/dT,1 d[d(d f/dT,1)/dT,2]/dT,3 d[d(d f/dT,1)/dT,3]/dT,2

d[d(d f/dT,2)/dT,1]/dT,3 d[d(d f/dT,2)/dT,3]/dT,1 d[d(d f/dT,3)/dT,1]/dT,2 d[d(d f/dT,3)/dT,2]/dT,1

T in each case is a timelike blend of various sorts of multiple time dimensions that appropriately and usefully enables derivation in one iteration.

We can go further to consider the following:

d f/dT̲,1 d(d f/dT̲,1)/dT̲,2 d(d f/dT̲,2)/dT̲,1 d[d(d f/dT̲,1)/dT̲,2]/dT̲,3 d[d(d f/dT̲,1)/dT̲,3]`/dT̲,2

d[d(d f/dT̲,2)/dT̲,1]/dT̲,3 d[d(d f/dT̲,2)/dT̲,3]/dT̲,1 d[d(d f/dT̲,3)/dT̲,1]/dT̲,2 d[d(d f/dT̲,3)/dT̲,2]/dT̲,1

T̲ in each case is a timelike blend of timelike blends of various sorts of multiple time dimensions that appropriately and usefully enables derivation in one iteration.

By the same token, we can also integrate functions for power, energy, velocity, momentum, gamma, and the like with respect to multiple time dimensions as follows:

∫ f(dt,1) ∫ ∫d f(dt,1)(dt,2) ∫ ∫d f(dt,2)(dt,1) ∭ f(dt,1)(dt,2)(dt,3) ∭ f(dt,1)(dt,3)(dt,2)

∭ f(dt,2)(dt,1)(dt,3) ∭ f(dt,2)(dt,3)(dt,1) ∭ f(dt,3)(dt,1)(dt,2) ∭ f(dt,3)(dt,2)(dt,1)

∫ f(dT,1) ∬ f(dT,1) (dT,2) ∬ f(dT,2) ((dT,1)) ∭ f(dT,1) (dT,2) (dT,3) ∭ f(dT,1) (dT,3) (dT,2)

∭ f(dT,2) (dT,1) (dT,3) ∭ f(dT,2) (dT,3) (dT,1) ∭ f(dT,3) (dT,1) (dT,2) ∭ f(dT,3) (dT,2) (dT,1)

We can go further to consider the following:

∫ f(dT̲,1) ∫ ∫ f(dT̲,1)(dT̲,2) ∫ ∫ f(dT̲,2)(dT̲,1) ∫ ∫ ∫ f(dT̲,1)(dT̲,2)(dT̲,3) ∫ ∫ ∫ f(dT̲,1)(dT̲,3)(dT̲,2)

∫ ∫ ∫ f(dT̲,2)(dT̲,1)(dT̲,3) ∫ ∫ ∫ f(dT̲,2)(dT̲,3)(dT̲,1) ∫ ∫ ∫ f(dT̲,3)(dT̲,1)(dT̲,2) ∫ ∫ ∫ f(dT̲,3)(dT̲,2)(dT̲,1)

We can continue to yet higher-order derivatives and integrals. Suffice it to say that the general patterns continue for higher numbers of dimensions.

We can also consider higher-order blends of time and thus are not limited to the blends defined by T and $\underline{T}$. We chose these two symbols for their resemblance to the letter t.

As for hyperspatial applications, we consider the following hyperspaces:

Such hyperspaces can be N-Time-M-Space, where N is an integer greater than or equal to 3, and M is any counting number.

Alternatively, N can be any rational number equal to 3 or greater, and M is any rational number greater than or equal to 1.

As another set of scenarios, N can be any irrational number equal to 3 or greater, and M is any irrational number greater than or equal to 1.

Alternatively, N can be any rational number greater than 0, and M is any rational number greater than 0.

As another set of scenarios, N can be any irrational number greater than 0, and M is any irrational number greater than 0.

The N-Space-M-Time may optionally be as follows:

> flat, positively curved, negatively curved, positively curved and torsioned at one or more scales in arbitrary patterns including, but not limited to, fractals, negatively curved and torsioned at one or more scales in arbitrary patterns including, but not limited to, fractals, positively super-curved, negatively super-curved, positively super-curved and torsioned at one or more scales in arbitrary patterns including, but not limited to, fractals, negatively super-curved and torsioned at one or more scales in arbitrary patterns including, but not limited to, fractals, positively super-curved and super-torsioned at one or more scales in arbitrary patterns including, but not limited to, fractals, negatively super-curved and super-torsioned at one or more scales in arbitrary patterns including, but not limited to, fractals, positively curved and positively torsioned at one or more scales in arbitrary patterns including, but not limited to, fractals, negatively curved and positively torsioned at one or more scales in arbitrary patterns including, but not limited to, fractals, positively super-curved, negatively super-curved, positively super-curved and positively torsioned at one or more scales in arbitrary patterns including, but not limited to, fractals, negatively super-curved and positively torsioned at one or more scales in arbitrary patterns including, but not limited to, fractals, positively super-

curved and positively super-torsioned at one or more scales in arbitrary patterns including, but not limited to, fractals, negatively super-curved and positively super-torsioned at one or more scales in arbitrary patterns including, but not limited to, fractals, positively curved and negatively torsioned at one or more scales in arbitrary patterns including, but not limited to, fractals, negatively curved and negatively torsioned at one or more scales in arbitrary patterns including, but not limited to, fractals, positively super-curved, negatively super-curved, positively super-curved and negatively torsioned at one or more scales in arbitrary patterns including, but not limited to, fractals, negatively super-curved and torsioned at one or more scales in arbitrary patterns including, but not limited to, fractals, positively super-curved and negatively super-torsioned at one or more scales in arbitrary patterns including, but not limited to, fractals, negatively super-curved and negatively super-torsioned at one or more scales in arbitrary patterns including, but not limited to, fractals, positively super-...-super-curved, negatively super-...-super-curved, positively super-...-super-curved and torsioned at one or more scales in arbitrary patterns including, but not limited to, fractals, negatively super-...-super-curved and torsioned at one or more scales in arbitrary patterns including, but not limited to, fractals, positively super-...-super-curved and super-...-super-torsioned at one or more scales in arbitrary patterns including, but not limited to, fractals, negatively super-...-super-curved and super-...-super-torsioned at one or more scales in arbitrary patterns including, but not limited to, fractals, positively curved and positively torsioned at one or more scales in arbitrary patterns including, but not limited to, fractals, negatively curved and positively torsioned at one or more scales in arbitrary patterns including, but not limited to, fractals, positively super-...-super-curved, negatively super-...-super-curved, positively super-...-super-curved and positively torsioned at one or more scales in arbitrary patterns including, but not limited to, fractals, negatively super-...-super-curved and positively torsioned at one or more scales in arbitrary patterns including, but not limited to, fractals, positively super-...-super-curved and positively super-...-super-torsioned at one or more scales in arbitrary patterns including, but not limited to, fractals, negatively super-...-super-curved and positively super-...-super-torsioned at one or more scales in arbitrary patterns including, but not limited to, fractals, positively curved and negatively torsioned at one or more scales in arbitrary patterns including, but not limited to, fractals, negatively curved and negatively torsioned at one or more scales in arbitrary patterns including, but not limited to, fractals, positively super-...-super-curved, negatively super-...-super-curved, positively super-...-super-curved and negatively torsioned at one or more scales in arbitrary patterns including, but not limited to, fractals, negatively super-...-super-curved and negatively torsioned at one or more scales in arbitrary patterns including, but not limited to, fractals, positively super-...-super-curved and negatively super-...-super-torsioned at one or more scales in arbitrary patterns including, but not limited to, fractals, negatively super-...-super-curved and negatively super-...-super-torsioned at one or more scales in arbitrary patterns including, but not limited to fractals.

ESSAY 40) Ascension into a Never-Ending Series of Omega Points and Eternities

Now we have considered scenarios for which a spacecraft traveling at the speed of light or ever so slightly faster than light would run out of future temporal room in the cosmic order of its origin.

Thus, we assumed that the spacecraft with infinite or super-infinite Lorentz factors would skip over into a future beyond the cosmic temporal order of initial travel and origin.

Such a spacecraft and her crew might existentially grow to huge proportions in parameters such as Lorentz factors, momentum, kinetic energy, experienced relativistic aberration, propulsion power, propulsion force, acceleration, and the like.

Accordingly, the craft and the crew may ontologically grow relative to the rest of the cosmic order to huge proportions, all the while not harming any sentient life-forms if the craft is suitably thermodynamically and causally cloaked from the environment through which it travels.

Essentially, the craft could grow from one level of immensity after another level in a never-ending series of levels of growth.

Now such a craft and crew might at first seem like energy and resource hogs. However, I like to present the argument that if no sentient life-forms are harmed, and every sentient life-form is on track to exist otherwise but still on track to exist with the extreme spacecraft scenarios, then no harm is done and the cosmic order actually grows as a result of the level-by-level growth of the spacecraft.

Having a spacecraft that can travel from one future eternity after another in minuscule ship-frame time intervals in a sense allows the spacecraft reference frame present to stitch together the space-time of origin and the serially entered future eternities.

Accordingly, the cosmic order may become more unified with each and every creature having a bigger role to play ontologically as the cosmic order progresses to a first omega point, then after one omega point or omega-like point after another in a never-ending series.

Now a first omega point would, in principle, arrive after the cosmic order reached maximum entropy according to the great Catholic priest Pierre Teilhard de Chardin, SJ (May 1, 1881–April 10, 1955). Teilhard was a renowned Jesuit priest, scientist, paleontologist, theologian, philosopher, and teacher. He was the author of several influential theological and philosophical books and was Darwinian in his approach.

I take Teilhard's work a step further to consider a never-ending series of omega-point–like scenarios.

After cosmic entropy reaches a maximum level, there can be no greater disorder to follow.

However, something else thermodynamically may occur for which we do not yet have the scientific language or understanding to mathematically evaluate.

We will refer to this statistical mechanical scenario as the increase in meta-entropy, or meta-1-entropy.

Accordingly, we will arrive at a cosmic state of meta-omega-point, or meta-1-omega-point.

Eventually, the cosmic order will then progress toward a state of maximum meta-meta-entropy or meta-2-entropy to thus arrive at a state of a meta-meta-omega-point or meta-2-omega-point.

Afterward, the cosmic order will then progress toward a state of maximum meta-meta-meta-entropy, or meta-3-entropy, to thus arrive at a state of a meta-meta-meta-omega-point, or meta-3-omega-point.

The serial pattern will continue forever, past meta-(Aleph 0)-omega-point, …, past meta-(Aleph 1)-omega-point, …, past meta-(Aleph 2)-omega-point, …, past meta-(Aleph 3)-omega-point, …, past meta-(Aleph Ω)-omega-point, …, and so on and so on.

Since each meta-k-omega-point starting at meta-1-omega-point involves going beyond maximum entropy. The implications are that new life or vital modes of physiological and psychological principles will evolve with the realization of new modes of experience, travel, technology being realized, and new scientific, cultural, medical, and religious aspects of associated civilizations coming into being.

Now as we progress farther and farther into the future, we in a sense, climb higher and higher into the cosmos. It is as if we are climbing a mountain with no summit.

For example, a spacecraft having achieved a Lorentz factor of Aleph 0 may well leave the current temporal eternity to travel into the proceeding temporal eternity.

We can use the analogy of the limitless mountain and refer to the upward climb as hyper-upward-climb-1 to denote the first future eternity beyond our current eternity. We indicate this realization symbolically as ↑,1.

Travel at a still greater infinite Lorentz factor, say at Aleph 1, may enable us to travel into a yet more distant future eternity, which we indicate as ↑,2.

Travel at a still greater infinite Lorentz factor, say at Aleph 2, may enable us to travel into a yet more distant future eternity, which we indicate as ↑,3.

We can consider the set of levels defined by the limitless range, ↑,k, where k is any finite or infinite positive integer.

All these ↑,k realizations are as if simply climbing ever more upward on a limitless proverbial mountain. Thus, all these scenarios so far are just another extreme of moving upward.

Now comes a zinger!

Imagine a construct of moving-up-beyond-up, which is to moving up as moving up is to staying in the same time period or being substantially held at one instant in the current time. We denote the construct of moving-up-beyond-up as Beyond-1-↑.

We can indicate the various levels of moving-up-beyond-up as Beyond-1-↑-k, where k is any finite or infinite positive integer.

Note that Beyond-1-↑-k's are not just a quantitative extreme beyond ↑,k's, but involve entirely new meanings and are thus qualitatively enhanced beyond ↑,k's.

We can indicate the various levels of moving-up-beyond-moving-up-beyond-up as Beyond-2-↑-k, where k is any finite or infinite positive integer.

Note that Beyond-2-↑-k's are not just a quantitative extreme beyond, Beyond-1-↑-k's, but involve entirely new meanings and are thus qualitatively enhanced beyond Beyond-1-↑-k's.

We can indicate the various levels of moving-up-beyond-moving-up-beyond-moving-up-beyond-up as Beyond-3-↑-k, where k is any finite or infinite positive integer.

Note that Beyond-3-↑-k's are not just a quantitative extreme beyond, Beyond-2-↑-k's, but involve entirely new meanings and are thus qualitatively enhanced beyond Beyond-2-↑-k's.

We can likewise consider ever more extreme scenarios, which we denote as the set of Beyond-j-↑-k's, where j and k may be either a finite or infinite positive integer and do not need to be equal, although they can be.

These notions should ignite your wanderlust for the future of space and/or time travel into the future.

Back in the day, while I was a teenager, I attended a private school. The school psychologist was a consecrated Catholic religious brother with dark hair and a dark beard, and he used to let me ride in his fancy Ford Thunderbird. The car had a black exterior and interior.

Well, at about the same time, the sitcom, *The Jeffersons* was popular, and the show's theme song had a refrain that went like, "Well we're movin' on up. / (Movin' on up). / To the east side."

Even back then I was interested in interstellar travel concepts.

To make a long story short, I associated with the school psychologist and rode in his Thunderbird with my internalized mantra of "Movin' on up," to the future, at near light-speed. Thus, I became more hooked on special relativistic space travel and time dilation. The fire of my imagination for near light-speed travel was lit just as assuredly as the black Ford Thunderbird resembled the black cosmic void. I knew then the ramifications of infinite time dilation and infinite forward time travel and distances through space mathematically plausible for light-speed impulse travel.

So if you have the courage to delve farther into this book, or even study only select portions thereof, you will likely, if you have not already, also become intrigued with Movin' on up into the future with special relativity. As we now have a space travel industry, we have set before us the seas of infinity.

In a way, we are all movin' on up into the future special relativistically. Such a future is unbounded in terms of forward temporal progression and the eternal duration of reality.

Even for spacecraft having attained light-speed in the least infinite Lorentz factor, there is no mathematical reason why the light-speed spacecraft cannot continue having increasing infinite Lorentz factors.

Essentially, we will always be movin' up into the future.

Now, within the observable universe, there are about 10 EXP 24 stars. This is about equal to the number of fine grains of table sugar that would cover the entire United States one hundred meters deep. The number of planets orbiting stars in our universe seems to be about ten times greater yet or equal to the number of fine grains of table sugar that would cover the entire United States one thousand meters deep. The number of moons, orbiting planets, orbiting stars is estimated to be ten times greater yet or equal to the number of fine grains of table sugar that would cover the entire United States ten thousand meters deep.

In theory, within the next 15 billion years, we could travel to a greater portion of these approximately 10 EXP 24 stars and settle on nice warm planets orbiting these stars. Since most of these stars are red dwarfs, they will naturally continue to burn at a more or less constant rate over a lifetime ranging from 3 trillion to 30 trillion years. Can you imagine how advanced a civilization could become over this time period!

Even right here in the Milky Way Galaxy, there are an estimated half of a trillion stars. This is about equal to the number of fine grains of table sugar that would fill a container that has an interior vacancy of ten meters by ten meters, and which is five meters tall. We can colonize the entire galaxy in only seventy thousand years in space-ark craft that achieve a small Lorentz factor of a mere five, which corresponds to about 98 percent of the speed of light.

And this says nothing about prospects for figuring out how to travel into any hyperspaces.

ESSAY 41) Another Take on Exotic Calculi and Light-Speed Travel

Now the arrow of time seems to travel in only one direction. This seems especially true from a statistical mechanical point of view with the mindset that entropy seems to increase overall with temporal progression. Accordingly, the universe seems to progress to maximize its entropy to attain a first omega point.

However, other cosmic events may happen in the meantime, such as phase-change events, symmetry-breaking events, and universal vacuum energy state decays or quantum mechanical tunnelings.

Currently, we measure time precisely where time is effected by special and general relativity in one reference frame relative to another observer. Temporal events, especially scheduled technological events and planned social communications, happen in highly predictable manners, especially with reference to their scheduling.

However, a spacecraft traveling at a suitably extreme finite or infinite Lorentz factor might only notice the arrow of time by experiencing the cosmic ticks of phase changes. Thus, each phase change would mark an outside event relative to the spacecraft reference frame. The smaller-scale external events relative to the spacecraft would not be noticeable because these events would happen in less than one Planck time unit ship-frame, a time unit theoretically the smallest possible time unit with any meaning.

We can define a new meaning of time by the following formulation:

d(Number of cosmic phase changes experienced by a suitably extreme finite or infinite Lorentz factor spacecraft)/dtship, where *tship* is ship-time.

We can define another meaning of time by the following formulation, and this formulation is a bit murky but profound:

d(Number of cosmic phase changes experienced by a suitably extreme finite or infinite Lorentz factor spacecraft)/dtbackground, where *tbackground* is effectively the background time.

Likewise, we can take higher-order time derivatives such as follows:

d[d(Number of cosmic phase changes experienced by a suitably extreme finite or infinite Lorentz factor spacecraft)/dtship]/dtship where *tship* is ship-time.

D[d(Number of cosmic phase changes experienced by a suitably extreme finite or infinite Lorentz factor spacecraft)/dtbackground]/ dtbackground, where *tbackground* is effectively the background time.

Likewise, we can take generally yet higher-order time derivatives such as follows:

d{...{d[d(Number of cosmic phase changes experienced by a suitably extreme finite or infinite Lorentz factor spacecraft)/dtship]/dtship}/...}/dtship, where *tship* is ship-time.

D{...{d[d(Number of cosmic phase changes experienced by a suitably extreme finite or infinite Lorentz factor spacecraft)/dtbackground]/ dtbackground}/...}/dtbackground, where *tbackground* is effectively the background time.

We can define analogues of time in new meanings by the following formulation:

d(Number of cosmic phase changes experienced by a suitably extreme finite or infinite Lorentz factor spacecraft)/dgamma.

D(Number of cosmic phase changes experienced by a suitably extreme finite or infinite Lorentz factor spacecraft)/dacceleration.

D(Number of cosmic phase changes experienced by a suitably extreme finite or infinite Lorentz factor spacecraft)/d Kinetic Energy.

Likewise, we can take higher-order derivatives such as follows:

d[d(Number of cosmic phase changes experienced by a suitably extreme finite or infinite Lorentz factor spacecraft)/dgamma]/dgamma.

D[d(Number of cosmic phase changes experienced by a suitably extreme finite or infinite Lorentz factor spacecraft)/dacceleration]/dacceleration.

D[d(Number of cosmic phase changes experienced by a suitably extreme finite or infinite Lorentz factor spacecraft)/d Kinetic Energy]/d kinetic energy.

Likewise, we can take generally yet higher-order time derivatives such as follows:

d{…{d[d(Number of cosmic phase changes experienced by a suitably extreme finite or infinite Lorentz factor spacecraft)/dgamma]/dgamma}/…}/dgamma.

D{…{d[d(Number of cosmic phase changes experienced by a suitably extreme finite or infinite Lorentz factor spacecraft)/dacceleration]/dacceleration}/…}/ dacceleration.

D{…{d[d(Number of cosmic phase changes experienced by a suitably extreme finite or infinite Lorentz factor spacecraft)/d Kinetic Energy]/d kinetic energy}/…}/d kinetic energy.

We can replace said cosmic phase changes by realized omega points to derive similar formulas.

Additionally, we may integrate to obtain the following analogues of velocity, acceleration, jerk, distance traveled and the like as follows:

∫(Number of cosmic phase changes experienced by a suitably extreme finite or infinite Lorentz factor spacecraft)dtship, where *tship* is ship-time.

∫(Number of cosmic phase changes experienced by a suitably extreme finite or infinite Lorentz factor spacecraft)dtbackground, where *tbackground* is effectively the background time.

Likewise, we can take higher-order time integrals such as follows:

∫ [∫ (Number of cosmic phase changes experienced by a suitably extreme finite or infinite Lorentz factor spacecraft)dtship] dtship, where *tship* is ship-time.

∫ [∫ (Number of cosmic phase changes experienced by a suitably extreme finite or infinite Lorentz factor spacecraft)dtbackground] dtbackground, where *tbackground* is effectively the background time.

Likewise, we can take generally yet higher-order time integrals as such as follows:

∫ {...{ ∫ [∫ (Number of cosmic phase changes experienced by a suitably extreme finite or infinite Lorentz factor spacecraft)dtship]dtship}...}dtship, where *tship* is ship-time.

∫ {...{ ∫ [∫ (Number of cosmic phase changes experienced by a suitably extreme finite or infinite Lorentz factor spacecraft)dtbackground] dtbackground}...}dtbackground, where *tbackground* is effectively the background time.

We can define analogues of time in new meanings by the following formulation.

∫ (Number of cosmic phase changes experienced by a suitably extreme finite or infinite Lorentz factor spacecraft)dgamma.

∫ (Number of cosmic phase changes experienced by a suitably extreme finite or infinite Lorentz factor spacecraft)dacceleration.

∫ (Number of cosmic phase changes experienced by a suitably extreme finite or infinite Lorentz factor spacecraft)d Kinetic Energy.

.

Likewise, we can take higher-order integrals such as follows,

∫ [∫ (Number of cosmic phase changes experienced by a suitably extreme finite or infinite Lorentz factor spacecraft)dgamma]dgamma.

∫ [∫ (Number of cosmic phase changes experienced by a suitably extreme finite or infinite Lorentz factor spacecraft)dacceleration]dacceleration.

∫ [∫ (Number of cosmic phase changes experienced by a suitably extreme finite or infinite Lorentz factor spacecraft)/ Kinetic Energy]d kinetic energy.

Likewise, we can take generally yet higher-order time integrals such as follows:

∫ {...{ ∫ [∫ (Number of cosmic phase changes experienced by a suitably extreme finite or infinite Lorentz factor spacecraft)dgamma]dgamma}...}dgamma.

∫ {...{ ∫ [∫ (Number of cosmic phase changes experienced by a suitably extreme finite or infinite Lorentz factor spacecraft)/dacceleration]dacceleration}...}dacceleration.

∫ {...{ ∫ [∫ (Number of cosmic phase changes experienced by a suitably extreme finite or infinite Lorentz factor spacecraft)d Kinetic Energy]d kinetic energy}...}d kinetic energy.

∫ {...{ ∫ [∫ (Number of cosmic phase changes experienced by a suitably extreme finite or infinite Lorentz factor spacecraft)dtship]dtbackground}...}dgamma.

∫ {...{ ∫ [∫ (Number of cosmic phase changes experienced by a suitably extreme finite or infinite Lorentz factor spacecraft)dtship]dtbackground}...}dacceleration.

∫ {...{ ∫ [∫ (Number of cosmic phase changes experienced by a suitably extreme finite or infinite Lorentz factor spacecraft)/d tship]d tbackground}...}d kinetic energy.

∫ {...{ ∫ [∫ (Number of cosmic phase changes experienced by a suitably extreme finite or infinite Lorentz factor spacecraft)dgamma]dacceleration}...}d kinetic energy.

∫ {...{ ∫ [∫ (Number of cosmic phase changes experienced by a suitably extreme finite or infinite Lorentz factor spacecraft)/dacceleration]dgamma}...} kinetic energy.

∫ {...{ ∫ [∫ (Number of cosmic phase changes experienced by a suitably extreme finite or infinite Lorentz factor spacecraft)d Kinetic Energy]d gamma}...}d acceleration.

∫ {...{ ∫ [∫ (Number of cosmic phase changes experienced by a suitably extreme finite or infinite Lorentz factor spacecraft)dgamma]dacceleration}...}dtship.

∫ {...{ ∫ [∫ (Number of cosmic phase changes experienced by a suitably extreme finite or infinite Lorentz factor spacecraft)/dacceleration]dgamma}...} dtship.

∫ {...{ ∫ [∫ (Number of cosmic phase changes experienced by a suitably extreme finite or infinite Lorentz factor spacecraft)d Kinetic Energy]d gamma}...}dtship.

∫ {...{ ∫ [∫ (Number of cosmic phase changes experienced by a suitably extreme finite or infinite Lorentz factor spacecraft)dgamma]dacceleration}...}dtbackground.

∫ {...{ ∫ [∫ (Number of cosmic phase changes experienced by a suitably extreme finite or infinite Lorentz factor spacecraft)/dacceleration]dgamma}...} dtbackground.

∫ {...{ ∫ [∫ (Number of cosmic phase changes experienced by a suitably extreme finite or infinite Lorentz factor spacecraft)d Kinetic Energy]d gamma}...}dtbackground.

∫ {...{ ∫ [∫ (Number of cosmic phase changes experienced by a suitably extreme finite or infinite Lorentz factor spacecraft)dgamma]dtship}...}d kinetic energy.

∫ {...{ ∫ [∫ (Number of cosmic phase changes experienced by a suitably extreme finite or infinite Lorentz factor spacecraft)/dacceleration]dtship}...} kinetic energy.

∫ {...{ ∫ [∫ (Number of cosmic phase changes experienced by a suitably extreme finite or infinite Lorentz factor spacecraft)d Kinetic Energy]d tship}...}d acceleration.

∫ {...{ ∫ [∫ (Number of cosmic phase changes experienced by a suitably extreme finite or infinite Lorentz factor spacecraft)dgamma]dtbackground}...}d kinetic energy.

∫ {...{ ∫ [∫ (Number of cosmic phase changes experienced by a suitably extreme finite or infinite Lorentz factor spacecraft)/dacceleration]dtbackground}...} kinetic energy.

∫ {...{ ∫ [∫ (Number of cosmic phase changes experienced by a suitably extreme finite or infinite Lorentz factor spacecraft)d Kinetic Energy]d tbackground}...}d acceleration.

We likewise can take derivatives. However, if the functions are not cyclical trigonometric functions expressible by Fourier series or Fourier transforms, then the functions may revert to a constant then 0, or the empty set after sufficiently high numbers of iterations of derivatives have been taken. Other conditions may also apply.

For integrals, we can perform any number of iterations even infinite numbers of iterations.

ESSAY 42) A Multifaceted Consideration of Infinite Spacecraft Lorentz Factors

The following lengthy digression is a multipart series for which each part essentially covers aspects of light-speed and infinite Lorentz factor travel. So I decided when compiling this text not to delimit the various parts by using the "Essay" nomenclature since the focus of the subject portions adds little distinction from the previous digressions preceding this final portion of the book. Instead, I indicate transitions in conceptual content by using the preface "Part."

PART 1

Hyper4(a, n) is equal to a tetrated n, or a raised to the power of itself n-1 times. The latter value is symbolically written as n subscript a.

For example,

3 EXP 4 = 81, but 4 subscript 3 is approximately equal to 10 EXP (1,000,000,000,000).

Alternatively,

4 subscript 2 = 2 EXP 2 EXP 2 EXP 2 = 2 EXP [2 EXP [2 EXP 2]] = 2 EXP (2 EXP 4) = 2 EXP 16 = 65,536.

For example,

Hyper5(4, 4)is equal to 4 tetrated 4 tetrated 4 tetrated 4. This value is commonly referred to as 4 pentated 4.

Hyper 6, (4,4) is 4 pentated 4 pentated 4 pentated 4 and is also referred to as 4 hexataed 4.

Hyper 7, (4,4) is 4 hexated 4 hexated 4 hexated 4 and so on.

Aleph 0 is the infinite number of integers.

Aleph 1 according to the perhaps unprovable, and thus unfalsifiable, continuum hypothesis, is the number of real numbers which is greater than Aleph 0 by a multiplicative factor of infinity.

Aleph 2 is similarly greater than Aleph 1.

Aleph 3 is similarly greater than Aleph 2.

Aleph 4 is similarly greater than Aleph 3.

And so on.

In general, Aleph n = 2 EXP [Aleph (n-1)].

The number Ω is commonly stated as the least infinite positive integer or ordinal.

Now here is a real zinger.

So we can produce the abstraction of [Hyper Aleph Ω (Aleph Ω, Aleph Ω)].

We can go to ever greater infinities.

So we can consider

> Hyper [Hyper Aleph Ω (Aleph Ω, Aleph Ω)] [[Hyper Aleph Ω (Aleph Ω, Aleph Ω)], [Hyper Aleph Ω (Aleph Ω, Aleph Ω)]] numbers

We can also consider uncountable infinities which that are often considered greater than countable infinities.

As with Aleph numbers, we can consider a large evolved uncountable infinity in the following form.

So we can consider

> Hyper [Hyper (large evolved uncountable infinity) [(large evolved uncountable infinity), (large evolved uncountable infinity)]] [[Hyper (large evolved uncountable infinity) [(large evolved uncountable infinity), (large evolved uncountable infinity)]], [Hyper (large evolved uncountable infinity) [(large evolved uncountable infinity), (large evolved uncountable infinity)]]] numbers

We can go to yet further extremes to produce the following number:

> Aleph {Hyper [Hyper (large evolved uncountable infinity) [(large evolved uncountable infinity), (large evolved uncountable infinity)]] [[Hyper (large evolved uncountable infinity) [(large evolved uncountable infinity), (large evolved uncountable infinity)]], [Hyper (large evolved uncountable infinity) [(large evolved uncountable infinity), (large evolved uncountable infinity)]]]}

Truly, there is no limit on the number of infinities, even in the standard real number system.

Now the number of integers is what it is, even if this number can be as great as one likes.

Likewise, the number of real numbers is what it is even if this number can be as great as one likes.

Now what I am about to present to you, hopefully, will fill your heart with delight.

We consider the concept of the unlimited.

Accordingly, the infinite number of integers is what it is regardless of whether this number can be as great as you like. The same goes for the size of the set of real numbers.

The concept of the unlimited is deep, broad, and literal.

The unlimited in my interpretation is greater than any specific infinite number.

A wonderful consideration is the following hyper-operator function:

> Hyper [Hyper (unlimited) [(unlimited), (unlimited)]] [[Hyper (unlimited) [(unlimited), (unlimited)]], [Hyper (unlimited) [(unlimited), (unlimited)]]] numbers

We can go to yet further extremes to produce the following number:

> Aleph {Hyper [Hyper (unlimited) [(unlimited), (unlimited)]] [[Hyper (unlimited) [(unlimited), (unlimited)]], [Hyper (unlimited) [(unlimited), (unlimited)]]]}

So we can consider

> Hyper [Hyper Aleph {Hyper [Hyper (unlimited) [(unlimited), (unlimited)]] [[Hyper (unlimited) [(unlimited), (unlimited)]], [Hyper (unlimited) [(unlimited), (unlimited)]]]} (Aleph {Hyper [Hyper (unlimited) [(unlimited), (unlimited)]] [[Hyper (unlimited) [(unlimited), (unlimited)]], [Hyper (unlimited) [(unlimited), (unlimited)]]]}, Aleph {Hyper [Hyper (unlimited) [(unlimited), (unlimited)]] [[Hyper (unlimited) [(unlimited), (unlimited)]], [Hyper (unlimited) [(unlimited), (unlimited)]]]})] [[Hyper Aleph {Hyper [Hyper (unlimited) [(unlimited), (unlimited)]] [[Hyper (unlimited) [(unlimited), (unlimited)]], [Hyper (unlimited) [(unlimited), (unlimited)]]]} (Aleph {Hyper [Hyper (unlimited) [(unlimited), (unlimited)]] [[Hyper (unlimited) [(unlimited), (unlimited)]], [Hyper (unlimited) [(unlimited), (unlimited)]]]}, Aleph {Hyper [Hyper (unlimited) [(unlimited), (unlimited)]] [[Hyper (unlimited) [(unlimited), (unlimited)]], [Hyper (unlimited) [(unlimited), (unlimited)]]]})], [Hyper Aleph {Hyper [Hyper (unlimited) [(unlimited), (unlimited)]] [[Hyper (unlimited) [(unlimited), (unlimited)]], [Hyper

(unlimited) [(unlimited), (unlimited)]]]} (Aleph {Hyper [Hyper (unlimited) [(unlimited), (unlimited)]] [[Hyper (unlimited) [(unlimited), (unlimited)]], [Hyper (unlimited) [(unlimited), (unlimited)]]]}, Aleph {Hyper [Hyper (unlimited) [(unlimited), (unlimited)]] [[Hyper (unlimited) [(unlimited), (unlimited)]], [Hyper (unlimited) [(unlimited), (unlimited)]]]})]] numbers

Might we dare contemplate Lorentz factors in light-speed travel equal to the above number. Or how about temperatures equal to this number:

A Lorentz factor of

> Hyper [Hyper Aleph {Hyper [Hyper (unlimited) [(unlimited), (unlimited)]] [[Hyper (unlimited) [(unlimited), (unlimited)]], [Hyper (unlimited) [(unlimited), (unlimited)]]]} (Aleph {Hyper [Hyper (unlimited) [(unlimited), (unlimited)]] [[Hyper (unlimited) [(unlimited), (unlimited)]], [Hyper (unlimited) [(unlimited), (unlimited)]]]}, Aleph {Hyper [Hyper (unlimited) [(unlimited), (unlimited)]] [[Hyper (unlimited) [(unlimited), (unlimited)]], [Hyper (unlimited) [(unlimited), (unlimited)]]]})] [[Hyper Aleph {Hyper [Hyper (unlimited) [(unlimited), (unlimited)]] [[Hyper (unlimited) [(unlimited), (unlimited)]], [Hyper (unlimited) [(unlimited), (unlimited)]]]} (Aleph {Hyper [Hyper (unlimited) [(unlimited), (unlimited)]] [[Hyper (unlimited) [(unlimited), (unlimited)]], [Hyper (unlimited) [(unlimited), (unlimited)]]]}, Aleph {Hyper [Hyper (unlimited) [(unlimited), (unlimited)]] [[Hyper (unlimited) [(unlimited), (unlimited)]], [Hyper (unlimited) [(unlimited), (unlimited)]]]})], [Hyper Aleph {Hyper [Hyper (unlimited) [(unlimited), (unlimited)]] [[Hyper (unlimited) [(unlimited), (unlimited)]], [Hyper (unlimited) [(unlimited), (unlimited)]]]} (Aleph {Hyper [Hyper (unlimited) [(unlimited), (unlimited)]] [[Hyper (unlimited) [(unlimited), (unlimited)]], [Hyper (unlimited) [(unlimited), (unlimited)]]]}, Aleph {Hyper [Hyper (unlimited) [(unlimited), (unlimited)]] [[Hyper (unlimited) [(unlimited), (unlimited)]], [Hyper (unlimited) [(unlimited), (unlimited)]]]})]]

This means that you will travel this number times faster than the speed of light, this number of light-years per year ship-time, this number of years into the future per year ship-time.

However, we have way too much fun in the present cosmic era to have first before we begin the eternal quest for infinite Lorentz factors and beyond. A glorious quest indeed.

As for some things we can do and use within the next 10 billion years, spacecraft Lorentz factor of 10 billion will enable us to travel about the current radius of the observable universe in about one-year ship-time. We can progress to ever greater Lorentz factors.

Now, within the observable universe, there are about 10 EXP 24 stars. This is about equal to the number of fine grains of table sugar that would cover the entire United States one hundred meters deep. The number of planets orbiting stars in our universe seems to be about ten times greater yet or equal to the number of fine grains of table sugar that would cover the entire United States one thousand meters deep. The number of moons, orbiting planets, orbiting stars is estimated to be ten times greater yet or equal to the number of fine grains of table sugar that would cover the entire United States ten thousand meters deep.

So let us go boldly into the future to explore the awesome realm that GOD has made. The next 13.78 billion years will keep us very busy as we travel distances from Earth equal to one current cosmic light-cone radius unit. However, our universe extends far beyond the observable portion.

For now, let us begin our interstellar travel by figuring out a mil-spec blue-printed nuclear fission reactor–powered starship that can reach a Lorentz factor of a mere 1.005 or a velocity of 0.1 c. This will get us to Alpha Centauri in about forty years and to Barnard's Star in about fifty-nine years.

Once we have mastered nuclear fission reactor–powered rockets, we can move on to nuclear fusion reactor–powered rockets to attain velocities of 0.5 c for a Lorentz factor of about 1.155.

Once we learn how to make lots of antimatter, we can have positronium-fueled rockets able to reach a Lorentz factor of 5 at a speed of 0.98 c. This would require a mass ratio of only about 10. For a mass ratio of 10 including purely antimatter fuel that obtains its normal matter partners from the background of space in a drag-recycling manner, the Lorentz factor that accrues is about 50 with a velocity equal to 0.9998 c. It only gets better from here.

PART 2

Now, again, consider hyper-operator notation.

Hyper4(a, n) is equal to a tetrated n, or a raised to the power of itself n-1 times. The latter value is symbolically written as n subscript a.

For example,

> 3 EXP 4 = 81, but 4 subscript 3 is approximately equal to 10 EXP (1,000,000,000,000).

Alternatively,

> 4 subscript 2 = 2 EXP 2 EXP 2 EXP 2 = 2 EXP [2 EXP [2 EXP 2]] = 2 EXP (2 EXP 4) = 2 EXP 16 = 65,536.

For example,

Hyper5(4, 4)is equal to 4 tetrated 4 tetrated 4 tetrated 4. This value is commonly referred to as 4 pentated 4.

Hyper 6, (4,4) is 4 pentated 4 pentated 4 pentated 4 and is also referred to as 4 hexataed 4.

Hyper 7, (4,4) is 4 hexated 4 hexated 4 hexated 4 and so on.

Aleph 0 is the infinite number of integers.

Aleph 1, according to the perhaps unprovable, and thus unfalsifiable, continuum hypothesis, is the number of real numbers which is greater than Aleph 0 by a multiplicative factor of infinity.

Aleph 2 is similarly greater than Aleph 1.

Aleph 3 is similarly greater than Aleph 2.

Aleph 4 is similarly greater than Aleph 3.

And so on.

In general, Aleph n = 2 EXP [Aleph (n-1)].

The number Ω is commonly stated as the least infinite positive integer or ordinal.

Now here is a real zinger.

So we can produce the abstraction of [Hyper Aleph Ω (Aleph Ω, Aleph Ω)].

We can go to ever greater infinities.

So we can consider

> Hyper [Hyper Aleph Ω (Aleph Ω, Aleph Ω)] [[Hyper Aleph Ω (Aleph Ω, Aleph Ω)], [Hyper Aleph Ω (Aleph Ω, Aleph Ω)]] numbers

We reconsider the concept of the unlimited.

Accordingly, the infinite number of integers is what it is regardless of whether this number can be as great as you like. The same goes for the size of the set of real numbers.

The concept of the unlimited is deep, broad, and literal.

The unlimited in my interpretation is greater than any specific infinite number.

Having considered the unlimited as greater than infinities, we can further consider constructs such as unlimited-2, which is to the unlimited as the unlimited is to the infinities. We will define the unlimited as unlimited-1.

We can then consider the unlimited-3 which is to the unlimited-2 as the unlimited-2 is to the unlimited-1 as the unlimited-1 is to the infinities.

We can then consider the unlimited-4 which is to the unlimited-3 as the unlimited-3 is to the unlimited-2 as the unlimited-2 is to the unlimited-1 as the unlimited-1 is to the infinities.

We can go on forever to consider

unlimited-{Hyper [Hyper Aleph Ω (Aleph Ω, Aleph Ω)] [[Hyper Aleph Ω (Aleph Ω, Aleph Ω)], [Hyper Aleph Ω (Aleph Ω, Aleph Ω)]]}

While we are at this, we can go to still more extremes to consider:

Hyper unlimited-{Hyper [Hyper Aleph Ω (Aleph Ω, Aleph Ω)] [[Hyper Aleph Ω (Aleph Ω, Aleph Ω)], [Hyper Aleph Ω (Aleph Ω, Aleph Ω)]]}, (Hyper unlimited-{Hyper [Hyper Aleph Ω (Aleph Ω, Aleph Ω)] [[Hyper Aleph Ω (Aleph Ω, Aleph Ω)], [Hyper Aleph Ω (Aleph Ω, Aleph Ω)]]}, Hyper unlimited-{Hyper [Hyper Aleph Ω (Aleph Ω, Aleph Ω)] [[Hyper Aleph Ω (Aleph Ω, Aleph Ω)], [Hyper Aleph Ω (Aleph Ω, Aleph Ω)]]})

While we are at this, we can go to still more extremes to consider:

Hyper unlimited-{Hyper [Hyper Aleph unlimited-Ω (Aleph unlimited-Ω, Aleph unlimited-Ω)] [[Hyper Aleph unlimited-Ω (Aleph unlimited-Ω, Aleph unlimited-Ω)], [Hyper Aleph unlimited-Ω (Aleph unlimited-Ω, Aleph unlimited-Ω)]]}, (Hyper unlimited-{Hyper [Hyper Aleph unlimited-Ω (Aleph unlimited-Ω, Aleph unlimited-Ω)] [[Hyper Aleph unlimited-Ω (Aleph unlimited-Ω, Aleph unlimited-Ω)], [Hyper Aleph unlimited-Ω (Aleph unlimited-Ω, Aleph unlimited-Ω)]]}, Hyper unlimited-{Hyper [Hyper Aleph unlimited-Ω (Aleph unlimited-Ω, Aleph unlimited-Ω)] [[Hyper Aleph unlimited-Ω (Aleph unlimited-Ω, Aleph unlimited-Ω)], [Hyper Aleph unlimited-Ω (Aleph unlimited-Ω, Aleph unlimited-Ω)]]})

We can go to still more extremes to consider:

Hyper Hyper unlimited-{Hyper [Hyper Aleph unlimited-Ω (Aleph unlimited-Ω, Aleph unlimited-Ω)] [[Hyper Aleph unlimited-Ω (Aleph unlimited-Ω, Aleph unlimited-Ω)], [Hyper Aleph unlimited-Ω (Aleph unlimited-Ω, Aleph unlimited-Ω)]]}, (Hyper unlimited-{Hyper [Hyper Aleph unlimited-Ω (Aleph unlimited-Ω, Aleph unlimited-Ω)] [[Hyper Aleph unlimited-Ω (Aleph unlimited-Ω, Aleph unlimited-Ω)], [Hyper Aleph unlimited-Ω (Aleph unlimited-Ω, Aleph unlimited-Ω)]]}, Hyper unlimited-{Hyper [Hyper Aleph unlimited-Ω (Aleph unlimited-Ω, Aleph unlimited-Ω)] [[Hyper Aleph unlimited-Ω (Aleph unlimited-Ω, Aleph unlimited-Ω)], [Hyper Aleph unlimited-Ω (Aleph unlimited-Ω, Aleph unlimited-Ω)]]}), (Hyper unlimited-{Hyper [Hyper Aleph unlimited-Ω (Aleph unlimited-Ω, Aleph unlimited-Ω)] [[Hyper Aleph unlimited-Ω (Aleph unlimited-Ω, Aleph unlimited-Ω)], [Hyper Aleph unlimited-Ω (Aleph unlimited-Ω, Aleph unlimited-Ω)]]}, (Hyper unlimited-{Hyper [Hyper Aleph unlimited-Ω (Aleph unlimited-Ω, Aleph unlimited-Ω)] [[Hyper Aleph unlimited-Ω (Aleph unlimited-Ω, Aleph unlimited-Ω)], [Hyper Aleph unlimited-Ω (Aleph unlimited-Ω, Aleph unlimited-Ω)]]}, Hyper unlimited-{Hyper [Hyper Aleph unlimited-Ω (Aleph unlimited-Ω, Aleph unlimited-Ω)] [[Hyper Aleph unlimited-Ω (Aleph unlimited-Ω, Aleph unlimited-Ω)], [Hyper Aleph unlimited-Ω (Aleph unlimited-Ω, Aleph unlimited-Ω)]]}), Hyper unlimited-{Hyper [Hyper Aleph unlimited-Ω (Aleph unlimited-Ω, Aleph unlimited-Ω)] [[Hyper Aleph unlimited-Ω (Aleph unlimited-Ω, Aleph unlimited-Ω)], [Hyper Aleph unlimited-Ω (Aleph unlimited-Ω, Aleph unlimited-Ω)]]}, (Hyper unlimited-{Hyper [Hyper Aleph unlimited-Ω (Aleph unlimited-Ω, Aleph unlimited-Ω)] [[Hyper Aleph unlimited-Ω (Aleph unlimited-Ω, Aleph unlimited-Ω)], [Hyper Aleph unlimited-Ω (Aleph unlimited-Ω, Aleph unlimited-Ω)]]}, Hyper unlimited-{Hyper [Hyper Aleph unlimited-Ω (Aleph unlimited-Ω, Aleph unlimited-Ω)] [[Hyper Aleph unlimited-Ω (Aleph unlimited-Ω, Aleph unlimited-Ω)], [Hyper Aleph unlimited-Ω (Aleph unlimited-Ω, Aleph unlimited-Ω)]]}))

Might we dare contemplate Lorentz factors in light-speed travel equal to the above number. Or how about temperatures equal to this number.

A Lorentz factor of the above number means that you will travel this number times faster than the speed of light, this number of light-years per year ship-time, this number of years into the future per year ship-time.

Now, our universe is almost certainly nowhere descriptively near definability by this huge number.

More than likely, our universe, if infinite, has extensity of only about Aleph 0 light-years. It is quite likely that our universe, although much greater than its visible portion to us, is finite.

However, there may be an Aleph 0 number of universes in a multiverse, an Aleph 0 number of multiverse in a forest, an Aleph 0 number of forests in a biosphere, all included in a hyperspace of at least four spatial dimensions.

The number of hyperspaces may also be Aleph 0 as may the number of bulks.

However, the process of creation and evolution will certainly go on forever. Thus we can at least have as an eternally distant goal Lorentz factors about equal to and even surpassing the above huge number.

PART 3

Now, it is a common belief that there is only one speed of light in a vacuum. Accordingly, this speed of light is associated with a Lorentz factor of infinity.

However, if Lorentz factors can be infinite, then there is no reason that the possible values and sizes of the infinities can vary.

Now, the velocity of light may plausibly come as a vanishingly small finite or infinitesimal width spectrum of velocities.

Another possibility not necessarily opposed to a spectral width of light-speed velocities is that reference frames traveling at variously infinite Lorentz factors may imply different types or species of light-speed.

A spectrum of light-speed velocities and/or a set of species of speeds of light may morph with respect to a suitably infinite Lorentz factor spacecraft as if a continuum or realm of travel all to its own.

It is possible that the set of light-speed velocities and/or light-speed species may be infinite in number and thus offer a huge realm of travel.

Travel in a realm composed of light-speed velocities and/or species of light-speed may have its own analogue of Lorentz factors and limiting velocities.

The set of limiting values and species of velocities for craft traveling in a realm composed of light-speed velocities and/or species of light-speed may itself serve as a realm higher, yet which may have its own limiting values of travel velocities and analogue(s) of Lorentz factors.

In reality, there may be a never-ending ascending series of analogues of realms according to extensions of the above patterns.

We can take the first and higher-order derivatives and integrals of velocities of travel in these higher realms, accelerations, Lorentz factor analogues, and the like, as well as the rate of progression from one realm to another with respect to Lorentz factors, velocities, accelerations, and the like.

So if you think light-speed impulse travel is boring, there is consolation in the incredibly rich set of aspects of light-speed travel.

PART 4

Now we consider a really cool concept.

The concept is what we will refer to as a Planck eye blink. Accordingly, a Planck eye blink is a process that occurs and which is started and completed within one Planck time unit. Thus, the process occurs in less than one Planck time unit.

Now, time is theoretically not defined on scales less than that of the Planck time unit. Thus, a timelike metric, which is completely distinct from time, is needed to explore Planck eye blink scale activities.

The activities completed in less than a Planck eye blink scale we will refer to as sub-Planck-eye-blink-1. These activities would likely need still another timelike metric which is completely distinct from the previously conjectured timelike metric, which is completely distinct from time that is needed to explore Planck eye blink scale activities.

The first such timelike metric we will refer to as timelike-1 and the second as timelike-2.

We can go on to likewise consider sub-Planck-eye-blink-k and timelike-(k-1), where k can be any finite or infinite counting number.

Analogues can be considered for distance scales associated with sub-Planck-eye-blink-k and timelike-(k-1).

So perhaps there are an infinite number of timelike and spacelike metrics or coordinate systems that are strongly too completely distinct from space and time.

Now, these systems are below the Planck space and time scales. So as a corollary by symmetry, we consider analogues of timelike and spacelike metrics or coordinate systems, which would be accessible at light-speed and once the least infinite Lorentz factors have been achieved for spacecraft. The reason for these plausible scenarios is such that a spacecraft having achieved even the least infinite Lorentz factor would run out of future finite time periods to travel forward in time in and thus would begin to have muted time conditions because time would become less and less meaningful. Once a craft achieved sufficiently great Lorentz factors somewhere between Aleph 0 and Aleph 1, the craft would leave time and enter one or more other timelike coordinate systems.

The first level of such infinite Lorentz factor timelike coordinate systems we refer to as ∞-γ-timelike-1.

Upon reaching still greater infinite Lorentz factors, a spacecraft and her crew will enter second level of such infinite Lorentz factor timelike coordinate systems we refer to as ∞-γ-timelike-2.

We can continue forever, thus conjecturing the set of ∞-γ-timelike-k, where k can be any finite or infinite counting number.

Now, we come to a yet more extreme concept. The concept is the notion of values that are less than 0 and have less-than-0 absolute value. Here, we are not simply considering arbitrary functions of infinity such as 0/∞ and the like but, instead, an entirely new and previously unrecognized set of objects that have absolute values qualitatively less than 0 and distinct from 0.

What is being proposed here also has nothing to do with negative numbers or imaginary or complex numbers, or tuples as in linear algebra.

Just as we can consider sub-Planck-eye-blink-k and timelike-(k-1), where k can be arbitrarily infinite. We can consider scales commensurately of less absolute value than 0. Thus, by symmetry, we can consider k to be so great that it is in the range of ∞/m, where $|m|$ qualitatively < 0. This also applies to ∞-γ-timelike-k.

So k can not only be greater than the first few least infinite values, but it can also be something so huge that the values of k are not currently philosophically intuitable and are even beyond the scope of present-era mathematics.

The set of ∞-γ-timelike-k also has spacelike analogues.

So there is a good moral lesson to be learned here. The lesson is to quite simply not fret over any light-speed limits. Light-speed in ascending infinite Lorentz factors enables travel so extreme that faster-than-light warp drive, time travel, and wormhole travel are like children playing with marbles compared to these light-speed impulse travel concepts I work on.

PART 5

Next we consider the notion that the speed of light has realm-like qualities, for which the speed of light has two or more dimensions.

Light-speed might not be just one exact monolithic value but, instead, may be a range or spectral width of velocities of vanishingly small to infinitesimal intervals. As such, light-speed may not only have an internal structure but, perhaps, also multiple dimensions.

To the extent that the speed of light is so ubiquitous in the various formulations of theoretical physics, we may conjecture that the speed of light may have an extremely large finite number of dimensions, if not an infinite number of dimensions. The argument for this assertion is the numerous ways the velocity of light is related to other physical parameters, as well as the perhaps infinite set of distinct reference frames for which the speed of light remains a constant. Note that a velocity of light having a vanishingly small spectral distribution of speed may still act and have an effectively constant velocity.

A fascinating prospect includes new meanings of travel for which the speed of light acts as if a traversable realm.

We can take the first and any higher-order derivatives and integrals of travel within the realm of light-speed with respect to any dimensions of light-speed, propagation rate within any coordinate system within the realm of light-speed, spacecraft Lorentz factor, spacecraft acceleration, spacecraft kinetic energy, ship-time, background time, and the like.

PART 6

Now, commonly, the values of $n/0$, where n is positive and finite, are commonly referred to as infinity.

Moreover, the absolute value of 0 is commonly held to be 0.

However, we can go further to conjecture that $n/[f(m,1)]$ may go the next step above infinite values, where the absolute value of $[f(m,1)]$ is less than zero where the less than zero is at level-1.

We can also go further yet to conjecture that $n/[f(m,2)]$ may go a next step above $n/[f(m,1)]$ values where the absolute value of $[f(m,2)]$ is less than 0, where the less than 0 is at level-2.

We can also go further yet to conjecture that $n/[f(m,3)]$ may go a next step above $n/[f(m,2)]$ values, where the absolute value of $[f(m,3)]$ is less than 0, where the less than 0 is at level-3.

We can also go still further yet to conjecture that n/[f(m,4)] may go a next step above n/[f(m,3)] values, where the absolute value of [f(m,4)] is less than 0 where the less than 0 is at level-4.

In general, we can conjecture that n/[f(m,k)] may go the next step above n/[f(m,k-1)] values, where the absolute value of [f(m,k)] is less than 0, where the less than 0 is at level-k.

Here, the value k can be any positive finite or infinite integer.

So can you imagine traveling at Lorentz factors of n/[f(m,k)] values where the absolute value of [f(m,k)] is less than 0, where the less than 0 is at level-k!

Here, we would effectively travel n/[f(m,k)] light-years through space in one-year ship-time, n/[f(m,k)] years into the future in one-year ship-time, and effectively a multiple n/[f(m,k)] of the speed of light.

All this may conceivably come to fruition at a time beginning in the eternally distant future. However, we have very large but finite Lorentz factors to achieve in the meantime. The process of developing spacecraft that can achieve progressively higher but finite Lorentz factors can continue for infinity scrapers numbers of years. An infinity scraper is a finite number so large that it cannot be represented by ordinary mathematical notation given all reasonably available writing templates.

Now, again consider hyper-operator notation:

Hyper4(a, n) is equal to a tetrated n, or a raised to the power of itself n-1 times. The latter value is symbolically written as n subscript a.

For example,

3 EXP 4 = 81, but 4 subscript 3 is approximately equal to 10 EXP (1,000,000,000,000).

Alternatively,

4 subscript 2 = 2 EXP 2 EXP 2 EXP 2 = 2 EXP [2 EXP [2 EXP 2]] = 2 EXP (2 EXP 4) = 2 EXP 16 = 65,536.

For example,

Hyper5(4, 4)is equal to 4 tetrated 4 tetrated 4 tetrated 4. This value is commonly referred to as 4 pentated 4.

Hyper 6, (4,4) is 4 pentated 4 pentated 4 pentated 4 and is also referred to as 4 hexataed 4.

Hyper 7, (4,4) is 4 hexated 4 hexated 4 hexated 4 and so on.

Aleph 0 is the infinite number of integers.

Aleph 1, according to the perhaps unprovable, and thus unfalsifiable, continuum hypothesis, is the number of real numbers which is greater than Aleph 0 by a multiplicative factor of infinity.

Aleph 2 is similarly greater than Aleph 1.

Aleph 3 is similarly greater than Aleph 2.

Aleph 4 is similarly greater than Aleph 3.

And so on.

In general, Aleph n = 2 EXP [Aleph (n-1)].

The number Ω is commonly stated as the least infinite positive integer or ordinal.

Now here is a real zinger.

So we can produce the abstraction of [Hyper Aleph Ω (Aleph Ω, Aleph Ω)].

We can go to ever greater infinities.

So we can consider

Hyper [Hyper Aleph Ω (Aleph Ω, Aleph Ω)] [[Hyper Aleph Ω (Aleph Ω, Aleph Ω)], [Hyper Aleph Ω (Aleph Ω, Aleph Ω)]] numbers

We reconsider the concept of the unlimited.

Accordingly, the infinite number of integers is what it is regardless of whether this number can be as great as you like. The same goes for the size of the set of real numbers.

So now we can consider

Hyper [Hyper Aleph n/[f(m,Ω)] (Aleph n/[f(m,Ω)], Aleph n/[f(m,Ω)])] [[Hyper Aleph n/[f(m,Ω)] (Aleph n/[f(m,Ω)], Aleph n/[f(m,Ω)])], [Hyper Aleph n/[f(m,Ω)] (Aleph n/[f(m,Ω)], Aleph n/[f(m,Ω)])]] numbers.

As for very large but still finite Lorentz factors, consider the following bad boy that we can obtain in the meantime:

[Hyper googol-plex (googol-plex, googol-plex)] [[Hyper googol-plex (googol-plex, googol-plex)], [Hyper googol-plex (googol-plex, googol-plex)]]

PART 7

Now, again, consider hyper-operator notation:

Hyper4(a, n) is equal to a tetrated n, or a raised to the power of itself n-1 times. The latter value is symbolically written as n subscript a.

For example,

> 3 EXP 4 = 81, but 4 subscript 3 is approximately equal to 10 EXP (1,000,000,000,000).

Alternatively,

> 4 subscript 2 = 2 EXP 2 EXP 2 EXP 2 = 2 EXP [2 EXP [2 EXP 2]] = 2 EXP (2 EXP 4) = 2 EXP 16 = 65,536.

For example,

Hyper5(4, 4)is equal to 4 tetrated 4 tetrated 4 tetrated 4. This value is commonly referred to as 4 pentated 4.

Hyper 6, (4,4) is 4 pentated 4 pentated 4 pentated 4 and is also referred to as 4 hexataed 4.

Hyper 7, (4,4) is 4 hexated 4 hexated 4 hexated 4 and so on.

Aleph 0 is the infinite number of integers.

Aleph 1 according to the perhaps unprovable, and thus unfalsifiable, continuum hypothesis, is the number of real numbers which is greater than Aleph 0 by a multiplicative factor of infinity.

Aleph 2 is similarly greater than Aleph 1.

Aleph 3 is similarly greater than Aleph 2.

Aleph 4 is similarly greater than Aleph 3.

And so on.

In general, Aleph n = 2 EXP [Aleph (n-1)].

The number Ω is commonly stated as the least infinite positive integer or ordinal.

Now here is a real zinger.

So we can produce the abstraction of [Hyper Aleph Ω (Aleph Ω, Aleph Ω)].

We can go to ever greater infinities.

So we can consider

Hyper [Hyper Aleph Ω (Aleph Ω, Aleph Ω)] [[Hyper Aleph Ω (Aleph Ω, Aleph Ω)], [Hyper Aleph Ω (Aleph Ω, Aleph Ω)]] numbers

Thus, we can understand that there is no limit to how big infinities come.

We can now consider the notion of a first level of undefinably large infinities. Such infinities are so large that they cannot be abstractly defined.

We can consider the following huge infinity, perhaps, which only GOD knows about in detail:

Hyper [Hyper Aleph (1st level of undefinable infinities) (Aleph 1st level of undefinable infinities, Aleph 1st level of undefinable infinities)] [[Hyper Aleph 1st level of undefinable infinities (Aleph 1st level of undefinable infinities, Aleph 1st level of undefinable infinities)], [Hyper Aleph 1st level of undefinable infinities (Aleph 1st level of undefinable infinities, Aleph 1st level of undefinable infinities)]]

We can go on to consider the next higher level of undefinable infinities, which we will refer as 2nd level of undefinable infinities:

We can develop the following huge numbers:

Hyper [Hyper Aleph (2nd level of undefinable infinities) (Aleph 2nd level of undefinable infinities, Aleph 2nd level of undefinable infinities)] [[Hyper Aleph 2nd level of undefinable infinities (Aleph 2nd level of undefinable infinities, Aleph 2nd level of undefinable infinities)], [Hyper Aleph 2nd level of undefinable infinities (Aleph 2nd level of undefinable infinities, Aleph 2nd level of undefinable infinities)]]

We can go on to consider the next higher level of undefinable infinities, which we will refer to as 3rd level of undefinable infinities.

We can develop the following huge numbers:

Hyper [Hyper Aleph (3rd level of undefinable infinities) (Aleph 3rd level of undefinable infinities, Aleph 3rd level of undefinable infinities)] [[Hyper Aleph 3rd level of undefinable infinities (Aleph 3rd level of undefinable infinities, Aleph 3rd level of undefinable infinities)], [Hyper Aleph 3rd level of undefinable infinities (Aleph 3rd level of undefinable infinities, Aleph 3rd level of undefinable infinities)]]

In fact, we can proceed to a 4th, 5th, 6th, and so on levels of undefinable infinities up to the (Aleph Ω)th level of undefinable infinities to produce the following huge numbers:

Hyper [Hyper Aleph ((Aleph Ω)th level of undefinable infinities) (Aleph (Aleph Ω)th level of undefinable infinities, Aleph (Aleph Ω)th level of undefinable infinities)] [[Hyper Aleph (Aleph Ω)th level of undefinable infinities (Aleph (Aleph Ω)th level of undefinable infinities, Aleph (Aleph Ω)th level of undefinable infinities)], [Hyper Aleph (Aleph Ω)th level of undefinable infinities (Aleph (Aleph Ω)th level of undefinable infinities, Aleph (Aleph Ω)th level of undefinable infinities)]]

We can go all the way up to and beyond the following huge numbers:

Hyper [Hyper Aleph (Aleph Ω level of Undefinable infinities of undefinable infinities) (Aleph Aleph Ω level of Undefinable infinities of undefinable infinities, Aleph Aleph Ω level of Undefinable infinities of undefinable infinities)] [[Hyper Aleph Aleph Ω level of Undefinable infinities of undefinable infinities (Aleph Aleph Ω level of Undefinable infinities of undefinable infinities, Aleph Aleph Ω level of Undefinable infinities of undefinable infinities)], [Hyper Aleph Aleph Ω level of Undefinable infinities of undefinable infinities (Aleph Aleph Ω level of Undefinable infinities of undefinable infinities, Aleph Aleph Ω level of Undefinable infinities of undefinable infinities)]]

Now can you imagine a fractal-verse so that it has the above numbers of universes, multiverses, forests, biospheres, and the like, or a bulk that has that many hyperspaces, some of which may have the above numbers of dimensions!

How about some impossibly distant future era for which spacecraft Lorentz factors can be as large as or exceed the above values!

How about Lorentz factors as large as the following expression and a fractal-verse so large that it has the following numbers of universes, multiverses, forests, biospheres, and the like or a bulk that has that many hyperspaces, some of which may have the following numbers of dimensions!

HYPER { Hyper [Hyper Aleph (Aleph Ω level of Undefinable infinities of undefinable infinities) (Aleph Aleph Ω level of Undefinable infinities of undefinable infinities, Aleph Aleph Ω level of Undefinable infinities of undefinable infinities)] [[Hyper Aleph Aleph Ω level of Undefinable infinities of undefinable infinities (Aleph Aleph Ω level of Undefinable infinities of undefinable infinities, Aleph Aleph Ω level of Undefinable infinities of undefinable infinities)], [Hyper Aleph Aleph Ω level of Undefinable infinities of undefinable infinities (Aleph Aleph Ω level of Undefinable infinities of undefinable infinities, Aleph Aleph Ω level of Undefinable infinities of undefinable infinities)]]} ({Hyper [Hyper Aleph (Aleph Ω level of Undefinable infinities of undefinable infinities) (Aleph Aleph Ω level of Undefinable infinities of undefinable infinities, Aleph Aleph Ω level of Undefinable infinities of undefinable infinities)] [[Hyper Aleph Aleph Ω level of Undefinable infinities of undefinable infinities (Aleph Aleph Ω level of Undefinable infinities of undefinable infinities, Aleph Aleph Ω level of Undefinable infinities of undefinable infinities)], [Hyper Aleph Aleph Ω level of Undefinable infinities of undefinable infinities (Aleph Aleph Ω level of Undefinable infinities of undefinable infinities, Aleph Aleph Ω level of Undefinable infinities of undefinable infinities)]]},{ Hyper [Hyper Aleph (Aleph Ω level of Undefinable infinities of undefinable infinities) (Aleph Aleph Ω level of Undefinable infinities of undefinable infinities, Aleph Aleph Ω level of Undefinable infinities of undefinable infinities)] [[Hyper Aleph Aleph Ω level of Undefinable infinities of undefinable infinities (Aleph Aleph Ω level of Undefinable infinities of undefinable infinities, Aleph Aleph Ω level of Undefinable infinities of undefinable infinities)], [Hyper Aleph Aleph Ω level of Undefinable infinities of undefinable infinities (Aleph Aleph Ω level of Undefinable infinities of undefinable infinities, Aleph Aleph Ω level of Undefinable infinities of undefinable infinities)]]})

We can ironically produce extended expressions by iterating the above mathematical process so great so as to require all storage media on Earth to denote these expressions and then go iterate the above mathematical process that would require the use of storage media made of a large fraction of all matter in our cosmic light-cone.

Such large Lorentz factors may not properly translate into meaningful travel at light-speed, perhaps because light-speed cannot be formulated from such huge Lorentz factors or be associated with such huge Lorentz factors. In essence, such huge Lorentz factors may inexpressibly transcend the speed of light, and perhaps even require currently unknowable constructs other than the speed of light as explanations. Thus, there may be an undefinably vast sea of other travel constructs other than the speed of light that become available well before the above Lorentz factors would be obtained.

What's more, Lorentz factors slightly in excess of Aleph 0 (the number of integers which is one of the smallest infinities), may morph into parameters other than Lorentz factors themselves. So travel otherwise definable at these above huge undefinably infinite Lorentz factors may not be associated with Lorentz factors at all, perhaps not even with infinite Lorentz factors. Thus, there may be an undefinably infinite sea of parameters other than Lorentz factors that manifest journeys associated with otherwise such huge undefinably infinite Lorentz factors.

However, the above scenarios are so extreme that they best serve as an impossibly distant eternity beacon that beckons us forward with its dimly perceptible intellectual light to attain the next greatest but merely finite Lorentz factors.

PART 8

Now, we are all familiar with the notions of greater than, less than, or equal especially in math. Even kindergarten folks understand such concepts.

We can certainly understand the notion that infinity is greater than the finite in terms of integers and real numbers.

Now, again consider hyper-operator notation.

Hyper4(a, n) is equal to a tetrated n, or a raised to the power of itself n-1 times. The latter value is symbolically written as n subscript a.

For example,

> 3 EXP 4 = 81, but 4 subscript 3 is approximately equal to 10 EXP (1,000,000,000,000).

Alternatively,

> 4 subscript 2 = 2 EXP 2 EXP 2 EXP 2 = 2 EXP [2 EXP [2 EXP 2]] = 2 EXP (2 EXP 4) = 2 EXP 16 = 65,536.

For example,

Hyper5(4, 4)is equal to 4 tetrated 4 tetrated 4 tetrated 4. This value is commonly referred to as 4 pentated 4.

Hyper 6, (4,4) is 4 pentated 4 pentated 4 pentated 4 and is also referred to as 4 hexataed 4.

Hyper 7, (4,4) is 4 hexated 4 hexated 4 hexated 4 and so on.

Aleph 0 is the infinite number of integers.

Aleph 1 according to the perhaps unprovable and thus unfalsifiable, continuum hypothesis, is the number of real numbers which is greater than Aleph 0 by a multiplicative factor of infinity.

Aleph 2 is similarly greater than Aleph 1.

Aleph 3 is similarly greater than Aleph 2.

Aleph 4 is similarly greater than Aleph 3.

And so on.

In general, Aleph n = 2 EXP [Aleph (n-1)].

The number Ω is commonly stated as the least infinite positive integer or ordinal.

Now here is a real zinger.

So we can produce the abstraction of [Hyper Aleph Ω (Aleph Ω, Aleph Ω)].

We can go to ever greater infinities.

So we can consider Hyper [Hyper Aleph Ω (Aleph Ω, Aleph Ω)] [[Hyper Aleph Ω (Aleph Ω, Aleph Ω)], [Hyper Aleph Ω (Aleph Ω, Aleph Ω)]] numbers.

Thus, we can understand that there is no limit to how big infinities come.

So the concepts of greater than, less than, or equal to are well taken by just about every conscious person.

Now here is a fascinating construct. What about the concept of hyper-greater-than?

Here, I am not simply proposing the notion of extremely greater than such as the notion of how any infinity is extremely greater than 1. Instead, I am proposing a concept that is so extreme that it is to greater than as greater than is to less than or equal.

As such, we can intuit entities that are hyper-greater than any infinite values.

We refer to the first level of hyper-greater-than as hyper-1-greater-than.

Now we go on to propose a second term or hyper-hyper-greater-than, which we will denote as hyper-2-greater-than.

Hyper-2-greater-than is roughly to hyper-1-greater-than as hyper-1-greater-than is to greater than.

Now we go on to propose a third term, or hyper-hyper-hyper-greater-than, which we will denote as hyper-3-greater-than.

Hyper-3-greater-than is roughly to hyper-2-greater-than as hyper-2-greater-than is hyper-1-greater-than as hyper-1-greater-than is to greater than.

We can continue the series up to and eternally beyond the following value:

Hyper-{Hyper [Hyper Aleph Ω (Aleph Ω, Aleph Ω)] [[Hyper Aleph Ω (Aleph Ω, Aleph Ω)], [Hyper Aleph Ω (Aleph Ω, Aleph Ω)]]}-greater-than.

A spacecraft having a Lorentz factor of,

Hyper-{Hyper [Hyper Aleph Ω (Aleph Ω, Aleph Ω)] [[Hyper Aleph Ω (Aleph Ω, Aleph Ω)], [Hyper Aleph Ω (Aleph Ω, Aleph Ω)]]}-greater-than

Hyper-{Hyper [Hyper Aleph Ω (Aleph Ω, Aleph Ω)] [[Hyper Aleph Ω (Aleph Ω, Aleph Ω)], [Hyper Aleph Ω (Aleph Ω, Aleph Ω)]]}

would be amazing.

Such craft would travel

Hyper-{Hyper [Hyper Aleph Ω (Aleph Ω, Aleph Ω)] [[Hyper Aleph Ω (Aleph Ω, Aleph Ω)], [Hyper Aleph Ω (Aleph Ω, Aleph Ω)]]}-greater-than

Hyper-{Hyper [Hyper Aleph Ω (Aleph Ω, Aleph Ω)] [[Hyper Aleph Ω (Aleph Ω, Aleph Ω)], [Hyper Aleph Ω (Aleph Ω, Aleph Ω)]]}

light-years in one-year ship-time

Hyper-{Hyper [Hyper Aleph Ω (Aleph Ω, Aleph Ω)] [[Hyper Aleph Ω (Aleph Ω, Aleph Ω)], [Hyper Aleph Ω (Aleph Ω, Aleph Ω)]]}-greater-than

Hyper-{Hyper [Hyper Aleph Ω (Aleph Ω, Aleph Ω)] [[Hyper Aleph Ω (Aleph Ω, Aleph Ω)], [Hyper Aleph Ω (Aleph Ω, Aleph Ω)]]}

years into the future in one-year ship-time and an effective multiple of

Hyper-{Hyper [Hyper Aleph Ω (Aleph Ω, Aleph Ω)] [[Hyper Aleph Ω (Aleph Ω, Aleph Ω)], [Hyper Aleph Ω (Aleph Ω, Aleph Ω)]]}-greater-than

Hyper-{Hyper [Hyper Aleph Ω (Aleph Ω, Aleph Ω)] [[Hyper Aleph Ω (Aleph Ω, Aleph Ω)], [Hyper Aleph Ω (Aleph Ω, Aleph Ω)]]} the speed of light.

However, before we become carried away, it is best for us to focus on slightly relativistic spacecraft, say, ones that may obtain a Lorentz factor of 1.005, which is slightly greater than 1. A Lorentz factor of 1.005 corresponds to a velocity of about 10 percent of the speed of light.

Such velocities may be obtained by nuclear fusion–powered spacecraft with a relatively modest mass-ratio.

We can likely design, assemble, and fly such craft this century and by the end of this century have human-crewed vessels reaching the two nearest stars.

PART 9

Now, folks familiar with my writing have covered a wide variety of examples of countable infinities, have seen my digressions on uncountable infinities, perhaps read my digressions on undefinably great infinities and the like.

Evidently, the logical and abstract realm is suffused with infinities of innumerable sizes.

One thing all these infinities, and finite numbers as well, have in common is value.

Now, I would like to conjecture on infinities so large that they are not values because these infinities are so large they transcend values.

We consider the first level of infinities that transcend values as ∞-transcend-values-1.

A second level of infinities or the next more extreme set of infinities that transcend values we will refer to as ∞-transcend-values-2.

A third level of infinities or the next more extreme set of infinities yet that transcend values we will refer to as ∞-transcend-values-3.

Now, again consider hyper-operator notation.

Hyper4(a, n) is equal to a tetrated n, or a raised to the power of itself n-1 times. The latter value is symbolically written as n subscript a.

For example,

3 EXP 4 = 81, but 4 subscript 3 is approximately equal to 10 EXP (1,000,000,000,000).

Alternatively,

4 subscript 2 = 2 EXP 2 EXP 2 EXP 2 = 2 EXP [2 EXP [2 EXP 2]] = 2 EXP (2 EXP 4) = 2 EXP 16 = 65,536.

For example,

Hyper5(4, 4)is equal to 4 tetrated 4 tetrated 4 tetrated 4. This value is commonly referred to as 4 pentated 4.

Hyper 6, (4,4) is 4 pentated 4 pentated 4 pentated 4 and is also referred to as 4 hexataed 4.

Hyper 7, (4,4) is 4 hexated 4 hexated 4 hexated 4 and so on.

Aleph 0 is the infinite number of integers.

Aleph 1, according to the perhaps unprovable, and thus unfalsifiable, continuum hypothesis, is the number of real numbers which is greater than Aleph 0 by a multiplicative factor of infinity.

Aleph 2 is similarly greater than Aleph 1.

Aleph 3 is similarly greater than Aleph 2.

Aleph 4 is similarly greater than Aleph 3.

And so on

In general, Aleph n = 2 EXP [Aleph (n-1)].

The number Ω is commonly stated as the least infinite positive integer or ordinal.

Now here is a real zinger.

So we can produce the abstraction of [Hyper Aleph Ω (Aleph Ω, Aleph Ω)].

We can go to ever greater infinities.

So we can consider

Hyper [Hyper Aleph Ω (Aleph Ω, Aleph Ω)] [[Hyper Aleph Ω (Aleph Ω, Aleph Ω)], [Hyper Aleph Ω (Aleph Ω, Aleph Ω)]] numbers

Thus, we can understand that there is no limit to how big standard infinities come.

Now consider a set of levels of infinities or still more extreme sets of infinities yet that transcend values we will refer to generally as ∞-transcend-values-k.

We can set *k* equal to any integer no matter how infinitely large the integer is.

So given that the set of integers has Aleph 0 elements, we can likewise conjecture levels of still more extreme sets of infinities yet that transcend values all the way to ∞-transcend-values-(Aleph 0).

Since integers can be as infinite as we would like, in reality Aleph 0 can be as great as we would like in practice.

However, we can also consider integers on hyperextended number lines. For such number lines, the number of integers can be arbitrarily greater than Aleph 0.

So we can conjecture levels of still more extreme sets of infinities yet that transcend values all the way to ∞-transcend-values-[>(Aleph 0)].

Here is a real zinger for you. Imagine a spacecraft having a Lorentz factor of ∞-transcend-values-1, or perhaps ∞-transcend-values-2, or perhaps ∞-transcend-values-3, or perhaps even all the way up to ∞-transcend-values-(googol-plex) and so on.

A googol-plex is equal to (10 EXP 100) EXP (10 EXP 100). This is a finite number but extremely huge.

A spacecraft traveling at a Lorentz factor of ∞-transcend-values-(googol-plex) would travel ∞-transcend-values-(googol-plex) light-years in one-year ship-time, an ∞-transcend-values-(googol-plex) number of years into the future in one-year ship-time and an effective multiple of ∞-transcend-values-(googol-plex) of the speed of light.

Most reasonably, Lorentz factors and velocities and forward time travel would not even be defined for spacecraft otherwise traveling at a Lorentz factor of ∞-transcend-values-(googol-plex). The meaning to time dilation, Lorentz contraction, relativistic aberration, relativistic kinetic energy, velocity, momentum and the like would have long since lost their meaning even for craft able to achieve standard infinite values or well before otherwise Lorentz factors of ∞-transcend-values-1 would be attained.

We have all of future eternity to play around with spacecraft Lorentz factors lived out in unmasked kinetic energies of finite multiples of the energy equivalent of spacecraft invariant masses.

For example, standard mathematical logic would agree that logically, spacecraft Lorentz factors of (10 EXP 100) EXP (10 EXP 100) are possible. If you hold standard special relativity to the law, any finite spacecraft Lorentz factor is fair game. As I said, we have all huge finite time intervals in the future to learn how to achieve ever greater finite Lorentz factors.

So even the least infinite or smallest infinite Lorentz factor as a concept can beckon us all onward into eternity as if a guiding light or a mysteriously dim, ever-so-distant beacon, to provide us motivation to go the next step forward in achieving still yet greater finite but huge Lorentz factors.

PART 10

A fascinating construct would involve a sufficiently accelerated extremely relativistic or perhaps even a light-speed craft with infinite Lorentz factors simply morphing into a composition of heavier quarks, and/or heavier charged leptons.

Ideally, such a craft would manifest as a perpetual series of heavier quarks and charged leptons of species heavier than the top quark and the tauon, respectively.

Perhaps such a craft can take on a despaced and/or detimed configuration for which it progresses up the series of ever more massive quarks and charged leptons. So one or more analogues of Lorentz factors and velocities may define the craft's motion. By *despaced* and *detimed*, we are referring to the craft location outside of space and/or time.

A fascinating construct would involve a spacecraft that, accordingly, becomes despaced and/or detimed but emplaced nowhere else or located nowhere else.

These spacecraft may instead travel up an infinite series of quarks and charged leptons where both matter and antimatter forms are considered. Alternatively, these spacecraft may travel up an infinite series of species of gauge bosons. As another example, the craft may travel up along an infinite series of ever heavier supersymmetric particles in progressively higher levels of extended supersymmetry.

To some extent, the despaced and detimed craft alludes to despaced and/or detimed nuclear explosives of shaped charge configurations. The explosive jets may include cascades of particles with the formation of currently unknown species of quarks, charged leptons, neutrinos, but also supersymmetric bosons and fermions of extended levels of supersymmetry.

The formation of such particles may be engineered into the designs of the associated nuclear explosives or follow naturally in unplanned and unstructured explosive mechanisms.

Some of these currently unknown species of quarks, charged leptons, neutrinos, but also supersymmetric bosons and fermions of extended levels of supersymmetry may be associated with and/or the production of net positive energy.

For example, some created net energy may be sourced from the zero-point or vacuum fields.

As another example, the created net energy may be sourced from other hidden dimensions, other universes, other forests, other biospheres, and the like, as well as hyperspaces and bulks.

As yet another example, the created net energy may to at least some extent be so in a defacto manner. Accordingly, the energy would manifest as a result of information conservation or information creation.

We may affix the following operator to formulas for the exotic nuclear explosive effects to denote and functionalize the above despaced and/or detimed mechanisms as well as for associated energy production.

$$[\Sigma(ds, dt, \uparrow E)]$$

Here, *ds* stands for despaced, *dt* stands for detimed, and the last item in the triplet stands for energy increases or energy creation.

PART 11

Now, there are countable and uncountable infinities.

But before we go further, consider the hyper-operator notation that was designed to express huge values not otherwise expressible.

For example, note that Hyper4(a, n) is equal to a tetrated *n*, or a raised to the power of itself n-1 times. The latter value is symbolically written as *n* subscript *a*.

For example,

3 EXP 4 = 81, but 4 subscript 3 is approximately equal to 10 EXP (1,000,000,000,000).

Alternatively,

4 subscript 2 = 2 EXP 2 EXP 2 EXP 2 = 2 EXP [2 EXP [2 EXP 2]] = 2 EXP (2 EXP 4) = 2 EXP 16 = 65,536.

For example,

Hyper5(4, 4)is equal to 4 tetrated 4 tetrated 4 tetrated 4. This value is commonly referred to as 4 pentated 4.

Hyper 6, (4,4) is 4 pentated 4 pentated 4 pentated 4 and is also referred to as 4 hexataed 4.

Hyper 7, (4,4) is 4 hexated 4 hexated 4 hexated 4 and so on.

Aleph 0 is the infinite number of integers.

Aleph 1, according to the perhaps unprovable, and thus unfalsifiable, continuum hypothesis, is the number of real numbers that is greater than Aleph 0 by a multiplicative factor of infinity.

Aleph 2 is similarly greater than Aleph 1.

Aleph 3 is similarly greater than Aleph 2.

Aleph 4 is similarly greater than Aleph 3.

And so on.

In general, Aleph n = 2 EXP [Aleph (n-1)].

The number Ω is commonly stated as the least infinite positive integer or ordinal.

Now here is a real zinger.

So we can produce the abstraction of Hyper Aleph Ω (Aleph Ω, Aleph Ω).

How about,

> Hyper [Hyper Aleph Ω (Aleph Ω, Aleph Ω)](Hyper Aleph Ω (Aleph Ω, Aleph Ω), Hyper Aleph Ω (Aleph Ω, Aleph Ω))

and even

> Hyper [Hyper [Hyper Aleph Ω (Aleph Ω, Aleph Ω)](Hyper Aleph Ω (Aleph Ω, Aleph Ω), Hyper Aleph Ω (Aleph Ω, Aleph Ω))](Hyper [Hyper Aleph Ω (Aleph Ω, Aleph Ω)](Hyper Aleph Ω (Aleph Ω, Aleph Ω), Hyper Aleph Ω (Aleph Ω, Aleph Ω)), Hyper [Hyper Aleph Ω (Aleph Ω, Aleph Ω)](Hyper Aleph Ω (Aleph Ω, Aleph Ω), Hyper Aleph Ω (Aleph Ω, Aleph Ω))).

So we can produce the abstraction of Hyper Aleph Ω (Aleph Ω, Aleph Ω) we will label as follows:

[Hyper Aleph Ω (Aleph Ω, Aleph Ω)]-x,1,1.

Hyper [Hyper Aleph Ω (Aleph Ω, Aleph Ω)](Hyper Aleph Ω (Aleph Ω, Aleph Ω), Hyper Aleph Ω (Aleph Ω, Aleph Ω)) we will label as

[Hyper Aleph Ω (Aleph Ω, Aleph Ω)]-x,1,2.

Hyper [Hyper [Hyper Aleph Ω (Aleph Ω, Aleph Ω)](Hyper Aleph Ω (Aleph Ω, Aleph Ω), Hyper Aleph Ω (Aleph Ω, Aleph Ω))](Hyper [Hyper Aleph Ω (Aleph Ω, Aleph Ω)](Hyper Aleph Ω (Aleph Ω, Aleph Ω), Hyper Aleph Ω (Aleph Ω, Aleph Ω)), Hyper [Hyper Aleph Ω (Aleph Ω, Aleph Ω)](Hyper Aleph Ω (Aleph Ω, Aleph Ω), Hyper Aleph Ω (Aleph Ω, Aleph Ω))) we will label as

[Hyper Aleph Ω (Aleph Ω, Aleph Ω)]-x,1,3.

We can likewise consider

[Hyper Aleph Ω (Aleph Ω, Aleph Ω)]-x,1,4

[Hyper Aleph Ω (Aleph Ω, Aleph Ω)]-x,1,5

[Hyper Aleph Ω (Aleph Ω, Aleph Ω)]-x,1,6

all the way to

[Hyper Aleph Ω (Aleph Ω, Aleph Ω)]-x,1,[Hyper Aleph Ω (Aleph Ω, Aleph Ω)]

and even to

[Hyper Aleph Ω (Aleph Ω, Aleph Ω)]-x,1, [Hyper Aleph Ω (Aleph Ω, Aleph Ω)]-x,1.

We can continue this sequence forever.

Now we can consider measures that are not quantities such as

K-x,2.

Here, *K* is a qualitative infinity and can be any such value.

So we can form the following expression:

(Qualitative Infinity)-x,2,1

We can also form the following expressions:

(Qualitative Infinity)-x,2,2

(Qualitative Infinity)-x,2,3

(Qualitative Infinity)-x,2,4

And so on to

(Qualitative Infinity)-x,2,[[Hyper Aleph Ω (Aleph Ω, Aleph Ω)]-x,1, [Hyper Aleph Ω (Aleph Ω, Aleph Ω)]-x,1]

We can consider measures that are not quantities or qualities such as follows:

K-x,3; K-x,4; K-x,5; and indeed all the way to and beyond

K-x,{ [Hyper Aleph Ω (Aleph Ω, Aleph Ω)]-x,1, [Hyper Aleph Ω (Aleph Ω, Aleph Ω)]-x,1}

Can you imagine some almost-impossible distant future time where we might embark on the research and development of a spacecraft that can attain a Lorentz factor of [Hyper Aleph Ω (Aleph Ω, Aleph Ω)]-x,1, [Hyper Aleph Ω (Aleph Ω, Aleph Ω)]-x,1.

Such a spacecraft could travel [Hyper Aleph Ω (Aleph Ω, Aleph Ω)]-x,1, [Hyper Aleph Ω (Aleph Ω, Aleph Ω)]-x,1 light-years through space in one-year ship-time, and essentially the same number of years into the future in one-year ship-time, and effectively essentially the same number multiple of the speed of light.

We most assuredly will be forever climbing the gamma ladder to ever greater Lorentz factors.

PART 12

Now, there is still more I have to say about quantitative infinities.

We have two "known" types of infinities. These are referred to as countable infinities and uncountable infinities.

We will refer to countable infinities as ∞-1 and uncountable infinities as ∞-2.

However, we can move even further afield to conjecture

∞-3, ∞-4, ∞-5 and so on all the way up to and beyond ∞-{[Hyper Aleph Ω (Aleph Ω, Aleph Ω)]-x,1, [Hyper Aleph Ω (Aleph Ω, Aleph Ω)]-x,1}

Can you imagine Lorentz factors of ∞-{[Hyper Aleph Ω (Aleph Ω, Aleph Ω)]-x,1, [Hyper Aleph Ω (Aleph Ω, Aleph Ω)]-x,1}!

Such an imbued spacecraft would travel ∞-{[Hyper Aleph Ω (Aleph Ω, Aleph Ω)]-x,1, [Hyper Aleph Ω (Aleph Ω, Aleph Ω)]-x,1} light-years through space and the same number of years into the future in one-year ship-time and travel and effective multiple of the same number of the speed of light.

So do not fret over any light-speed limits imposed by nature. It works out all the better.

PART 13

Now, it may be considered that a suitably infinite spacecraft Lorentz factor would serve as a boundary condition.

However, I conjecture that such a light-speed boundary condition can be broken and exceeded.

Breaking any first infinite Lorentz factor-based light-speed boundary condition, ironically, may not be the same mechanism as traveling faster than light.

Instead of a spacecraft breaking a first light-speed boundary condition traveling faster-than-light, the boundary condition breaking may imply something far more exotic, for which we have yet to develop language and thought structure to consider.

We will denote a first broken infinite Lorentz factor light-speed boundary condition as ∞-c-broken-abstract-1.

The inclusion of the word "abstract" is a placeholder that denotes a real scenario but one which we have not yet the science and mathematics developed to logically explore and formulate in workable detail.

Upon undergoing a ∞-c-broken-abstract-1, we may pump super-abounding infinite kinetic energy into a finite invariant mass spacecraft to achieve yet another boundary condition breakage which we will refer to as ∞-c-broken-abstract-2.

Upon undergoing a ∞-c-broken-abstract-2, we may pump super-abounding infinite kinetic energy into a finite invariant mass spacecraft to achieve yet another boundary condition breakage which we will refer to as ∞-c-broken-abstract-3.

We may be able to go all the way to and beyond,

∞-c-broken-abstract-{[Hyper Aleph Ω (Aleph Ω, Aleph Ω)]-x,1, [Hyper Aleph Ω (Aleph Ω, Aleph Ω)]-x,1} and perhaps even beyond,

$$\infty\text{-c-broken-abstract-}\{\infty\text{-}\{[\text{Hyper Aleph } \Omega \ (\text{Aleph } \Omega, \text{ Aleph } \Omega)]\text{-x,1, } [\text{Hyper Aleph } \Omega \ (\text{Aleph } \Omega, \text{ Aleph } \Omega)]\text{-x,1}\}\}$$

The moral of the story is to never fret about impulse light-speed travel and whatever comes later.

The opportunities for impulse travel are boundless!

Note by breaking boundary conditions, I am not merely implying that such a spacecraft would leave this universe or realm of travel because it runs out of future temporal room to travel forward in time in, for example, our universe. Instead, I am implying much more extreme scenarios where mathematically functionally infinite boundary conditions are broken through. As such, the breakthrough would be trying transcending mathematical and logical infinite boundary conditions.

So do not fret over impulse travel.

PART 14.

Light-speed travel is a gateway to entering exotic realms.

Now we may validly intuit that a spacecraft having attained the velocity of light thus being imbued with infinite Lorentz factors and time dilations may run out of future time to travel forward in time. Accordingly, the spacecraft would be eternalized with respect to the background and may pop out of the space-time of travel to enter a larger, broader hyperspace-time, and perhaps even one or more levels of meta-eternities. As another option, the spacecraft may realize a state of eternal nows and meta-eternal-nows of ascending levels as its Lorentz factor grows to ever greater infinities.

Now, what I propose is that such a spacecraft may enter states and/or places that are strongly distinct from hyper-futures, levels of hyper-futures, eternities, levels of eternities, nows, levels of nows, meta-eternal-nows, levels of meta-eternal nows.

Now, we compose the following ordinated set for the proceeding conjectures.

1) hyper-futures,
2) levels of hyper-futures,

3) eternities,
4) levels of eternities,
5) nows,
6) levels of nows,
7) meta-eternal-nows,
8) levels of meta-eternal nows

Moreover, the number of elements in each of the above 8 classes can be infinite as the following expressions indicate.

1) $\Sigma(j = 1; j \geq \infty)$☺hyper-futures),j,
2) $\Sigma(j = 1; j \geq \infty)$:(levels of hyper-futures),j,
3) $\Sigma(j = 1; j \geq \infty)$:(eternities),j,
4) $\Sigma(j = 1; j \geq \infty)$:(levels of eternities),j,
5) $\Sigma(j = 1; j \geq \infty)$:(nows),j,
6) $\Sigma(j = 1; j \geq \infty)$:(levels of nows),j,
7) $\Sigma(j = 1; j \geq \infty)$:(meta-eternal-nows),j,
8) $\Sigma(j = 1; j \geq \infty)$:(levels of meta-eternal nows),j

Now, the number of ordinated constructs that are distinct from each other and from the above eight items is greater than or equal to infinity. These constructs are essentially realms potentially available to initially light-speed spacecraft with suitably mounting infinite Lorentz factors or spacecraft with kinetic energies of suitably infinite multiples of the spacecraft invariant mass-energies.

So the opportunities for light-speed travel and para-light-speed travel are limitless.

We may go further to consider realms of travel that are neither the same nor different than the elements of the two sets of eight ordinated items, and indeed, neither the same nor different than the set consisting of an infinite extension beyond the eight ordinated items.

Thus, these states and/or places of travel would require an infinite number of ordinated terms, which would require an infinite number of positive integers as list item indices.

We may take the ship-time, spacecraft kinetic energy, spacecraft acceleration, and so on, single and higher-order derivatives and integrals of the rate of spacecraft travel in each of the above realms, as well as the rate of transition from one such realm to another.

Moreover, it may be possible for spacecraft to undergo quantum teleportation and/or tunneling within a given realm of travel or from one realm of travel to another.

Additionally, wormholes or analogues thereof may convey a spacecraft from one location of a given realm to another location within the same realm or from one realm to another realm.

PART 15

We can all appreciate the concept that travel at the velocity of light is associated with an infinite Lorentz factor. Mathematically, the Lorentz factor of light-speed travel should be at least equal to the least transfinite number.

Ironically, the Lorentz factor of a light-speed spacecraft may continue to increase to ever greater infinite values.

However, it is mainly held paradigms that state that light-speed impulse travel is associated with the greatest possible velocity.

So we have some mathematical paradoxes. One such paradox is how a light-speed spacecraft may have continually increasing but already infinite Lorentz factors while limited to the velocity of light.

I propose a resolution to the latter paradox by suggesting that after a spacecraft would attain light-speed in the least transfinite number, it would gain incremental virtual velocity components that can commensurately grow in magnitude to progressively greater infinite magnitudes.

Accordingly, the conventional velocity of spacecraft with progressively greater infinite Lorentz factors would be limited to precisely the speed of light or c. However, the full velocity of the spacecraft would be equal to {c + [(k)(virtual c)]}. Thus, the conventional velocity of the spacecraft would be limited precisely to c while the full velocity of the spacecraft would be greater than c.

The component [(k)(virtual c)] is a vector that may have multiple interpretations.

For example, one interpretation would be [(k)(virtual c)] oriented in higher dimensional space most likely ranging in relation to c from orthogonality to parallel orientation.

A second example is such that [(k)(virtual c)] would be oriented in scalar fields as the precursor material out of which mass-energy-space-time compositions are formed.

A third example is such that [(k)(virtual c)] would be oriented in bulks, whereby bulks, I mean realms for which space-times and hyperspace-times manifest and which underwrite or contain space-times and hyperspace-times.

A fourth example is such that [(k)(virtual c)] would act as if one or more hidden variables yet then so in ways that have objective meaning and that are metaphysically real.

A fifth example is such that [(k)(virtual c)] would have no extension but instead be deposited in realms or states not defined by geometric dimensions.

A sixth example is such that [(k)(virtual c)] would have an objective extension but instead be deposited in realms or states not defined by geometric dimensions.

Other scenarios additionally are possible.

For many scenarios for which the spacecraft Lorentz factor would be limited to approximately a range of (Aleph 0) to [m(Aleph 0)], where *m* is finitely greater than 1 but vanishingly small positive fractions greater than 1, *k* should likewise be similarly vanishingly small extents greater than 1.

As a spacecraft would attain a Lorentz factor of around [n(Aleph 0)], where *n* is significantly greater than 1 but finite and small, *k* would increase in scale from values that are small but significantly greater than 1 to values at least equal to infinity scrapers.

As a spacecraft would attain a Lorentz factor of around [o(Aleph 0)], were *o* is very large and finite, *k* should increase in scale from pre-infinite values to values of [m(Aleph 0)], where *m* is finitely greater than 1 but vanishingly small positive fractions greater than 1.

As a spacecraft would attain a Lorentz factor of around [p(Aleph 0)], where *p* is [a(Aleph 0)], where *a* is finitely greater than 1 but vanishingly small positive fractions greater than 1, *k* should increase in scale from pre-infinite values to values of [m(Aleph 0)] where *m* ranges from values greater than 1 to large finite values.

As a spacecraft would attain a Lorentz factor of around [q(Aleph 0)], where *q* is [a(Aleph 0)], where *a* is greater than 1 to very large finite values, *k* should increase in scale to [m(Aleph 0)], where *m* ranges in values from pre-finite numbers to small superunitary real multiples of (Aleph 0).

As a spacecraft would attain a Lorentz factor of around [r(Aleph 0)], where *r* is [a(Aleph 0)], where *a* ranges from pre-finite values to small superunitary multiples of (Aleph 0), *k* should increase in scale to [(Aleph 0) EXP m], where *m* ranges in values that are about equal to 2 to large finite values.

For many scenarios for which the spacecraft Lorentz factor would be limited to approximately a range of (Aleph 1) to [m(Aleph 1)], where *m* is finitely greater than 1 but vanishingly small positive fractions greater than 1, *k* should likewise be similarly vanishingly small extents greater than 1.

As a spacecraft would attain a Lorentz factor of around [n(Aleph 1)], where *n* is significantly greater than 1 but finite and small, *k* would increase in scale from values which are small but significantly greater than 1 to values at least equal to infinity scrapers.

As a spacecraft would attain a Lorentz factor of around [o(Aleph 1)], where *o* is very large and finite, *k* should increase in scale from pre-infinite values to values of [m(Aleph 1)], where *m* is finitely greater than 1, but vanishingly small positive fractions greater than 1.

As a spacecraft would attain a Lorentz factor of around [p(Aleph 1)], where *p* is [a(Aleph 1)], where *a* is finitely greater than 1 but vanishingly small positive fractions greater than 1, *k* should increase in scale from pre-infinite values to values of [m(Aleph 1)], where *m* ranges from values greater than1 to large finite values.

As a spacecraft would attain a Lorentz factor of around [q(Aleph 1)], where *q* is [a(Aleph 1)], where *a* is greater than 1 to very large finite values, *k* should increase in scale to [m(Aleph 1)], where *m* ranges in values from pre-finite numbers to small superunitary real multiples of (Aleph 1).

As a spacecraft would attain a Lorentz factor of around [r(Aleph 1)], where *r* is [a(Aleph 1)], where *a* ranges from pre-finite values to small superunitary multiples of (Aleph 1), *k* should increase in scale to [(Aleph 1) EXP m], where *m* ranges in values that are about equal to 2 to large finite values.

For many scenarios for which the spacecraft Lorentz factor would be limited to approximately a range of (Aleph 2) to [m(Aleph 2)], where *m* is finitely greater than 1 but vanishingly small positive fractions greater than 1, *k* should likewise be similarly vanishingly small extents greater than 1.

As a spacecraft would attain a Lorentz factor of around [n(Aleph 2)], where *n* is significantly greater than 1 but finite and small, *k* would increase in scale from values that are small but significantly greater than 1 to values at least equal to infinity scrapers.

As a spacecraft would attain a Lorentz factor of around [o(Aleph 2)], where *o* is very large and finite, *k* should increase in scale from pre-infinite values to values of [m(Aleph 2)], where *m* is finitely greater than 1 but vanishingly small positive fractions greater than 1.

As a spacecraft would attain a Lorentz factor of around [p(Aleph 2)], where *p* is [a(Aleph 2)], where *a* is finitely greater than 1 but vanishingly small positive fractions greater than

1, *k* should increase in scale from pre-infinite values to values of [m(Aleph 2)], where *m* ranges from values greater than 1 to large finite values.

As a spacecraft would attain a Lorentz factor of around [q(Aleph 2)], where *q* is [a(Aleph 2)], where *a* is greater than 1 to very large finite values, *k* should increase in scale to [m(Aleph 2)], where *m* ranges in values from pre-finite numbers to small superunitary real multiples of (Aleph 2).

As a spacecraft would attain a Lorentz factor of around [r(Aleph 2)], where *r* is [a(Aleph 2)], where *a* ranges from pre-finite values to small superunitary multiples of (Aleph 2), *k* should increase in scale to [(Aleph 2) EXP m], where *m* ranges in values which are about equal to 2 to large finite values.

For many scenarios for which the spacecraft Lorentz factor would be limited to approximately a range of (Aleph 3) to [m(Aleph 3)], where *m* is finitely greater than 1 but vanishingly small positive fractions greater than 1, *k* should likewise be similarly vanishingly small extents greater than 1.

As a spacecraft would attain a Lorentz factor of around [n(Aleph 3)], where *n* significantly greater than 1 but finite and small, *k* would increase in scale from values that are small but significantly greater than 1 to values at least equal to infinity scrapers.

As a spacecraft would attain a Lorentz factor of around [o(Aleph 3)], where *o* is very large and finite, *k* should increase in scale from pre-infinite values to values of [m(Aleph 3)], where *m* is finitely greater than 1 but vanishingly small positive fractions greater than 1.

As a spacecraft would attain a Lorentz factor of around [p(Aleph 3)], where *p* is [a(Aleph 3)], where *a* is finitely greater than 1 but vanishingly small positive fractions greater than 1, *k* should increase in scale from pre-infinite values to values of [m(Aleph 3)], where *m* ranges from values greater than 1 to large finite values.

As a spacecraft would attain a Lorentz factor of around [q(Aleph 3)], where *q* is [a(Aleph 3)], where *a* is greater than 1 to very large finite values, *k* should increase in scale to [m(Aleph 3)], where *m* ranges in values from pre-finite numbers to small superunitary real multiples of (Aleph 3).

As a spacecraft would attain a Lorentz factor of around [r(Aleph 3)], where *r* is [a(Aleph 3)], where *a* ranges from pre-finite values to small superunitary multiples of (Aleph 3), *k* should increase in scale to [(Aleph 3) EXP m], where *m* ranges in values that are about equal to 2 to large finite values.

We can likewise continue on with functional arguments of Aleph 4, Aleph 5, Aleph 6, and so on. Generally, we can consider functional arguments of Aleph g, where g is any finite or infinite counting number.

PART 16

Regarding how big infinities can be, consider the following digression that is repeated in this book.

Now, there are countable and uncountable infinities.

But before we go further, consider the hyper-operator notation that was designed to express huge values not otherwise expressible.

For example, note that Hyper4(a, n) is equal to a tetrated n, or a raised to the power of itself n-1 times. The latter value is symbolically written as n subscript a.

For example,

> 3 EXP 4 = 81, but 4 subscript 3 is approximately equal to 10 EXP (1,000,000,000,000).

Alternatively,

> 4 subscript 2 = 2 EXP 2 EXP 2 EXP 2 = 2 EXP [2 EXP [2 EXP 2]] = 2 EXP (2 EXP 4) = 2 EXP 16 = 65,536.

For example,

Hyper5(4, 4)is equal to 4 tetrated 4 tetrated 4 tetrated 4. This value is commonly referred to as 4 pentated 4.

Hyper 6, (4,4) is 4 pentated 4 pentated 4 pentated 4 and is also referred to as 4 hexataed 4.

Hyper 7, (4,4) is 4 hexated 4 hexated 4 hexated 4 and so on.

Aleph 0 is the infinite number of integers.

Aleph 1, according to the perhaps unprovable, and thus unfalsifiable, continuum hypothesis, is the number of real numbers which is greater than Aleph 0 by a multiplicative factor of infinity.

Aleph 2 is similarly greater than Aleph 1.

Aleph 3 is similarly greater than Aleph 2.

Aleph 4 is similarly greater than Aleph 3.

And so on.

In general, Aleph n = 2 EXP [Aleph (n-1)].

The number Ω is commonly stated as the least infinite positive integer or ordinal.

Now here is a real zinger.

So we can produce the abstraction of Hyper Aleph Ω (Aleph Ω, Aleph Ω).

How about,

>Hyper [Hyper Aleph Ω (Aleph Ω, Aleph Ω)](Hyper Aleph Ω (Aleph Ω, Aleph Ω), Hyper Aleph Ω (Aleph Ω, Aleph Ω))

and even

>Hyper [Hyper [Hyper Aleph Ω (Aleph Ω, Aleph Ω)](Hyper Aleph Ω (Aleph Ω, Aleph Ω), Hyper Aleph Ω (Aleph Ω, Aleph Ω))](Hyper [Hyper Aleph Ω (Aleph Ω, Aleph Ω)](Hyper Aleph Ω (Aleph Ω, Aleph Ω), Hyper Aleph Ω (Aleph Ω, Aleph Ω)), Hyper [Hyper Aleph Ω (Aleph Ω, Aleph Ω)](Hyper Aleph Ω (Aleph Ω, Aleph Ω), Hyper Aleph Ω (Aleph Ω, Aleph Ω))).

So we can produce the abstraction of Hyper Aleph Ω (Aleph Ω, Aleph Ω), which we will label as follows:

>[Hyper Aleph Ω (Aleph Ω, Aleph Ω)]-x,1,1.

>Hyper [Hyper Aleph Ω (Aleph Ω, Aleph Ω)](Hyper Aleph Ω (Aleph Ω, Aleph Ω), Hyper Aleph Ω (Aleph Ω, Aleph Ω)) we will label as

>[Hyper Aleph Ω (Aleph Ω, Aleph Ω)]-x,1,2.

>Hyper [Hyper [Hyper Aleph Ω (Aleph Ω, Aleph Ω)](Hyper Aleph Ω (Aleph Ω, Aleph Ω), Hyper Aleph Ω (Aleph Ω, Aleph Ω))](Hyper [Hyper Aleph Ω (Aleph Ω, Aleph Ω)](Hyper Aleph Ω (Aleph Ω, Aleph Ω), Hyper Aleph Ω (Aleph Ω, Aleph Ω)), Hyper [Hyper Aleph Ω (Aleph Ω, Aleph Ω)](Hyper Aleph Ω (Aleph Ω, Aleph Ω), Hyper Aleph Ω (Aleph Ω, Aleph Ω))) we will label as

>[Hyper Aleph Ω (Aleph Ω, Aleph Ω)]-x,1,3.

We can likewise consider

[Hyper Aleph Ω (Aleph Ω, Aleph Ω)]-x,1,4

[Hyper Aleph Ω (Aleph Ω, Aleph Ω)]-x,1,5

[Hyper Aleph Ω (Aleph Ω, Aleph Ω)]-x,1,6

all the way to

[Hyper Aleph Ω (Aleph Ω, Aleph Ω)]-x,1,[Hyper Aleph Ω (Aleph Ω, Aleph Ω)]

and even to

[Hyper Aleph Ω (Aleph Ω, Aleph Ω)]-x,1, [Hyper Aleph Ω (Aleph Ω, Aleph Ω)]-x,1.

We can continue this sequence forever.

So velocities of {c + [(k)(virtual c)]} may be associated, with *k* having any infinite value based on the above infinity building hyper-operator notation as also may spacecraft Lorentz factors.

We may likewise take the ship-time, background time, Lorentz factor, spacecraft kinetic energy, ship-frame acceleration, background frame acceleration, and the like associated with the spacecraft, single and higher-order derivatives and integrals of the rate of change of {c + [(k)(virtual c)]} and of [(k)(virtual c)].

In short, I beg you not to fret about any light-speed limits or apparent light-speed limits nature imposes on us because the opportunities for exotic travel scenarios under such limits are far, far more extreme than any superluminal warp drive(s) or wormhole travel would enable.

Perhaps at each point in various abstract spaces, there are seas of abstract hyperspaces that directly connect to the point.

Accordingly, the sea of abstract spaces that connect to said point in spaces of three dimensions or less range in dimensionality from 1 to uncountable infinities.

Additionally, the number of spaces in the said sea of abstract spaces may be uncountably infinite. It may be the case where mathematical systems can be worked for which the number of abstract spaces in the said sea of abstract spaces of each number of dimensions is infinite.

It may even be the case for which the sea of abstract spaces contains arbitrary infinite numbers of curved topologies, per each number of dimensions.

The sea of abstract spaces may include spaces of any number of dimensions that do not encompass the spaces for which a point is considered as well as abstract hyperspaces that envelop the abstract spaces for which a point is considered.

Moreover, scenarios for which each point in the above digression of abstract spaces may have analogues in volumes of the least dimensionality greater than 0 through unboundable numbers of dimensions. Additionally, such greater than 0 spatial volumes may extend into infinities along each dimension. Ironically, the latter scenario would include all spaces of dimensionality greater than 0 but less than 1.

Regarding real physical analogues, perhaps the above sea of abstract spaces has analogues in physical space-times, even in cases where the space-times are finitely quantized, continuous, or super-continuous. In cases where the real space-times enveloping a reference point or higher dimensional space-time volume where the number of dimensions of the volume includes no more than one temporal dimension, and no more than three spatial dimensions, we refer to the enveloping space-times as hyperspaces or more properly hyper-space-times. Accordingly, such real enveloping space-times may be selected from the following set.

Such hyperspaces can be N-Time-M-Space, where N is an integer greater than or equal to 3, and M is any counting number.

Alternatively, N can be any rational number equal to 3 or greater and M is any rational number greater than or equal to 1.

As another set of scenarios, N can be any irrational number equal to 3 or greater and M is any irrational number greater than or equal to 1.

Alternatively, N can be any rational number greater than 0, and M is any rational number greater than 0.

As another set of scenarios, N can be any irrational number greater than 0, and M is any irrational number greater than 0.

The N-Space-M-Time may optionally be as follows:

> Flat, positively curved, negatively curved, positively curved and torsioned at one or more scales in arbitrary patterns including, but not limited to, fractals, negatively curved and torsioned at one or more scales in arbitrary patterns including, but not limited to, fractals, positively super-curved, negatively super-curved, positively super-curved and torsioned at one or more scales in arbitrary patterns including, but not limited to, fractals, negatively super-curved and torsioned at one or more scales in arbitrary patterns including but not limited to fractals, positively super-curved and super-torsioned at one or

more scales in arbitrary patterns including, but not limited to, fractals, negatively super-curved and super-torsioned at one or more scales in arbitrary patterns including, but not limited to, fractals, positively curved and positively torsioned at one or more scales in arbitrary patterns including, but not limited to, fractals, negatively curved and positively torsioned at one or more scales in arbitrary patterns including, but not limited to, fractals, positively super-curved, negatively super-curved, positively super-curved and positively torsioned at one or more scales in arbitrary patterns including, but not limited to, fractals, negatively super-curved and positively torsioned at one or more scales in arbitrary patterns including, but not limited to, fractals, positively super-curved and positively super-torsioned at one or more scales in arbitrary patterns including, but not limited to, fractals, negatively super-curved and positively super-torsioned at one or more scales in arbitrary patterns including, but not limited to, fractals, positively curved and negatively torsioned at one or more scales in arbitrary patterns including, but not limited to, fractals, negatively curved and negatively torsioned at one or more scales in arbitrary patterns including, but not limited to, fractals, positively super-curved, negatively super-curved, positively super-curved and negatively torsioned at one or more scales in arbitrary patterns including, but not limited to, fractals, negatively super-curved and negatively torsioned at one or more scales in arbitrary patterns including, but not limited to, fractals, positively super-curved and negatively super-torsioned at one or more scales in arbitrary patterns including, but not limited to, fractals, negatively super-curved and negatively super-torsioned at one or more scales in arbitrary patterns including, but not limited to, fractals, positively super-...-super-curved, negatively super-...-super-curved, positively super-...-super-curved and torsioned at one or more scales in arbitrary patterns including, but not limited to, fractals, negatively super-...-super-curved and torsioned at one or more scales in arbitrary patterns including, but not limited to, fractals, positively super-...-super-curved and super-...-super-torsioned at one or more scales in arbitrary patterns including, but not limited to, fractals, negatively super-...-super-curved and super-...-super-torsioned at one or more scales in arbitrary patterns including, but not limited to, fractals, positively curved and positively torsioned at one or more scales in arbitrary patterns including, but not limited to, fractals, negatively curved and positively torsioned at one or more scales in arbitrary patterns including, but not limited to, fractals, positively super-...-super-curved, negatively super-...-super-curved, positively super-...-super-curved and positively torsioned at one or more scales in arbitrary patterns including, but not limited to, fractals, negatively super-...-super-curved and positively torsioned at one or more scales in arbitrary patterns including, but not limited to, fractals, positively super-...-super-curved and positively super-...-super-torsioned at one or more scales in arbitrary patterns including, but not limited to, fractals, negatively super-...-super-curved and positively super-...-super-torsioned at one or more scales in arbitrary patterns including, but not limited to, fractals, positively curved and negatively torsioned at one or more scales in arbitrary patterns including, but not limited to, fractals, negatively curved and negatively torsioned at one or more scales in arbitrary patterns including, but not limited to, fractals, positively super-...-super-curved, negatively super-...-super-curved, positively super-...-super-curved and negatively torsioned at one or more scales in arbitrary patterns including, but not limited to, fractals, negatively super-...-super-curved and negatively torsioned at one or more scales in arbitrary patterns including, but not limited to, fractals, positively super-...-super-curved and negatively super-...-super-torsioned at one or more scales in arbitrary patterns including, but not limited to, fractals, negatively super-...-super-curved and negatively super-...-super-torsioned at one or more scales in arbitrary patterns including, but not limited to, fractals.

PART 17

Now, there are countable and uncountable infinities.

But before we go further, consider the hyper-operator notation that was designed to express huge values not otherwise expressible.

For example, note that Hyper4(a, n) is equal to a tetrated *n*, or a raised to the power of itself n-1 times. The latter value is symbolically written as *n* subscript *a*.

For example,

> 3 EXP 4 = 81, but 4 subscript 3 is approximately equal to 10 EXP (1,000,000,000,000).

Alternatively,

> 4 subscript 2 = 2 EXP 2 EXP 2 EXP 2 = 2 EXP [2 EXP [2 EXP 2]] = 2 EXP (2 EXP 4) = 2 EXP 16 = 65,536.

For example,

Hyper5(4, 4)is equal to 4 tetrated 4 tetrated 4 tetrated 4. This value is commonly referred to as 4 pentated 4.

Hyper 6, (4,4) is 4 pentated 4 pentated 4 pentated 4 and is also referred to as 4 hexataed 4.

Hyper 7, (4,4) is 4 hexated 4 hexated 4 hexated 4 and so on.

Aleph 0 is the infinite number of integers.

Aleph 1, according to the perhaps unprovable, and thus unfalsifiable, continuum hypothesis, is the number of real numbers which is greater than Aleph 0 by a multiplicative factor of infinity.

Aleph 2 is similarly greater than Aleph 1.

Aleph 3 is similarly greater than Aleph 2.

Aleph 4 is similarly greater than Aleph 3.

And so on.

In general, Aleph n = 2 EXP [Aleph (n-1)].

The number Ω is commonly stated as the least infinite positive integer or ordinal.

Now here is a real zinger.

So we can produce the abstraction of Hyper Aleph Ω (Aleph Ω, Aleph Ω).

We can produce greater infinities as great as we want even using infinite numbers of letters and symbols.

Now consider an extremely evolved spacecraft Lorentz factor of Hyper Aleph Ω (Aleph Ω, Aleph Ω). Such a craft would travel about Hyper Aleph Ω (Aleph Ω, Aleph Ω) years through space in a one-year ship-frame, Hyper Aleph Ω (Aleph Ω, Aleph Ω) years into the future in one-year ship-frame, and an effective multiple of Hyper Aleph Ω (Aleph Ω, Aleph Ω) of the speed of light.

Now we can consider the mathematical limits of infinities and refer to these values as the Final Limits or Final-Limits-1.

Now here is a real zinger! What about values abstractly greater than Final-Limits-1, which we will refer to as Final-Limits-2->?

What about values abstractly greater than Final-Limits-2->, which we will refer to as Final-Limits-3->?

What about values abstractly greater than Final-Limits-3->, which we will refer to as Final-Limits-4->?

What about values abstractly greater than Final-Limits-4->, which we will refer to as Final-Limits-5->?

We can likewise progress to Final-Limits-[Hyper Aleph Ω (Aleph Ω, Aleph Ω)]-> and continue on forever and ever.

GOD can actually imagine and comprehend in a perfect way something larger than Himself. How do I know this to be true, you might ask? The answer is rather quite elementary, my dear Watson! GOD, who knows all, knows that He is not the sum total of existent reality. Existent reality is the set consisting of GOD and all creatures, substantial, accidental, and both substantial and accidental. What is more is that there are no limits to GOD's creative might, and He can, and I believe He will, continue the process of creation forever. Even though not widely accepted by the Catholic Church, I believe as a Catholic that the process of human procreation will continue forever, thus implying no upper boundary to the potential number of living human persons. However, these human persons will need living space, so I am driven to believe that the size of human-occupied creation must also increase utterly forever. What this means for Lorentz factors achievable is at present is a mystery and a wonderful one at that.

Essentially, our spaceships will keep getting faster and faster. Ironically, even if we reach the speed of light. Ideal ultimate light-speed is as if a beacon, so impossibly distant and glimmering, so impossibly dim, that can serve to beckon us forward with its mystique. May we ever achieve the first ideal ultimate light-speed? Perhaps, and this may take all eternity to accomplish. However, perhaps just beyond the first ideal ultimate light-speed or ideal-ultimate-light-speed-1 lies a faster light-speed or ultimate-light-speed-2. Beyond ultimate-light-speed-2 perhaps lies a greater velocity of light or ultimate-light-speed-3.

We may progress to ultimate-light-speed-[Hyper Aleph Ω (Aleph Ω, Aleph Ω)] and continue forever onward from there.

None of this will happen proverbially overnight. However, right here and now, we can start the journey under nuclear power in nuclear fission–powered spacecraft able to attain velocities of about 20 percent of the speed of light. I believe such craft can and will be built this century to begin the process of human exploration and colonization of the Milky Way Galaxy. However, we are morally obliged not to take over or steal planets and moons already inhabited by extraterrestrial peoples. We likely should not even take over or occupy planets with infra-personal life-forms such as animals and the like. That being said, the quantity of available interstellar hydrogen and other elements is as limitless as our universe. These resources can be used by us humans to construct vast space colonies.

Now, within the observable universe, there are about 10 EXP 24 stars. This is about equal to the number of fine grains of table sugar that would cover the entire United States one hundred meters deep. The number of planets orbiting stars in our universe seems to be about ten times greater yet or equal to the number of fine grains of table sugar that would cover the entire United States one thousand meters deep. The number of moons, orbiting planets, orbiting stars is estimated to be ten times greater yet or equal to the number of fine grains of table sugar that would cover the entire United States ten thousand meters deep. This portion of space is huge but likely only a tiny fraction of our universe.

The opportunities are boundless for human evolution, and technological, scientific, medical, social, emotional, psychological, economic, political, and spiritual progress.

Do we have the courage to set sail into the Seas of Forever? I think we do.

PART 18

Now, there are countable and uncountable infinities.

But before we go further, consider the hyper-operator notation that was designed to express huge values not otherwise expressible.

For example, note that Hyper4(a, n) is equal to a tetrated n, or a raised to the power of itself n-1 times. The latter value is symbolically written as n subscript a.

For example,

> 3 EXP 4 = 81, but 4 subscript 3 is approximately equal to 10 EXP (1,000,000,000,000).

Alternatively,

> 4 subscript 2 = 2 EXP 2 EXP 2 EXP 2 = 2 EXP [2 EXP [2 EXP 2]] = 2 EXP (2 EXP 4) = 2 EXP 16 = 65,536.

For example,

Hyper5(4, 4)is equal to 4 tetrated 4 tetrated 4 tetrated 4. This value is commonly referred to as 4 pentated 4.

Hyper 6, (4,4) is 4 pentated 4 pentated 4 pentated 4 and is also referred to as 4 hexataed 4.

Hyper 7, (4,4) is 4 hexated 4 hexated 4 hexated 4 and so on.

Aleph 0 is the infinite number of integers.

Aleph 1, according to the perhaps unprovable, and thus unfalsifiable, continuum hypothesis, is the number of real numbers that is greater than Aleph 0 by a multiplicative factor of infinity.

Aleph 2 is similarly greater than Aleph 1.

Aleph 3 is similarly greater than Aleph 2.

Aleph 4 is similarly greater than Aleph 3.

And so on.

In general, Aleph n = 2 EXP [Aleph (n-1)].

The number Ω is commonly stated as the least infinite positive integer or ordinal.

Now here is a real zinger.

So we can produce the abstraction of Hyper Aleph Ω (Aleph Ω, Aleph Ω).

So we can likewise form abstractions such as the following:

[Hyper Aleph Ω (Aleph Ω, Aleph Ω)]([Hyper Aleph Ω (Aleph Ω, Aleph Ω)], [Hyper Aleph Ω (Aleph Ω, Aleph Ω)])

Or,

{[Hyper Aleph Ω (Aleph Ω, Aleph Ω)]([Hyper Aleph Ω (Aleph Ω, Aleph Ω)], [Hyper Aleph Ω (Aleph Ω, Aleph Ω)])}({[Hyper Aleph Ω (Aleph Ω, Aleph Ω)]([Hyper Aleph Ω (Aleph Ω, Aleph Ω)], [Hyper Aleph Ω (Aleph Ω, Aleph Ω)])}, {[Hyper Aleph Ω (Aleph Ω, Aleph Ω)]([Hyper Aleph Ω (Aleph Ω, Aleph Ω)], [Hyper Aleph Ω (Aleph Ω, Aleph Ω)])})

Or,

{{[Hyper Aleph Ω (Aleph Ω, Aleph Ω)]([Hyper Aleph Ω (Aleph Ω, Aleph Ω)], [Hyper Aleph Ω (Aleph Ω, Aleph Ω)])}({[Hyper Aleph Ω (Aleph Ω, Aleph Ω)]([Hyper Aleph Ω (Aleph Ω, Aleph Ω)], [Hyper Aleph Ω (Aleph Ω, Aleph Ω)])}, {[Hyper Aleph Ω (Aleph Ω, Aleph Ω)]([Hyper Aleph Ω (Aleph Ω, Aleph Ω)], [Hyper Aleph Ω (Aleph Ω, Aleph Ω)])})}({{[Hyper Aleph Ω (Aleph Ω, Aleph Ω)]([Hyper Aleph Ω (Aleph Ω, Aleph Ω)], [Hyper Aleph Ω (Aleph Ω, Aleph Ω)])}({[Hyper Aleph Ω (Aleph Ω, Aleph Ω)]([Hyper Aleph Ω (Aleph Ω, Aleph Ω)], [Hyper Aleph Ω (Aleph Ω, Aleph Ω)])}, {[Hyper Aleph Ω (Aleph Ω, Aleph Ω)]([Hyper Aleph Ω (Aleph Ω, Aleph Ω)], [Hyper Aleph Ω (Aleph

Ω)])})}, {{[Hyper Aleph Ω (Aleph Ω, Aleph Ω)]([Hyper Aleph Ω (Aleph Ω, Aleph Ω)], [Hyper Aleph Ω (Aleph Ω, Aleph Ω)])}({[Hyper Aleph Ω (Aleph Ω, Aleph Ω)]([Hyper Aleph Ω (Aleph Ω, Aleph Ω)], [Hyper Aleph Ω (Aleph Ω, Aleph Ω)])}, {[Hyper Aleph Ω (Aleph Ω, Aleph Ω)]([Hyper Aleph Ω (Aleph Ω, Aleph Ω)], [Hyper Aleph Ω (Aleph Ω, Aleph Ω)])}})}}

And so on.

Now, what if GOD is the creator or generator of infinities but has not yet exhausted the creation of infinities?

What if GOD will forever create or generate ever greater infinities as abstract eternal forms?

GOD will for certain continue to create and wants us humans, any extraterrestrials, any ultraterrestrials, and any etherians to have ever greater delight and wanderlust in the ever-growing scope of infinities from now on without end.

I do not know if extraterrestials, ultraterrestrials, and/or etherians exist, but the notions of such personal life-forms are well ingrained in the fringe UFO space alien literature and UFO new age religions.

Regarding spacecraft able to attain the speed of light, if special relativity remains true to its abstract formulations, such spacecraft would at least be able to travel Ω light-years in about one-year ship-time, an Ω number of years into the future in one-year ship-time, and an effective travel velocity of Ω times the speed of light.

Perhaps a light-speed spacecraft Lorentz factor can grow to ever greater infinities such as values of Ω EXP 2, Ω EXP 3, Ω EXP 4, and so on to Aleph 2, then Aleph 3, then Aleph 4, and so on. You get the picture.

As for spacecraft able to attain infinite Lorentz factors of infinities that GOD has not yet created or manifested, we have the depths of future eternity to see these Lorentz factors become possible for spacecraft.

I hope the above digression fills you with wonderment and joy. I feel joy every time I think about this stuff.

PART 19

Now, there are numerous ways to manifest light-speed travel in theory.

First, there is mass-less travel of photons in an ideal vacuum at the speed of light.

Second, gravitational waves also propagate at the speed of light.

Third, gluons or particles of the strong nuclear force travel at the speed of light.

Fourth, quantum mechanical state teleportation is limited to the speed of light.

Fifth, theoretical quantum mechanical teleportation of energy or material objects is limited to the speed of light.

Sixth, electrical currents in a conductor or superconductor travel at the speed of light.

Seventh, quantum tunneling should not occur faster than light but at speeds as great as those of light.

Eighth, reducing the inertial mass of an object in motion to a magnitude of 0 should induce the velocity of the object to become that of light-speed.

Ninth, the conversion of an object or person's body to pure light-energy and/or gravitational energy followed by beaming the object or person should enable light-speed transport. Here, the object or person transported would be rematerialized at the destination.

Tenth, the conversion of an object or person's body to an electrical pulse through a conductor or superconductor should enable light-speed transport. Again, the object or person transported would be rematerialized at a distant location.

Other meanings of light-speed travel are also plausible. However, we do not know what they are.

Definitely, as we develop physics, engineering, and language, more meanings of light-speed travel will manifest.

Nature may protect herself by not allowing superluminal transport. However, even light-speed travel would enable infinite numbers of light-years traveled in one-second ship-time, infinite numbers of years of travel into the future in one-second ship-time, and effective travel of infinite multiples of the speed of light.

Thus, much territory can be accessed at light-speed for spacecraft so traveling.

Regarding travel to new destinations, we have our entire cosmic light-cone to explore and perhaps even more.

First, we have stars, galaxies, planets, and moons to travel to.

Second, we have about six times more cold dark matter in our cosmic light-cone than we have Standard Model baryonic matter. Thus, we have plausibly sixfold additional realms to explore in our cosmic light-cone.

Third, we have theoretical dark energy embodiments within our cosmic light-cone. Dark energy is theoretically in some models as an imbuement of energy associated with space-time. However, the dark energy might be distributed in lumps or other portions, but in ways such that its effects on Standard Model mass and energy and cold dark matter would be very uniform, and thus homogenous.

Now, ordinary Standard Model matter and energy make up about 4 percent of the mass of our universe whereas theoretically, cold dark matter makes up about 24 percent of the mass in our universe. So there is six times more cold dark matter than ordinary matter and energy. However, dark energy theoretically comprises about 72 percent of the total mass-energy of our universe. If dark energy and cold dark matter are divided into classes proportional to the Standard Model matter and energy, there may be effectively 100/4 or 25 parallel universes for us to explore. These we may be able to access in our cosmic light-cone and beyond per universal expansion processes.

So we have a large number of opportunities with light-speed travel, and even travel less than but close to the speed of light.

Additionally, it may be possible that local travel, ever so slightly faster than light, is possible but limited to speeds that do not enable backward time travel and do not violate the laws of special relativity. If the excess velocity components in superluminal travel are infinitesimal fractions of light-speed, then we have a fascinating scenario in which we are limited to light-speed yet can travel infinitesimally faster than light, where the chronological protection conjecture still holds. The number of ways infinitesimally faster-than-light travel can manifest and meanings thereof is also likely very large.

PART 20

Note that in this digression, we refer to the subtraction process by the word "minus" instead of using the symbol -. This is because the latter symbol is used for indexing, as you will see upon further reading.

Now, there are countable and uncountable infinities.

But before we go further, consider the hyper-operator notation that was designed to express huge values not otherwise expressible.

For example, note that Hyper4(a, n) is equal to a tetrated *n*, or a raised to the power of itself n minus 1 times. The latter value is symbolically written as *n* subscript *a*.

For example,

3 EXP 4 = 81, but 4 subscript 3 is approximately equal to 10 EXP (1,000,000,000,000).

Alternatively,

4 subscript 2 = 2 EXP 2 EXP 2 EXP 2 = 2 EXP [2 EXP [2 EXP 2]] = 2 EXP (2 EXP 4) = 2 EXP 16 = 65,536.

For example,

Hyper5(4, 4)is equal to 4 tetrated 4 tetrated 4 tetrated 4. This value is commonly referred to as 4 pentated 4.

Hyper 6, (4,4) is 4 pentated 4 pentated 4 pentated 4 and is also referred to as 4 hexataed 4.

Hyper 7, (4,4) is 4 hexated 4 hexated 4 hexated 4 and so on.

Aleph 0 is the infinite number of integers.

Aleph 1, according to the perhaps unprovable, and thus unfalsifiable, continuum hypothesis, is the number of real numbers which is greater than Aleph 0 by a multiplicative factor of infinity.

Aleph 2 is similarly greater than Aleph 1.

Aleph 3 is similarly greater than Aleph 2.

Aleph 4 is similarly greater than Aleph 3.

And so on.

In general, Aleph n = 2 EXP [Aleph (n-1)].

The number Ω is commonly stated as the least infinite positive integer or ordinal.

Now here is a real zinger.

So we can produce the abstraction of Hyper Aleph Ω (Aleph Ω, Aleph Ω).

Now I present my spin on hyper-operator notation.

Hyper-E-1 Aleph Ω (Aleph Ω, Aleph Ω)

I state is equal to

Hyper [Aleph Ω (Aleph Ω, Aleph Ω)]([Aleph Ω (Aleph Ω, Aleph Ω)], [Aleph Ω (Aleph Ω, Aleph Ω)])

Hyper-E-2 Aleph Ω (Aleph Ω, Aleph Ω)

I state is equal to

Hyper [Hyper [Aleph Ω (Aleph Ω, Aleph Ω)]([Aleph Ω (Aleph Ω, Aleph Ω)], [Aleph Ω (Aleph Ω, Aleph Ω)])]([Hyper [Aleph Ω (Aleph Ω, Aleph Ω)]([Aleph Ω (Aleph Ω, Aleph Ω)], [Aleph Ω (Aleph Ω, Aleph Ω)])], [Hyper [Aleph Ω (Aleph Ω, Aleph Ω)]([Aleph Ω (Aleph Ω, Aleph Ω)], [Aleph Ω (Aleph Ω, Aleph Ω)])])

Hyper-E-3 Aleph Ω (Aleph Ω, Aleph Ω)

I state is equal to

Hyper [Hyper [Hyper [Aleph Ω (Aleph Ω, Aleph Ω)]([Aleph Ω (Aleph Ω, Aleph Ω)], [Aleph Ω (Aleph Ω, Aleph Ω)])]([Hyper [Aleph Ω (Aleph Ω, Aleph Ω)]([Aleph Ω (Aleph Ω, Aleph Ω)], [Aleph Ω (Aleph Ω, Aleph Ω)])], [Hyper [Aleph Ω (Aleph Ω, Aleph Ω)]([Aleph Ω (Aleph Ω, Aleph Ω)], [Aleph Ω (Aleph Ω, Aleph Ω)])])]([Hyper [Hyper [Aleph Ω (Aleph Ω, Aleph Ω)]([Aleph Ω (Aleph Ω, Aleph Ω)], [Aleph Ω (Aleph Ω, Aleph Ω)])]([Hyper [Aleph Ω (Aleph Ω, Aleph Ω)]([Aleph Ω (Aleph Ω, Aleph Ω)], [Aleph Ω (Aleph Ω, Aleph Ω)])], [Hyper [Aleph Ω (Aleph Ω, Aleph Ω)]([Aleph Ω (Aleph Ω, Aleph Ω)], [Aleph Ω (Aleph Ω, Aleph Ω)])])], [Hyper [Hyper [Aleph Ω (Aleph Ω, Aleph Ω)]([Aleph Ω (Aleph Ω, Aleph Ω)], [Aleph Ω (Aleph Ω, Aleph Ω)])]([Hyper [Aleph Ω (Aleph Ω, Aleph Ω)]([Aleph Ω (Aleph Ω, Aleph Ω)], [Aleph Ω (Aleph Ω, Aleph Ω)])], [Hyper [Aleph Ω (Aleph Ω, Aleph Ω)]([Aleph Ω (Aleph Ω, Aleph Ω)], [Aleph Ω (Aleph Ω, Aleph Ω)])])])

Hyper-E-4 Aleph Ω (Aleph Ω, Aleph Ω)

I state is equal to

Hyper [Hyper [Hyper [Hyper [Aleph Ω (Aleph Ω, Aleph Ω)]([Aleph Ω (Aleph Ω, Aleph Ω)], [Aleph Ω (Aleph Ω, Aleph Ω)])]([Hyper [Aleph Ω (Aleph Ω, Aleph Ω)]([Aleph Ω (Aleph Ω, Aleph Ω)], [Aleph Ω (Aleph Ω, Aleph Ω)])], [Hyper [Aleph Ω (Aleph Ω, Aleph Ω)]([Aleph Ω (Aleph Ω, Aleph Ω)], [Aleph Ω (Aleph Ω, Aleph

Ω)])])]([Hyper [Hyper [Aleph Ω (Aleph Ω, Aleph Ω)]([Aleph Ω (Aleph Ω, Aleph Ω)], [Aleph Ω (Aleph Ω, Aleph Ω)])]([Hyper [Aleph Ω (Aleph Ω, Aleph Ω)]([Aleph Ω (Aleph Ω, Aleph Ω)], [Aleph Ω (Aleph Ω, Aleph Ω)])], [Hyper [Aleph Ω (Aleph Ω, Aleph Ω)]([Aleph Ω (Aleph Ω, Aleph Ω)], [Aleph Ω (Aleph Ω, Aleph Ω)])])], [Hyper [Hyper [Aleph Ω (Aleph Ω, Aleph Ω)]([Aleph Ω (Aleph Ω, Aleph Ω)], [Aleph Ω (Aleph Ω, Aleph Ω)])]([Hyper [Aleph Ω (Aleph Ω, Aleph Ω)]([Aleph Ω (Aleph Ω, Aleph Ω)], [Aleph Ω (Aleph Ω, Aleph Ω)])], [Hyper [Aleph Ω (Aleph Ω, Aleph Ω)]([Aleph Ω (Aleph Ω, Aleph Ω)], [Aleph Ω (Aleph Ω, Aleph Ω)])])])]([Hyper [Hyper [Hyper [Aleph Ω (Aleph Ω, Aleph Ω)]([Aleph Ω (Aleph Ω, Aleph Ω)], [Aleph Ω (Aleph Ω, Aleph Ω)])]([Hyper [Aleph Ω (Aleph Ω, Aleph Ω)]([Aleph Ω (Aleph Ω, Aleph Ω)], [Aleph Ω (Aleph Ω, Aleph Ω)])], [Hyper [Aleph Ω (Aleph Ω, Aleph Ω)]([Aleph Ω (Aleph Ω, Aleph Ω)], [Aleph Ω (Aleph Ω, Aleph Ω)])])]([Hyper [Hyper [Aleph Ω (Aleph Ω, Aleph Ω)]([Aleph Ω (Aleph Ω, Aleph Ω)], [Aleph Ω (Aleph Ω, Aleph Ω)])]([Hyper [Aleph Ω (Aleph Ω, Aleph Ω)]([Aleph Ω (Aleph Ω, Aleph Ω)], [Aleph Ω (Aleph Ω, Aleph Ω)])], [Hyper [Aleph Ω (Aleph Ω, Aleph Ω)]([Aleph Ω (Aleph Ω, Aleph Ω)], [Aleph Ω (Aleph Ω, Aleph Ω)])])], [Hyper [Hyper [Aleph Ω (Aleph Ω, Aleph Ω)]([Aleph Ω (Aleph Ω, Aleph Ω)], [Aleph Ω (Aleph Ω, Aleph Ω)])]([Hyper [Aleph Ω (Aleph Ω, Aleph Ω)]([Aleph Ω (Aleph Ω, Aleph Ω)], [Aleph Ω (Aleph Ω, Aleph Ω)])], [Hyper [Aleph Ω (Aleph Ω, Aleph Ω)]([Aleph Ω (Aleph Ω, Aleph Ω)], [Aleph Ω (Aleph Ω, Aleph Ω)])])])], [Hyper [Hyper [Hyper [Aleph Ω (Aleph Ω, Aleph Ω)]([Aleph Ω (Aleph Ω, Aleph Ω)], [Aleph Ω (Aleph Ω, Aleph Ω)])]([Hyper [Aleph Ω (Aleph Ω, Aleph Ω)]([Aleph Ω (Aleph Ω, Aleph Ω)], [Aleph Ω (Aleph Ω, Aleph Ω)])], [Hyper [Aleph Ω (Aleph Ω, Aleph Ω)]([Aleph Ω (Aleph Ω, Aleph Ω)], [Aleph Ω (Aleph Ω, Aleph Ω)])])]([Hyper [Hyper [Aleph Ω (Aleph Ω, Aleph Ω)]([Aleph Ω (Aleph Ω, Aleph Ω)], [Aleph Ω (Aleph Ω, Aleph Ω)])]([Hyper [Aleph Ω (Aleph Ω, Aleph Ω)]([Aleph Ω (Aleph Ω, Aleph Ω)], [Aleph Ω (Aleph Ω, Aleph Ω)])], [Hyper [Aleph Ω (Aleph Ω, Aleph Ω)]([Aleph Ω (Aleph Ω, Aleph Ω)], [Aleph Ω (Aleph Ω, Aleph Ω)])])], [Hyper [Hyper [Aleph Ω (Aleph Ω, Aleph Ω)]([Aleph Ω (Aleph Ω, Aleph Ω)], [Aleph Ω (Aleph Ω, Aleph Ω)])]([Hyper [Aleph Ω (Aleph Ω, Aleph Ω)]([Aleph Ω (Aleph Ω, Aleph Ω)], [Aleph Ω (Aleph Ω, Aleph Ω)])], [Hyper [Aleph Ω (Aleph Ω, Aleph Ω)]([Aleph Ω (Aleph Ω, Aleph Ω)], [Aleph Ω (Aleph Ω, Aleph Ω)])])])])])

Now can you imagine this:

Hyper-E-{Aleph Ω (Aleph Ω, Aleph Ω)} Aleph Ω (Aleph Ω, Aleph Ω).

Hyper-E-1-1 Aleph Ω (Aleph Ω, Aleph Ω)

I refer to as,

Hyper-E-{Hyper-E-{Aleph Ω (Aleph Ω, Aleph Ω)}} Aleph Ω (Aleph Ω, Aleph Ω)

Hyper-E-1-2 Aleph Ω (Aleph Ω, Aleph Ω)

I refer to as,

Hyper-E-{Hyper-E-1-1 Aleph Ω (Aleph Ω, Aleph Ω)} Aleph Ω (Aleph Ω, Aleph Ω)

Hyper-E-1-3 Aleph Ω (Aleph Ω, Aleph Ω)

I refer to as,

Hyper-E-{Hyper-E-1-2 Aleph Ω (Aleph Ω, Aleph Ω)} Aleph Ω (Aleph Ω, Aleph Ω)

Hyper-E-1-4 Aleph Ω (Aleph Ω, Aleph Ω)

I refer to as,

Hyper-E-{Hyper-E-1-3 Aleph Ω (Aleph Ω, Aleph Ω)} Aleph Ω (Aleph Ω, Aleph Ω)

Hyper-E-1-5 Aleph Ω (Aleph Ω, Aleph Ω)

I refer to as,

Hyper-E-{Hyper-E-1-4 Aleph Ω (Aleph Ω, Aleph Ω)} Aleph Ω (Aleph Ω, Aleph Ω)

Hyper-E-1-{Aleph Ω (Aleph Ω, Aleph Ω)} Aleph Ω (Aleph Ω, Aleph Ω)

I refer to as,

Hyper-E-1-{{Hyper-E-1-{Aleph Ω (Aleph Ω, Aleph Ω)} Aleph Ω (Aleph Ω, Aleph Ω)} minus1} Aleph Ω (Aleph Ω, Aleph Ω)

Hyper-E-1-{Aleph Ω (Aleph Ω, Aleph Ω)}-1 Aleph Ω (Aleph Ω, Aleph Ω)

I refer to as,

Hyper-E-1-{Hyper-E-1-{Aleph Ω (Aleph Ω, Aleph Ω)} Aleph Ω (Aleph Ω, Aleph Ω)} Aleph Ω (Aleph Ω, Aleph Ω)

Hyper-E-1-{Aleph Ω (Aleph Ω, Aleph Ω)}-2 Aleph Ω (Aleph Ω, Aleph Ω)

I refer to as,

Hyper-E-{Hyper-E-1-{Aleph Ω (Aleph Ω, Aleph Ω)}}-1 Aleph Ω (Aleph Ω, Aleph Ω)

Hyper-E-1-{Aleph Ω (Aleph Ω, Aleph Ω)}-3 Aleph Ω (Aleph Ω, Aleph Ω)

I refer to as,

Hyper-E-{Hyper-E-1-{Aleph Ω (Aleph Ω, Aleph Ω)}}-2 Aleph Ω (Aleph Ω, Aleph Ω)

Hyper-E-1-{Aleph Ω (Aleph Ω, Aleph Ω)}-4 Aleph Ω (Aleph Ω, Aleph Ω)

I refer to as,

Hyper-E-{Hyper-E-1-{Aleph Ω (Aleph Ω, Aleph Ω)}}-3 Aleph Ω (Aleph Ω, Aleph Ω)

Hyper-E-1-{Aleph Ω (Aleph Ω, Aleph Ω)}-5 Aleph Ω (Aleph Ω, Aleph Ω)

I refer to as,

Hyper-E-{Hyper-E-1-{Aleph Ω (Aleph Ω, Aleph Ω)}}-4 Aleph Ω (Aleph Ω, Aleph Ω)

Hyper-E-1-{Aleph Ω (Aleph Ω, Aleph Ω)}-{Aleph Ω (Aleph Ω, Aleph Ω)} Aleph Ω (Aleph Ω, Aleph Ω)

Hyper-E-1-{{Hyper-E-1-{Aleph Ω (Aleph Ω, Aleph Ω)}-{Aleph Ω (Aleph Ω, Aleph Ω)}} minus 1} Aleph Ω (Aleph Ω, Aleph Ω)

Now, can you imagine a spacecraft Lorentz factor of

Hyper-E-1-{{Hyper-E-1-{Aleph Ω (Aleph Ω, Aleph Ω)}-{Aleph Ω (Aleph Ω, Aleph Ω)}} minus 1} Aleph Ω (Aleph Ω, Aleph Ω)

Accordingly, the spacecraft would travel the above number of light-years in one-year ship-time, the above number of years into the future in one-year ship-time, and travel an effective multiple of light-speed equal to the above number.

No one need stay behind. Perhaps whole planets, galaxies, and cosmic light-cones can be moved with such Lorentz factors in light-speed travel to another location in our universe, multiverse, forest, biosphere, and the like, or in hyperspaces.

PART 21

Here we consider functions for which a number increasingly and expansively operates on itself.

For example, we use the following terminology:

↑ googol-plex $^{\text{googol-plex}}$ indicates an exponentiation function that continuously is iterated in an abstract atemporal manner.

↑ googol-plex $^{\text{googol-plex}}$ indicates an exponentiation function that is continuously iterated in an abstract temporal manner, where the time may be quantized, continuous, or super-continuous.

PART 22

Light-speed seems to be an inviolable limit of travel through space. Perhaps nature does not permit faster-than-light travel because she has evolved to a stable state where events cannot be removed from history. Otherwise, chaos could reign and a stable or meta-stable universe could never exist. Reality would thus become a complicated chaos.

However, light-speed is associated with infinite Lorentz factors that in turn enable infinite time dilation, infinite Lorentz contraction, and infinite relativistic aberration.

Perhaps light-speed is associated with many additional asymptotic limits that we have yet to formulate or discover.

Another fascinating set of conjectures involves different types of velocity that also have a limiting value.

Accordingly, some of these velocities may have magnitudes of *c* or the same as that of the velocity of light. Other velocities may have different meanings than the metric of distance/time but also be associated with limiting values for which additional parameters dependent on these alternative velocities grow asymptotically to infinite magnitudes in the limits of these velocities.

Now some of the analogue light-speed velocity terms may have dimensional variations that are almost the conventional meaning of velocity in variables.

For example, there may be pseudo-velocity limits expressed as [dx/d(t EXP (1 ± e))], where *e* ranges from the infinitesimal to the infinite.

Associated pseudo-Lorentz factors may be expressible as follows:

{1/{1 − [[[dx/d(t EXP (1 ± e))]/c] EXP 2]}} EXP (1/2)

or perhaps as

{1/{1 − [[v/[dx/d(t EXP (1 ± e))]] EXP 2]}} EXP (1/2)

or perhaps as

{1/{1 − [[[dx/d(t EXP (1 ± e))]/c] EXP (2 EXP f)]}} EXP (1/2)

or perhaps as

{1/{1 − [[v/[dx/d(t EXP (1 ± e))]] EXP (2 EXP f)]}} EXP (1/2)

or perhaps as

{1/{1 − [[[dx/d(t EXP (1 ± e))]/c] EXP 2]}} EXP [(1/2) EXP g]

or perhaps as,

{1/{1 − [[v/[dx/d(t EXP (1 ± e))]] EXP 2]}} EXP [(1/2) EXP g]

or perhaps as

{1/{1 − [[[dx/d(t EXP (1 ± e))]/c] EXP (2 EXP f)]}} EXP [(1/2) EXP g]

or perhaps as

{1/{1 − [[v/[dx/d(t EXP (1 ± e))]] EXP (2 EXP f)]}} EXP [(1/2) EXP g]

Here, *f* and *g* may be a real number ranging from the infinitesimal to the infinite.

We may also consider ship-frame acceleration analogues such as

d[dx/d(t EXP (1 ± e))]/dt

d[dx/d(t EXP (1 ± e))]/d(t EXP (1 ± e))

and even jerk, or the time rate of change of acceleration, such as

d{d[dx/d(t EXP (1 ± e))]/dt}/dt

d{d[dx/d(t EXP (1 ± e))]/d(t EXP (1 ± e))}/d(t EXP (1 ± e))

Likewise we can take arbitrary orders of timelike derivatives and/or integrals of ship-frame acceleration, and thus of velocity, as well as first or higher-order derivatives and/or integrals of ship-frame acceleration, and thus of velocity with respect to analogues of gamma, acceleration, kinetic energy, and any other relevant parameters.

In the limits as the formulaic physics for the above alternative parameters approaches Newtonian and special relativistic forms, e, f, and g approach 0.

PART 23

Now, there are countable and uncountable infinities.

But before we go further, consider the hyper-operator notation that was designed to express huge values not otherwise expressible.

For example, Note that Hyper4(a, n) is equal to a tetrated n, or a raised to the power of itself n minus 1 times. The latter value is symbolically written as n subscript a.

For example,

3 EXP 4 = 81, but 4 subscript 3 is approximately equal to 10 EXP (1,000,000,000,000).

Alternatively,

4 subscript 2 = 2 EXP 2 EXP 2 EXP 2 = 2 EXP [2 EXP [2 EXP 2]] = 2 EXP (2 EXP 4) = 2 EXP 16 = 65,536.

For example,

Hyper5(4, 4)is equal to 4 tetrated 4 tetrated 4 tetrated 4. This value is commonly referred to as 4 pentated 4.

Hyper 6, (4,4) is 4 pentated 4 pentated 4 pentated 4 and is also referred to as 4 hexataed 4.

Hyper 7, (4,4) is 4 hexated 4 hexated 4 hexated 4 and so on.

Aleph 0 is the infinite number of integers.

Aleph 1, according to the perhaps unprovable, and thus unfalsifiable, continuum hypothesis, is the number of real numbers that is greater than Aleph 0 by a multiplicative factor of infinity.

Aleph 2 is similarly greater than Aleph 1.

Aleph 3 is similarly greater than Aleph 2.

Aleph 4 is similarly greater than Aleph 3.

And so on.

In general, Aleph n = 2 EXP [Aleph (n-1)].

The number Ω is commonly stated as the least infinite positive integer or ordinal.

Now here is a real zinger.

So we can produce the abstraction of Hyper Aleph Ω (Aleph Ω, Aleph Ω).

We can also produce the abstraction of Hyper Aleph Extreme Uncountable Infinity (Aleph Extreme Uncountable Infinity, Aleph Extreme Uncountable Infinity).

We can consider scenarios of numbers that are greater than infinities.

In extreme circumstances, we can state that not only are there abstractions greater than infinities, but even greater abstractions that are "infinitier" than infinities.

Likewise, we can come up with abstractions that are infinitier than greater than infinity abstractions.

Additionally, we can contemplate abstractions that are infinitier than infinitier than greater than infinity abstractions.

Additionally, we can contemplate abstractions that are infinitier than infinitier than infinitier than greater than infinity abstractions. This latest abstraction we will label infinitier-3->∞.

Abstractions that are infinitier than infinitier than infinitier than infinitier than greater than infinity abstractions we will label infinitier-4->∞.

Abstractions that are infinitier than infinitier than infinitier than infinitier than infinitier than greater than infinity abstractions we will label infinitier-5->∞.

We can likewise consider infinitier-6->∞, infinitier-7->∞, infinitier-8->∞, …, infinitier-[Hyper Aleph Ω (Aleph Ω, Aleph Ω)]->∞ and even eventually infinitier-[infinitier-[Hyper Aleph Ω (Aleph Ω, Aleph Ω)]->∞]->∞, and so on.

Now a light-speed craft traveling at a Lorentz factor of infinitier-[infinitier-[Hyper Aleph Ω (Aleph Ω, Aleph Ω)]->∞]->∞ would travel an infinitier-[infinitier-[Hyper Aleph Ω (Aleph Ω, Aleph Ω)]->∞]->∞ light-years through space in one-year ship-time, an infinitier-[infinitier-[Hyper Aleph Ω (Aleph Ω, Aleph Ω)]->∞]->∞ years into the future in one-year ship-time, and an effective travel velocity of infinitier-[infinitier-[Hyper Aleph Ω (Aleph Ω, Aleph Ω)]->∞]->∞ times the speed of light.

Now, we in modern Standard Model physics, quantum mechanics, and special relativity have enshrined the speed of light as of utmost importance. Naturally, we give a simple name to the speed of light as denoted by the letter c. For light-speed, we will assign the expression $c,1$. However, as our spacecraft Lorentz factor climbs beyond Aleph 0, we likely need another constant, $c,2$, to define an analogue of light-speed. As our spacecraft Lorentz factor climbs to, say, $[(m)(Aleph\ 0)]$, where m is finite but very large, we likely need yet another constant, $c,3$, to define an analogue of c,2 and of c,1.

Oddly enough, perhaps we can continue to go all the way to c { infinitier-[infinitier-[Hyper Aleph Ω (Aleph Ω, Aleph Ω)]->∞]->∞} and continue from there.

We may affix the following operator to formulas for gamma and related parameters to denote the effects of travel at Lorentz factors from

> Aleph 0 to {infinitier-[infinitier-[Hyper Aleph Ω (Aleph Ω, Aleph Ω)]->∞]->∞} and greater and analogues of light-speed ranging from c,1 through c,{ infinitier-[infinitier-[Hyper Aleph Ω (Aleph Ω, Aleph Ω)]->∞]->∞} and greater

> {Σ{f{γ of Aleph 0 to {infinitier-[infinitier-[Hyper Aleph Ω (Aleph Ω, Aleph Ω)]->∞]->∞} and greater and analogues of light-speed ranging from c,1 through c,{ infinitier-[infinitier-[Hyper Aleph Ω (Aleph Ω, Aleph Ω)]->∞]->∞} and greater}}}

PART 24

Now, there are countable and uncountable infinities.

But before we go further, consider the hyper-operator notation that was designed to express huge values not otherwise expressible.

For example, Note that Hyper4(a, n) is equal to a tetrated n, or a raised to the power of itself n minus 1 times. The latter value is symbolically written as n subscript a.

For example,

> 3 EXP 4 = 81, but 4 subscript 3 is approximately equal to 10 EXP (1,000,000,000,000).

Alternatively,

> 4 subscript 2 = 2 EXP 2 EXP 2 EXP 2 = 2 EXP [2 EXP [2 EXP 2]] = 2 EXP (2 EXP 4) = 2 EXP 16 = 65,536.

For example,

Hyper5(4, 4)is equal to 4 tetrated 4 tetrated 4 tetrated 4. This value is commonly referred to as 4 pentated 4.

Hyper 6, (4,4) is 4 pentated 4 pentated 4 pentated 4 and is also referred to as 4 hexataed 4.

Hyper 7, (4,4) is 4 hexated 4 hexated 4 hexated 4 and so on.

Aleph 0 is the infinite number of integers.

Aleph 1, according to the perhaps unprovable, and thus unfalsifiable, continuum hypothesis, is the number of real numbers that is greater than Aleph 0 by a multiplicative factor of infinity.

Aleph 2 is similarly greater than Aleph 1.

Aleph 3 is similarly greater than Aleph 2.

Aleph 4 is similarly greater than Aleph 3.

And so on.

In general, Aleph n = 2 EXP [Aleph (n-1)].

The number Ω is commonly stated as the least infinite positive integer or ordinal.

Now here is a real zinger.

So we can produce the abstraction of Hyper Aleph Ω (Aleph Ω, Aleph Ω).

We can also produce the abstraction of Hyper Aleph Extreme Uncountable Infinity (Aleph Extreme Uncountable Infinity, Aleph Extreme Uncountable Infinity).

To the extent that the above infinities are bounded in their statements of abstract existence, they are commensurate with boundary conditions. This makes some sense when considering the scope of individual universes, multiverses, forests, biospheres, and the like, and hyperspaces or hyperspace-times.

Even in some cases where the above realms would be infinite, they likely are still grounded in bulks and are existentially underwritten by bulks. Bulks are likely bounded and underwritten by hyper-bulks and so on.

Now we consider the truly unbounded infinities of a first level we will refer to as unbounded-∞-1.

The next and higher level of unbounded infinities we will refer to as unbounded-∞-2.

The next and higher level of unbounded infinities we will refer to as unbounded-∞-3.

We can go all the way to and beyond unbounded-∞-[Hyper Aleph Ω (Aleph Ω, Aleph Ω)]

and to

We can go all the way to and beyond unbounded-∞-[Hyper Aleph Extreme Uncountable Infinity (Aleph Extreme Uncountable Infinity, Aleph Extreme Uncountable Infinity)].

We by definition will never reach the year AD Aleph 0 because such a future time is always ahead. Even reaching variously low infinite Lorentz factors in light-speed travel will never get us to the year AD Aleph 0.

However, we can look forward to AD Aleph 0 and, indeed, glimpse toward this date as if looking forward to an eternally distant awesomely mystical dim beacon of light.

After AD Aleph 0, we can take whimsical awe and amazement that there is as if a year to be penned by Mother Nature of AD Aleph 1.

After AD Aleph 1, we can take whimsical awe and amazement that there is as if a year to be penned by Mother Nature of AD Aleph 2.

After AD Aleph 2, we can take whimsical awe and amazement that there is as if a year to be penned by Mother Nature of 3 AD Aleph.

We can likewise ponder in awe about the ever-yet-to-arrive AD [Hyper Aleph Ω (Aleph Ω, Aleph Ω)] and ponder what more still lies beyond.

These extreme dates may not ever be attained while at the same time, perhaps in some impossibly distant future eternity, actually arrived at.

We may affix the following operator to formulas for gamma and related parameters to indicate going all the way to and beyond unbounded-∞-[Hyper Aleph Ω (Aleph Ω, Aleph Ω)] and beyond unbounded-∞-[Hyper Aleph Extreme Uncountable Infinity (Aleph Extreme Uncountable Infinity, Aleph Extreme Uncountable Infinity)] as forever distant open opportunities for travel distance through space and forward in time and in effective multiples of light-speed travel and the way these scenarios can effect gamma or the meaning of gamma:

> [w[Beyond unbounded-∞-[Hyper Aleph Ω (Aleph Ω, Aleph Ω)] and beyond unbounded-∞-[Hyper Aleph Extreme Uncountable Infinity (Aleph Extreme Uncountable Infinity, Aleph Extreme Uncountable Infinity)]]]

PART 25

Now we consider magnified light-speeds at sufficiently infinite Lorentz factors.

We also consider adjunct light-speed at sufficiently infinite Lorentz factors.

Accordingly, we elaborate on extreme finite to infinite multiples of light-speed in a spacecraft tearing through one realm after another, where sail realms may be cosmic light-cone extensities, domains, universes, multiverses, forests, biospheres, and the like and/or hyperspaces. Here, the spacecraft effects are limited to light-speed, magnified light-speeds, or adjunct light-speed in a given realm over classical scale distances. The spacecraft enters and leaves a realm in less than one Planck time unit in the background reference frame or non-unitary real positive real number multiple of one Planck time unit provided the realm of travel is commensurately quantized in space-time.

We may affix the following operator to formulas for gamma and related parameters to indicate the above light-speed, magnified light-speeds, and/or adjunct light-speeds in a given realm over quantum and/or classical scale distances.

We also consider ship-frame time, background frame time, gamma, acceleration, jerk, kinetic energy, etc., differentiation and integration of arbitrary orders thereof.

> [w(Light-speed, magnified light-speeds, and/or adjunct light-speeds in a given realm over quantum and/or classical scale distances and ship-frame time, background frame time, gamma, acceleration, jerk, kinetic energy, etc., differentiation and integration of arbitrary orders thereof)]

PART 26

Now, there are countable and uncountable infinities.

But before we go further, consider the hyper-operator notation that was designed to express huge values not otherwise expressible.

For example, Note that Hyper4(a, n) is equal to a tetrated n, or a raised to the power of itself n-1 times. The latter value is symbolically written as n subscript a.

For example,

> 3 EXP 4 = 81, but 4 subscript 3 is approximately equal to 10 EXP (1,000,000,000,000).

Alternatively,

> 4 subscript 2 = 2 EXP 2 EXP 2 EXP 2 = 2 EXP [2 EXP [2 EXP 2]] = 2 EXP (2 EXP 4) = 2 EXP 16 = 65,536.

For example,

Hyper5(4, 4)is equal to 4 tetrated 4 tetrated 4 tetrated 4. This value is commonly referred to as 4 pentated 4.

Hyper 6, (4,4) is 4 pentated 4 pentated 4 pentated 4 and is also referred to as 4 hexataed 4.

Hyper 7, (4,4) is 4 hexated 4 hexated 4 hexated 4 and so on.

Aleph 0 is the infinite number of integers.

Aleph 1, according to the perhaps unprovable, and thus unfalsifiable, continuum hypothesis, is the number of real numbers which is greater than Aleph 0 by a multiplicative factor of infinity.

Aleph 2 is similarly greater than Aleph 1.

Aleph 3 is similarly greater than Aleph 2.

Aleph 4 is similarly greater than Aleph 3.

And so on.

In general, Aleph n = 2 EXP [Aleph (n-1)].

The number Ω is commonly stated as the least infinite positive integer or ordinal.

Now here is a real zinger.

So we can produce the abstraction of Hyper Aleph Ω (Aleph Ω, Aleph Ω).

So we can imagine the expression such as

Hyper [Hyper Aleph Ω (Aleph Ω, Aleph Ω)](Hyper Aleph Ω (Aleph Ω, Aleph Ω), Hyper Aleph Ω (Aleph Ω, Aleph Ω))

and

Hyper [Hyper [Hyper Aleph Ω (Aleph Ω, Aleph Ω)](Hyper Aleph Ω (Aleph Ω, Aleph Ω), Hyper Aleph Ω (Aleph Ω, Aleph Ω))](Hyper [Hyper Aleph Ω (Aleph Ω, Aleph Ω)](Hyper Aleph Ω (Aleph Ω, Aleph Ω), Hyper Aleph Ω (Aleph Ω, Aleph Ω)), Hyper [Hyper Aleph Ω (Aleph Ω, Aleph Ω)](Hyper Aleph Ω (Aleph Ω, Aleph Ω), Hyper Aleph Ω (Aleph Ω, Aleph Ω))).

Now we come to a fascinating construct. We can imagine constructing infinities with a number Hyper Aleph Ω (Aleph Ω, Aleph Ω) of symbols using the above hyper-operator notation.

So a light-speed craft may conceivably travel any of the above conjectured and constructed infinities of light-years through space in one-second ship-time and essentially the same number of years into the future in one-second ship-time, and travel through space essentially and effectively the same number multiple of the speed of light.

So a light-speed craft may conceivably travel any of the above conjectured and constructed infinities of light-years through space in one-second ship-time and essentially the same number of years into the future in one-second ship-time, and travel through space essentially and effectively the same number multiple of the speed of light.

And no one need stay behind. Perhaps entire planets, solar systems, galaxies, and the like can be transported through the greater space at the above Lorentz factors.

Now we can consider hyper-operator notation with the argument Aleph Ω used with a high degree of mathematical sophistication, where the number of hyper-operator nestings is equal to Aleph Ω.

For example, Hyper Aleph Ω (Aleph Ω, Aleph Ω)

has one nesting.

Hyper Hyper Aleph Ω (Aleph Ω, Aleph Ω)(Hyper Aleph Ω (Aleph Ω, Aleph Ω), Hyper Aleph Ω (Aleph Ω, Aleph Ω))

has two nestings.

Hyper Hyper Hyper Aleph Ω (Aleph Ω, Aleph Ω)(Hyper Aleph Ω (Aleph Ω, Aleph Ω), Hyper Aleph Ω (Aleph Ω, Aleph Ω))(Hyper Hyper Aleph Ω (Aleph Ω, Aleph Ω)(Hyper Aleph Ω (Aleph Ω, Aleph Ω), Hyper Aleph Ω (Aleph Ω, Aleph Ω)), Hyper Hyper Aleph Ω (Aleph Ω, Aleph Ω)(Hyper Aleph Ω (Aleph Ω, Aleph Ω), Hyper Aleph Ω (Aleph Ω, Aleph Ω)))

has three nestings.

Hyper Hyper Hyper Hyper Aleph Ω (Aleph Ω, Aleph Ω)(Hyper Aleph Ω (Aleph Ω, Aleph Ω), Hyper Aleph Ω (Aleph Ω, Aleph Ω))(Hyper Hyper Aleph Ω (Aleph Ω, Aleph Ω)(Hyper Aleph Ω (Aleph Ω, Aleph Ω), Hyper Aleph Ω (Aleph Ω, Aleph Ω)), Hyper Hyper Aleph Ω (Aleph Ω, Aleph Ω)(Hyper Aleph Ω (Aleph Ω, Aleph Ω), Hyper Aleph Ω (Aleph Ω, Aleph Ω)))(Hyper Hyper Hyper Aleph Ω (Aleph Ω, Aleph Ω)(Hyper Aleph Ω (Aleph Ω, Aleph Ω), Hyper Aleph Ω (Aleph Ω, Aleph Ω))(Hyper Hyper Aleph Ω (Aleph Ω, Aleph Ω)(Hyper Aleph Ω (Aleph Ω, Aleph Ω), Hyper Aleph Ω (Aleph Ω, Aleph Ω)), Hyper Hyper Aleph Ω (Aleph Ω, Aleph Ω)(Hyper Aleph Ω (Aleph Ω, Aleph Ω), Hyper Aleph Ω (Aleph Ω, Aleph Ω))), Hyper Hyper Hyper Aleph Ω (Aleph Ω, Aleph Ω)(Hyper Aleph Ω (Aleph Ω, Aleph Ω), Hyper Aleph Ω (Aleph Ω, Aleph Ω))(Hyper Hyper Aleph Ω (Aleph Ω, Aleph Ω)(Hyper Aleph Ω (Aleph Ω, Aleph Ω), Hyper Aleph Ω (Aleph Ω, Aleph Ω)), Hyper Hyper Aleph Ω (Aleph Ω, Aleph Ω)(Hyper Aleph Ω (Aleph Ω, Aleph Ω), Hyper Aleph Ω (Aleph Ω, Aleph Ω))))

has four nestings.

So can you imagine an infinite Lorentz factor so great that it has

Hyper Hyper Hyper Hyper Aleph Ω (Aleph Ω, Aleph Ω)(Hyper Aleph Ω (Aleph Ω, Aleph Ω), Hyper Aleph Ω (Aleph Ω, Aleph Ω))(Hyper Hyper Aleph Ω (Aleph Ω, Aleph Ω)(Hyper Aleph Ω (Aleph Ω, Aleph Ω), Hyper Aleph Ω (Aleph Ω, Aleph Ω)), Hyper Hyper Aleph Ω (Aleph Ω, Aleph Ω)(Hyper Aleph Ω (Aleph Ω, Aleph Ω), Hyper Aleph Ω (Aleph Ω, Aleph Ω)))(Hyper Hyper Hyper Aleph Ω (Aleph Ω, Aleph Ω)(Hyper Aleph Ω (Aleph Ω, Aleph Ω), Hyper Aleph Ω (Aleph Ω, Aleph Ω))(Hyper Hyper Aleph Ω (Aleph Ω, Aleph Ω)(Hyper Aleph Ω (Aleph Ω, Aleph Ω), Hyper Aleph Ω (Aleph Ω, Aleph Ω)), Hyper Hyper Aleph Ω (Aleph Ω, Aleph

Ω)(Hyper Aleph Ω (Aleph Ω, Aleph Ω), Hyper Aleph Ω (Aleph Ω, Aleph Ω))), Hyper Hyper Hyper Aleph Ω (Aleph Ω, Aleph Ω)(Hyper Aleph Ω (Aleph Ω, Aleph Ω), Hyper Aleph Ω (Aleph Ω, Aleph Ω))(Hyper Hyper Aleph Ω (Aleph Ω, Aleph Ω)(Hyper Aleph Ω (Aleph Ω, Aleph Ω), Hyper Aleph Ω (Aleph Ω, Aleph Ω)), Hyper Hyper Aleph Ω (Aleph Ω, Aleph Ω)(Hyper Aleph Ω (Aleph Ω, Aleph Ω), Hyper Aleph Ω (Aleph Ω, Aleph Ω))))

nestings.

Such a spacecraft would travel about that number of light-years through space in one-year ship-time, the same number of years into the future, and effective travel velocity equal to about that number multiple of light-speed in a vacuum.

PART 27

Regarding how big infinities can be, consider the following digression that is repeated in this book.

Now, there are countable and uncountable infinities.

But before we go further, consider the hyper-operator notation that was designed to express huge values not otherwise expressible.

For example, note that Hyper4(a, n) is equal to a tetrated n, or a raised to the power of itself n-1 times. The latter value is symbolically written as n subscript a.

For example,

3 EXP 4 = 81, but 4 subscript 3 is approximately equal to 10 EXP (1,000,000,000,000).

Alternatively,

4 subscript 2 = 2 EXP 2 EXP 2 EXP 2 = 2 EXP [2 EXP [2 EXP 2]] = 2 EXP (2 EXP 4) = 2 EXP 16 = 65,536.

For example,

Hyper5(4, 4)is equal to 4 tetrated 4 tetrated 4 tetrated 4. This value is commonly referred to as 4 pentated 4.

Hyper 6, (4,4) is 4 pentated 4 pentated 4 pentated 4 and is also referred to as 4 hexataed 4.

Hyper 7, (4,4) is 4 hexated 4 hexated 4 hexated 4 and so on.

Aleph 0 is the infinite number of integers.

Aleph 1, according to the perhaps unprovable, and thus unfalsifiable, continuum hypothesis, is the number of real numbers that is greater than Aleph 0 by a multiplicative factor of infinity.

Aleph 2 is similarly greater than Aleph 1.

Aleph 3 is similarly greater than Aleph 2.

Aleph 4 is similarly greater than Aleph 3.

And so on.

In general, Aleph n = 2 EXP [Aleph (n-1)].

The number Ω is commonly stated as the least infinite positive integer or ordinal.

Now here is a real zinger.

So we can produce the abstraction of Hyper Aleph UNCOUNTABLE∞ (Aleph UNCOUNTABLE∞, Aleph UNCOUNTABLE∞).

So we can imagine expression such as the following:

Hyper [Hyper Aleph UNCOUNTABLE∞ (Aleph UNCOUNTABLE∞, Aleph UNCOUNTABLE∞)](Hyper Aleph UNCOUNTABLE∞ (Aleph UNCOUNTABLE∞, Aleph UNCOUNTABLE∞), Hyper Aleph UNCOUNTABLE∞ (Aleph UNCOUNTABLE∞, Aleph UNCOUNTABLE∞))

and

Hyper [Hyper [Hyper Aleph UNCOUNTABLE∞ (Aleph UNCOUNTABLE∞, Aleph UNCOUNTABLE∞)](Hyper Aleph UNCOUNTABLE∞ (Aleph UNCOUNTABLE∞, Aleph UNCOUNTABLE∞), Hyper Aleph UNCOUNTABLE∞ (Aleph UNCOUNTABLE∞, Aleph UNCOUNTABLE∞))](Hyper [Hyper Aleph UNCOUNTABLE∞ (Aleph UNCOUNTABLE∞, Aleph UNCOUNTABLE∞)](Hyper Aleph

UNCOUNTABLE∞ (Aleph UNCOUNTABLE∞, Aleph UNCOUNTABLE∞), Hyper Aleph UNCOUNTABLE∞ (Aleph UNCOUNTABLE∞, Aleph UNCOUNTABLE∞)), Hyper [Hyper Aleph UNCOUNTABLE∞ (Aleph UNCOUNTABLE∞, Aleph UNCOUNTABLE∞)](Hyper Aleph UNCOUNTABLE∞ (Aleph UNCOUNTABLE∞, Aleph UNCOUNTABLE∞), Hyper Aleph UNCOUNTABLE∞ (Aleph UNCOUNTABLE∞, Aleph UNCOUNTABLE∞))).

Now we come to a fascinating construct. We can imagine constructing infinities with a number Hyper Aleph UNCOUNTABLE∞ (Aleph UNCOUNTABLE∞, Aleph UNCOUNTABLE∞) of symbols using the above hyper-operator notation.

So a light-speed craft may conceivably travel any of the above conjectured and constructed infinities of light-years through space in one-second ship-time and essentially the same number of years into the future in one-second ship-time and travel through space essentially and effectively the same number multiple of the speed of light.

So a light-speed craft may conceivably travel any of the above conjectured and constructed infinities of light-years through space in one-second ship-time and essentially the same number of years into the future in one-second ship-time, and travel through space essentially and effectively the same number multiple of the speed of light.

And no one need stay behind. Perhaps entire planets, solar systems, galaxies, and the like can be transported through the greater space at the above Lorentz factors.

Now we can consider hyper-operator notation with the argument Aleph UNCOUNTABLE∞ used with a high degree of mathematical sophistication, where the number of hyper-operator nestings is equal to Aleph UNCOUNTABLE∞.

For example,

Hyper Aleph UNCOUNTABLE∞ (Aleph UNCOUNTABLE∞, Aleph UNCOUNTABLE∞)

has one nesting.

Hyper Hyper Aleph UNCOUNTABLE∞ (Aleph UNCOUNTABLE∞, Aleph UNCOUNTABLE∞)(Hyper Aleph UNCOUNTABLE∞ (Aleph UNCOUNTABLE∞,

Aleph UNCOUNTABLE∞), Hyper Aleph UNCOUNTABLE∞ (Aleph UNCOUNTABLE∞, Aleph UNCOUNTABLE∞))

has two nestings.

Hyper Hyper Hyper Aleph UNCOUNTABLE∞ (Aleph UNCOUNTABLE∞, Aleph UNCOUNTABLE∞)(Hyper Aleph UNCOUNTABLE∞ (Aleph UNCOUNTABLE∞, Aleph UNCOUNTABLE∞), Hyper Aleph UNCOUNTABLE∞ (Aleph UNCOUNTABLE∞, Aleph UNCOUNTABLE∞))(Hyper Hyper Aleph UNCOUNTABLE∞ (Aleph UNCOUNTABLE∞, Aleph UNCOUNTABLE∞)(Hyper Aleph UNCOUNTABLE∞ (Aleph UNCOUNTABLE∞, Aleph UNCOUNTABLE∞), Hyper Aleph UNCOUNTABLE∞ (Aleph UNCOUNTABLE∞, Aleph UNCOUNTABLE∞)), Hyper Hyper Aleph UNCOUNTABLE∞ (Aleph UNCOUNTABLE∞, Aleph UNCOUNTABLE∞)(Hyper Aleph UNCOUNTABLE∞ (Aleph UNCOUNTABLE∞, Aleph UNCOUNTABLE∞), Hyper Aleph UNCOUNTABLE∞ (Aleph UNCOUNTABLE∞, Aleph UNCOUNTABLE∞)))

has three nestings.

Hyper Hyper Hyper Hyper Aleph UNCOUNTABLE∞ (Aleph UNCOUNTABLE∞, Aleph UNCOUNTABLE∞)(Hyper Aleph UNCOUNTABLE∞ (Aleph UNCOUNTABLE∞, Aleph UNCOUNTABLE∞), Hyper Aleph UNCOUNTABLE∞ (Aleph UNCOUNTABLE∞, Aleph UNCOUNTABLE∞))(Hyper Hyper Aleph UNCOUNTABLE∞ (Aleph UNCOUNTABLE∞, Aleph UNCOUNTABLE∞)(Hyper Aleph UNCOUNTABLE∞ (Aleph UNCOUNTABLE∞, Aleph UNCOUNTABLE∞), Hyper Aleph UNCOUNTABLE∞ (Aleph UNCOUNTABLE∞, Aleph UNCOUNTABLE∞)), Hyper Hyper Aleph UNCOUNTABLE∞ (Aleph UNCOUNTABLE∞, Aleph UNCOUNTABLE∞)(Hyper Aleph UNCOUNTABLE∞ (Aleph UNCOUNTABLE∞, Aleph UNCOUNTABLE∞), Hyper Aleph UNCOUNTABLE∞ (Aleph UNCOUNTABLE∞, Aleph UNCOUNTABLE∞)))(Hyper Hyper Hyper Aleph UNCOUNTABLE∞ (Aleph UNCOUNTABLE∞, Aleph UNCOUNTABLE∞)(Hyper Aleph

UNCOUNTABLE∞ (Aleph UNCOUNTABLE∞, Aleph UNCOUNTABLE∞), Hyper Aleph UNCOUNTABLE∞ (Aleph UNCOUNTABLE∞, Aleph UNCOUNTABLE∞))(Hyper Hyper Aleph UNCOUNTABLE∞ (Aleph UNCOUNTABLE∞, Aleph UNCOUNTABLE∞)(Hyper Aleph UNCOUNTABLE∞ (Aleph UNCOUNTABLE∞, Aleph UNCOUNTABLE∞), Hyper Aleph UNCOUNTABLE∞ (Aleph UNCOUNTABLE∞, Aleph UNCOUNTABLE∞)), Hyper Hyper Aleph UNCOUNTABLE∞ (Aleph UNCOUNTABLE∞, Aleph UNCOUNTABLE∞)(Hyper Aleph UNCOUNTABLE∞ (Aleph UNCOUNTABLE∞, Aleph UNCOUNTABLE∞), Hyper Aleph UNCOUNTABLE∞ (Aleph UNCOUNTABLE∞, Aleph UNCOUNTABLE∞))), Hyper Hyper Hyper Aleph UNCOUNTABLE∞ (Aleph UNCOUNTABLE∞, Aleph UNCOUNTABLE∞)(Hyper Aleph UNCOUNTABLE∞ (Aleph UNCOUNTABLE∞, Aleph UNCOUNTABLE∞), Hyper Aleph UNCOUNTABLE∞ (Aleph UNCOUNTABLE∞, Aleph UNCOUNTABLE∞))(Hyper Hyper Aleph UNCOUNTABLE∞ (Aleph UNCOUNTABLE∞, Aleph UNCOUNTABLE∞)(Hyper Aleph UNCOUNTABLE∞ (Aleph UNCOUNTABLE∞, Aleph UNCOUNTABLE∞), Hyper Aleph UNCOUNTABLE∞ (Aleph UNCOUNTABLE∞, Aleph UNCOUNTABLE∞)), Hyper Hyper Aleph UNCOUNTABLE∞ (Aleph UNCOUNTABLE∞, Aleph UNCOUNTABLE∞)(Hyper Aleph UNCOUNTABLE∞ (Aleph UNCOUNTABLE∞, Aleph UNCOUNTABLE∞), Hyper Aleph UNCOUNTABLE∞ (Aleph UNCOUNTABLE∞, Aleph UNCOUNTABLE∞))))

has four nestings.

So, can you imagine an infinite Lorentz factor so great that it has

Hyper Hyper Hyper Hyper Aleph UNCOUNTABLE∞ (Aleph UNCOUNTABLE∞, Aleph UNCOUNTABLE∞)(Hyper Aleph UNCOUNTABLE∞ (Aleph UNCOUNTABLE∞, Aleph UNCOUNTABLE∞), Hyper Aleph UNCOUNTABLE∞ (Aleph UNCOUNTABLE∞, Aleph UNCOUNTABLE∞))(Hyper Hyper Aleph UNCOUNTABLE∞ (Aleph UNCOUNTABLE∞, Aleph UNCOUNTABLE∞)(Hyper Aleph UNCOUNTABLE∞ (Aleph UNCOUNTABLE∞, Aleph UNCOUNTABLE∞), Hyper Aleph UNCOUNTABLE∞ (Aleph UNCOUNTABLE∞, Aleph UNCOUNTABLE∞)), Hyper Hyper Aleph UNCOUNTABLE∞ (Aleph UNCOUNTABLE∞, Aleph UNCOUNTABLE∞)(Hyper Aleph UNCOUNTABLE∞ (Aleph UNCOUNTABLE∞, Aleph UNCOUNTABLE∞), Hyper Aleph UNCOUNTABLE∞ (Aleph UNCOUNTABLE∞, Aleph

UNCOUNTABLE∞)))(Hyper Hyper Hyper Aleph UNCOUNTABLE∞ (Aleph UNCOUNTABLE∞, Aleph UNCOUNTABLE∞)(Hyper Aleph UNCOUNTABLE∞ (Aleph UNCOUNTABLE∞, Aleph UNCOUNTABLE∞), Hyper Aleph UNCOUNTABLE∞ (Aleph UNCOUNTABLE∞, Aleph UNCOUNTABLE∞))(Hyper Hyper Aleph UNCOUNTABLE∞ (Aleph UNCOUNTABLE∞, Aleph UNCOUNTABLE∞)(Hyper Aleph UNCOUNTABLE∞ (Aleph UNCOUNTABLE∞, Aleph UNCOUNTABLE∞), Hyper Aleph UNCOUNTABLE∞ (Aleph UNCOUNTABLE∞, Aleph UNCOUNTABLE∞)), Hyper Hyper Aleph UNCOUNTABLE∞ (Aleph UNCOUNTABLE∞, Aleph UNCOUNTABLE∞)(Hyper Aleph UNCOUNTABLE∞ (Aleph UNCOUNTABLE∞, Aleph UNCOUNTABLE∞), Hyper Aleph UNCOUNTABLE∞ (Aleph UNCOUNTABLE∞, Aleph UNCOUNTABLE∞))), Hyper Hyper Hyper Aleph UNCOUNTABLE∞ (Aleph UNCOUNTABLE∞, Aleph UNCOUNTABLE∞)(Hyper Aleph UNCOUNTABLE∞ (Aleph UNCOUNTABLE∞, Aleph UNCOUNTABLE∞), Hyper Aleph UNCOUNTABLE∞ (Aleph UNCOUNTABLE∞, Aleph UNCOUNTABLE∞))(Hyper Hyper Aleph UNCOUNTABLE∞ (Aleph UNCOUNTABLE∞, Aleph UNCOUNTABLE∞)(Hyper Aleph UNCOUNTABLE∞ (Aleph UNCOUNTABLE∞, Aleph UNCOUNTABLE∞), Hyper Aleph UNCOUNTABLE∞ (Aleph UNCOUNTABLE∞, Aleph UNCOUNTABLE∞)), Hyper Hyper Aleph UNCOUNTABLE∞ (Aleph UNCOUNTABLE∞, Aleph UNCOUNTABLE∞)(Hyper Aleph UNCOUNTABLE∞ (Aleph UNCOUNTABLE∞, Aleph UNCOUNTABLE∞), Hyper Aleph UNCOUNTABLE∞ (Aleph UNCOUNTABLE∞, Aleph UNCOUNTABLE∞))))

nestings.

Such a spacecraft would travel about that number of light-years through space in one-year ship-time, the same number of years into the future, and effective travel velocity equal to about that number multiple of light-speed in a vacuum.